W. J. Beal

Grasses of North America

W. J. Beal

Grasses of North America

ISBN/EAN: 9783743344815

Manufactured in Europe, USA, Canada, Australia, Japa

Cover: Foto ©berggeist007 / pixelio.de

Manufactured and distributed by brebook publishing software (www.brebook.com)

W. J. Beal

Grasses of North America

GRASSES

OF

NORTH AMERICA

BY

W. J. BEAL, M.A., M.S., Ph.D.

Professor of Botany and Forestry in Michigan Agricultural College

IN TWO VOLUMES

VOL. I

CHAPTERS ON THE PHYSIOLOGY, COMPOSITION, SELECTION, IMPROVING, AND CULTIVATION OF GRASSES; MANAGEMENT OF GRASS LANDS; ALSO CHAPTERS ON CLOVERS, INJURIOUS INSECTS, AND FUNGI

NEW YORK

HENRY HOLT AND COMPANY

1896

PREFACE.

This volume may appear fragmentary and disconnected and may contain repetitions, and there may be important omissions. This is partially owing to the fact that much of the work has been performed at odd hours, sometimes with many days intervening. A full index will enable any one to find the topics which are mentioned.

It is hoped that the farmer or general reader who has never studied botany, will find much to interest and help him, while it is believed the agricultural student will find still more.

While many points are mentioned and illustrations freely used, no attempt has been made to write a complete account of the structure and physiology of grasses.

Although not grasses, after some misgivings of the author, a chapter on clovers was added, because the farmer would doubtless be disappointed if he did not find one.

An effort has been made to give credit to everyone who has been of much assistance in contributing to the volume, first, because it is due to such persons, and second, because if accurately quoted, it makes them responsible for the statements which are made.

In the chapter on bibliography will be found most of the sources of information, aside from the studies and experiments of the author. I have freely used, without quotation marks, my own contributions made at various times during the past seventeen years, to the Rural New Yorker, Philadelphia Press, New York Tribune, Prairie Farmer, the Clover Leaf, also using my reports as printed in the Michigan Board of Agriculture. Liberal quotations have been made from the publications of Sir J. B. Lawes, Charles Darwin, Maxwell T. Masters, George Bentham, Wm. Caruthers, Prof. James Buckman, Dr. A. Voelcker, of England; M. J. Duval-Jouvé, of France; Ernest Hackel, of Austria; J. S. Gould, C. W. Howard, Dr. D. L. Phares, Joseph Harris and J. B. Killebrew, of our own country.

My colleague, Prof. A. J. Cook, contributed the long and valuable chapter on Insects Injurious to Grasses and Clovers, while Prof. William Trelease, D. Sc., contributed that on Fungi injurious to the same orders of plants.

My friend, Prof. F. Lamson Scribner, has shown much interest in the work, and to him I am especially indebted for many of the excellent drawings of grasses. Mr. G. B. Sudworth also made many drawings and copied others.

The Levytype Company, of Philadelphia, and the Lewis Engraving Company, of Boston, prepared most of the plates from the drawings illustrating the work.

Frequent quotations have been made from the Rural New Yorker, Country Gentleman, New York Tribune, Prairie Farmer, Farmers' Review, and many agricultural reports of this country and of Europe.

The following persons also prepared more or less manuscript for these pages, and credit is given on the pages where the contributions appear: Prof. E. M. Shelton, of Kansas; Dr. H. P. Armsby, of Wisconsin; Prof. C. G. Pringle, of Vermont; Dr. C. E. Bessey, of Nebraska; Prof. J. J. Thomas, Major H. E. Alvord, Prof. I. P. Roberts, Dr. E. L. Sturtevant and J. S. Woodward, of New York; A. W. Cheever, of Massachusetts; Prof. J. W. Sanborn, of Missouri; Prof. F.

A. Gulley, of Mississippi; Professors Latta and Troop, of Purdue University, Indiana.

The following list should be added as freely quoted: Prof. F. L. Scribner, Clifford Richardson and Dr. G. Vasey, of Washington; Dr. R. C. Kedzie, of Michigan; Dr. A. Gray, Prof. N. S. Shaler, and Prof. F. L. Storer, of Massachusetts; Secretary W. I. Chamberlain, now president of Iowa Agricultural College; Mr. L. N. Bonham, of Ohio; E. S. Carman, of New York; Prof. Wm. Brown, of Ontario; Prof. G. E. Morrow, of Illinois, and Mr. De Laune, of England.

For reading portions of proof I am indebted to my colleagues, Prof. A. J. Cook, Dr. Lewis McLouth, Prof. E. J. MacEwan, Prof. L. H. Bailey, jr., also to Prof. V. M. Spaulding, of Ann Arbor, and Prof. F. L. Scribner, of Washington. Thorp & Godfrey, of Lansing, Mich., are credited with the mechanical part of the work.

A second volume is in preparation. This is to contain the description of all known grasses of North America, 700 or more species, with illustrations of one species in each genus, and in some cases more than one. Full notes in regard to their value for cultivation will be given. A chapter on Geographical Distribution and other information likely to interest the student of grasses will be presented.

W. J. BEAL.

BOTANICAL LABORATORY,
AGRICULTURAL COLLEGE, MICH.,
JANUARY, 1887.

PREFACE TO THE SECOND EDITION.

IN the revision it has not seemed important to replace a few of the generic names by others that they might be uniform with those of the second volume, which was published in 1896. I refer to the following: *Cynodon*, *Deyeuxia*, *Eragrostis*, *Leersia*, *Gly-*

ceria, *Setaria*, and a few others. Nor have I revised the places of residence of those who assisted in the preparation of this volume. G. C. Davis has made a few minor changes and corrections in the chapter which was chiefly prepared by Prof. A. J. Cook.

W. J. BEAL.

AGRICULTURAL COLLEGE, MICH.,
DECEMBER 2, 1897.

CONTENTS.

CHAPTER I.

THE STRUCTURE, FORM AND DEVELOPMENT OF THE GRASSES.

CHAPTER II.

THE POWER OF MOTION IN PLANTS.

CHAPTER III.

PLANT GROWTH.

CHAPTER IV.

CLASSIFYING, NAMING, DESCRIBING, COLLECTING, STUDYING.

CHAPTER V.

NATIVE GRAZING LANDS.

CHAPTER VI.

GRASSES FOR CULTIVATION.

CHAPTER VII.

EARLY ATTEMPTS TO CULTIVATE GRASSES.

CHAPTER VIII.

TESTING SEEDS, SOME COMMON WEEDS.

CHAPTER IX.

GRASSES FOR PASTURES AND MEADOWS.

CHAPTER X.

PREPARATION OF THE SOIL AND SEEDING.

CHAPTER XI.

CARE OF GRASS LANDS.

CHAPTER XII.

MAKING HAY.

CHAPTER XIII.

LOOK THE WORLD OVER FOR BETTER GRASSES AND IMPROVE THOSE WE NOW HAVE.

CHAPTER XIV.

GRASSES FOR THE LAWN, THE GARDEN, AND FOR DECORATION.

CHAPTER XV.

THE LEGUMINOSÆ, PULSE FAMILY.

CHAPTER XVI.

THE ENEMIES OF GRASSES AND CLOVERS.

CHAPTER XVII.

THE FUNGI OF FORAGE PLANTS.

CHAPTER I.

THE STRUCTURE, FORM AND DEVELOPMENT OF THE GRASSES.

Protoplasm is the living portion of a plant. It is sensitive to heat and cold and is the essential part without which the cell cannot live, take in or assimilate food or make any growth. Protoplasm is a soft-solid, generally containing a multitude of small granules, and when everything is favorable it is in unceasing motion. Delicate currents, often changing in direction and rapidity, are traced by the granules which they carry as they gracefully glide from one part of the cell to the other. Under the microscope this motion may be seen in the sting of a nettle, hair of a pumpkin vine, style of Indian corn, or a hair at the tip of a young kernel of wheat and in many other parts of plants. Protoplasm is most abundant in the newer or younger portions of the roots, stems, leaves, buds and seeds, and constitutes most of the nourishment as food for herbiverous animals. Very young cells are filled with protoplasm, while the older ones contain less, little, or none.

Cells. All parts of plants, except a few very small one-celled species, are composed of cells which are generally microscopic. When any part of a plant is soft and can be easily crushed or broken in any direction, the cell walls are thin; when it is hard the cell walls are thick; when tough like the fibre of flax, the cell walls are quite long and have thick walls.

Chlorophyll. All the green parts of a plant are so colored by a little green substance called *chlorophyll*, without which the plant is unable to assimilate any thing or to make any real progress in growth.

Roots. Although popularly so considered, it is by no means the case that all parts of plants which grow beneath the surface of the ground are roots. There are many stems beneath the surface and many roots above. Roots have no leaves, and are otherwise simpler than stems. They elongate by a rapid multiplication and growth of the cells a very short distance (perhaps the one-sixteenth of an inch in case of Indian corn) back of the extreme tip end. At such place, called "primary meristem," the cells rapidly increase by division, some of which continue to remain small and keep on dividing.

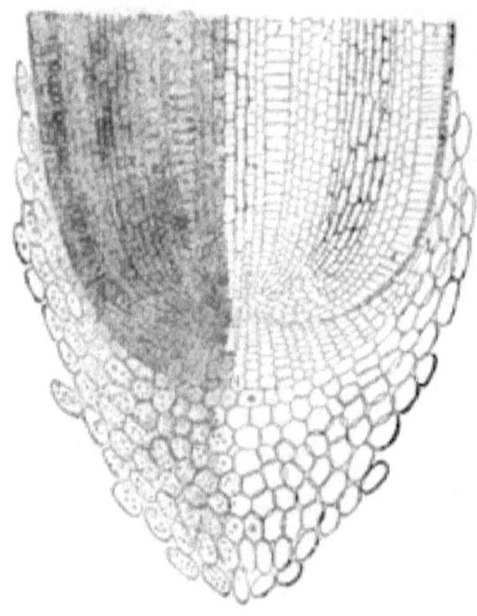

FIG. 1.—Longitudinal section through the apex of a root of Indian corn, half of which represents the cells as empty; *a, a,* outer and older portion of the root-cap; above this is the younger portion, just above which are very small cells that divide and make new cells for increasing the length of the root and replenishing the root-cap.—(Sachs.)

A portion of stem, on the other hand, usually produces leaves, buds, and when young elongates by a multiplication and growth of the cells for a considerable portion of its length. The tender, growing tip of a root is protected as it pushes along through the soil by a *root-cap* consisting of some older and harder cells. As these cells wear off, others crowd forward and take their places. In grasses the growth of the primary root is soon overtaken and is scarcely distinguishable from the secondaries or their branches.

Roots perforate the leaf-sheaths or rudimentary leaves and spring freely from the nodes or joints of underground stems of June grass, quack grass, and in some instances they grow from the nodes or joints of the stems above ground, especially where they are moist and well shaded. All the secondary roots—branch roots from roots and stems—originate from an internal layer of tissue where there are fibro-vascular bundles and break through the external portions of the root or stem.

The soil has much to do with the length and number of roots. In light, poor soil, in a dry time, we have found the roots of June grass to extend over four feet below the surface of the soil.

The roots of grasses are numerous, long, and fibrous, and when young the slender and delicate tips have a feeble power of moving from side to side, which enables them to find and penetrate the places of least resistance in the soil.

Although they are so small, it is estimated that in most farm crops, while growing, the aggregate surface of the roots is equal to that of the stems and leaves above ground. In hard clay subsoil in Central Michigan, oats pushed down their roots three feet four inches, and those of barley went down three feet nine inches. In mellow, sandy soil the roots of oats extended four feet two inches below the surface and those of barley five feet six inches. The famous buffalo grass (*Buchloë*) is often mentioned as having very short roots, but one of my students found in Kansas that they went down seven feet. The roots grow best where the best food is to be found, provided there is sufficient heat and moisture. They extend more or less in every direction; if one finds food it flourishes and enlarges and sends out numerous branches, and they in turn send out others. If rich earth or manure is placed above the roots they will grow upwards as well as downwards. In rich earth the roots of grasses will be densely matted; in sterile soil they will be longer, with fewer branches. Where the food is best, there we shall find the most roots. Roots cannot be accredited with any faculty which enables them to search for food as an animal hunts its prey.

The roots of all the grasses and most other flowering plants while in a growing condition are well supplied with **Trichomes or root-hairs** which vastly increase their surface.

Root-hairs are continuations of some of the outer cells of the younger roots and are brought into very close contact with the particles of soil. Their number depends much on the nature of the medium in which the roots are grown. Where the soil is rich, moist and porous, root-hairs are abundant. They are very short-lived, often lasting only for a few days, new hairs from other rootlets taking their places.

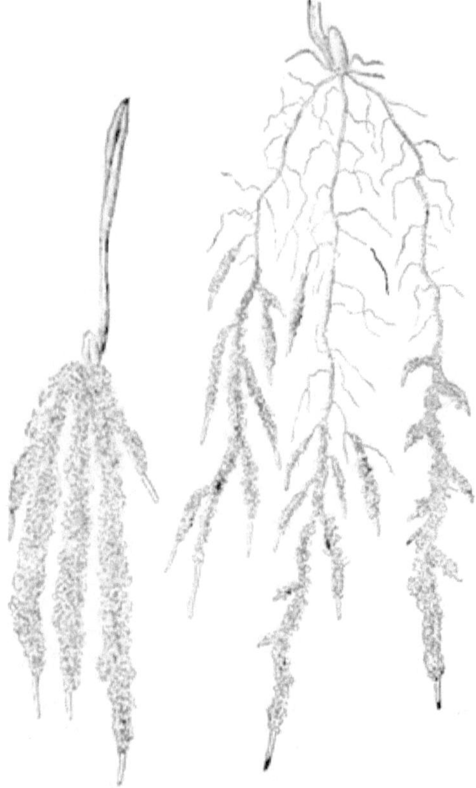

FIG. 2. Roots of young wheat plant lifted from the soil, holding soil by the root-hairs excepting near the apex where the hairs have not yet been produced.

FIG. 3. Plant a little older with soil clinging to the younger parts, but not to the older parts as there the root-hairs have perished.

The upper and older portions of the roots merely serve to hold the plant in position and act as conductors for the transmission of matter to the leaves of the plant and some of it back again to the newer roots. The reader should consult figure 2, representing a young wheat plant carefully lifted with the sand which is held fast by its close contact with the root-hairs. The tips of the roots have not put forth hairs and hence they are still naked.

Figure 3 represents the roots of a

wheat plant still older than the one shown in the previous figure. Here the root-tips are naked and the older roots fail to retain the particles of soil because the hairs have perished. It will be seen that the root-hairs are confined to the younger portions of the roots, beginning a little back from the tip.

These hairs look somewhat like mould or a mass of spider's webs and can be easily seen where Indian corn or wheat is sprouted between folds of damp cloth or paper. They are the chief agents for absorbing water and gases from the soil.

Root-hairs not only take up substances held in solution, but through their acid act on solid substances and render them soluble.

They also obtain nitrogen in the form of nitrates, which to some extent are formed in the soil through the action of bacteria, the lowest and simplest and smallest of plants.

The root-hairs nearly or quite all perish when a plant is at rest or ceases to grow, but when growth begins again it sends out new rootlets which produce new root-hairs.

Trichomes are usually found to a greater or less extent on stems, leaves, and even on some parts of the flower.

The Stem. The ascending axis or stem of a grass is called the *culm*. Some grasses produce stems on the surface of the ground or beneath it; these are called *rhizomes* or *root-stocks*. They often bear roots and sheathing scales, or rudimentary leaves with good buds, as is seen in June grass and quack grass.

The full grown culms of nearly all grasses are hollow, with solid or knotted joints called *nodes*. When very young the *internodes* or spaces between the nodes are solid, and even when full grown they are solid in most root-stocks, and in the culms above ground of such grasses as Indian corn, broom corn, sorghum, and sugar-cane. In case of *Phleum pratense* (Timothy), *Poa bulbosa*, *Arrhenatherum avenaceum* (tall oat grass), some of the lower short internodes become enlarged and contain a store of nourishment. Such grasses are called *bulbous*, though the term *tuber* or *corm* would be more nearly accurate.

The culms of most grasses produce branches, especially from the lower nodes near the ground. This branching is popularly called *tillering*, or *stooling*, or *mooting*, and is familiar in the case of wheat, oats, and rye, where one kernel not unfrequently produces twenty or more culms. Tillering is favored by shallow, thin seeding. Grasses are generally erect, though some are trailing; one or more climb over trees 100 feet high; others, like *Leersia* (rice cut-grass), are feeble climbers or sustain themselves on plants by means of numerous hooked prickles on their leaves.

Buds are undeveloped leaf or flower branches, and one or more may be looked for at every node. The apex of the young stem is covered by the young leaves.

The nodes are usually swollen or larger than the internodes, but seldom have a length very much greater than their diameter. The nodes remain short when the culm is erect, but if by any accident or otherwise the culm is tipped over, the nodes at once become longer on the lower side, and this curves the culm towards an erect position. In this way, to some extent, lodged wheat or other grasses can again partially regain their former position. At least, in most cases, the blossoms may be turned up from the ground.

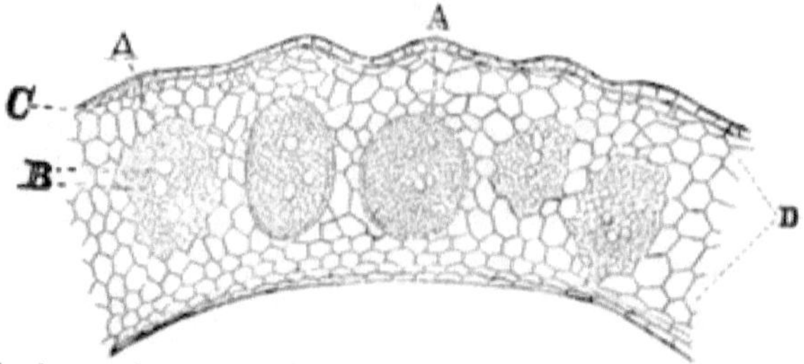

FIG. 4.—A part of a cross-section of wheat straw. A, fibro-vascular bundles; D, fundamental tissue made of thin-walled cells with hexagonal outlines. × 75.—(Mrs. L. R. Stowell.)

When quite young each internode elongates, by the multiplication and enlargement of the cells throughout its whole length, but as it gets older elongation for a considerable portion of the internode ceases, and finally there comes a time when the culm is incapable of further elongation. If taken in hand when young, and properly shaded, a stem may be made to grow to an almost indefinite length. The lower portion of an internode of most grasses, the part within the leaf-sheath, remains soft and continues to grow for a considerable time after the upper and main portion has lost this power.

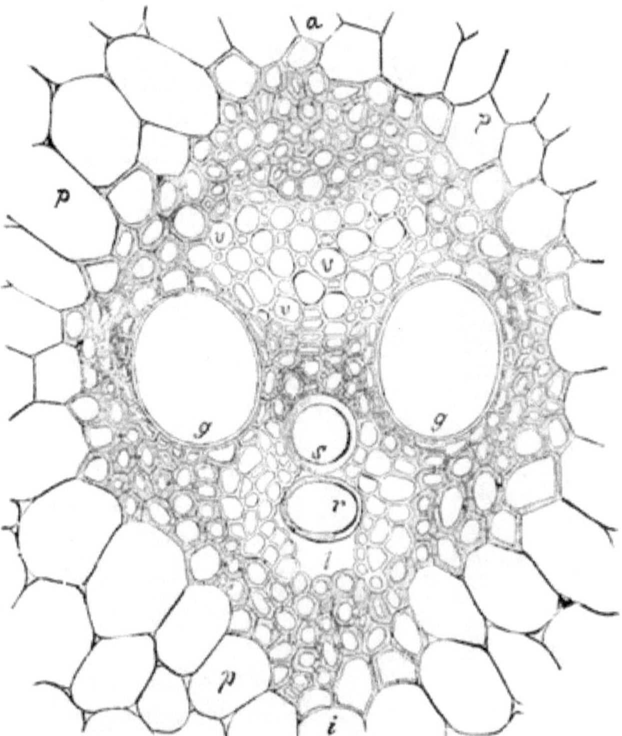

Fig. 5.—A cross-section of fibro-vascular bundle of Indian corn: *a*, side of bundle looking toward the circumference of the stem; *i*, side of bundle toward the center of stem; *p*, thin-walled cells of fundamental tissues of stem; *g*, *g*, large pitted vessels; *s*, spiral vessel; *r*, one ring of an annular vessel; *l*, air cavity formed by the breaking apart of the surrounding cells; *v*, *v*, latticed cells, or soft bast, a form of sieve tissue. × 550.—(After Sachs; notes after Bessey.)

The young stem of a grass when cut across will be found to contain numerous threads (*fibro-vascular bundles*) scattered from the center to the circumference. An epidermis covers the whole. In many instances, as the stem enlarges, the inside is ruptured and a hollow is formed.

Neither roots nor leaves could last long without each other.

Fig. 6.—Represents a young stem of *Festuca* as it branches at the base. (Hackel.)

The slender branches of the panicles of *Sporobolus heterolepis*, a grass common on the prairies of the west, are covered in places with a gummy excretion which entraps small insects. Dr. Bessey in the American Naturalist, p. 420, 1884, suggests that they may serve the same purpose as the similar sticky belts in *Silene* or catch-fly, viz.: to entrap crawling insects and prevent them from reaching the flowers which they are incapable of fertilizing.

The naked portions of the internodes of *Tragus racemosus* var. *occidentalis*, a wild grass of Arizona, are furnished with a sticky substance. The specimens examined are covered with many particles of sand and dust.

The main uses of the stem appear to be to convey the sap to or from the leaves, to support the leaves and extend them in every direction, giving each its share of room and exposure to light and air, and to bear the flowers and seeds.

"The stem, in fact, is the agency by which the work of individual leaves is combined and concentrated for the general benefit of the plant. Each separate leaf, like each separate cell, has a life of its own, and to some extent is independent of every other leaf; but if they are to be of any use to the plant as a whole, there must be coöperation." (Masters' Plant Life.)

The explanation for the ascent and movement of the "sap" in plants is by no means simple and easy. The swaying of the stems, branches, and leaves by the wind renders some assistance. The chemical changes going on within the plant cause some movements of the liquid materials. The evaporation from the leaves helps "draw" the water and gases from below to ascend and fill the spaces which would otherwise be vacant. "There is no continuous tube or set of tubes, and there is no fluid of uniformly the same composition throughout. Near the root the juice of the plant has one composition, near the leaf another. The word 'sap,' then, though convenient, must not be used or conceived of as indicating the existence of a current absolutely fixed in its direction or uniform in its composition." (Masters' Plant Life.)

The Leaf. Springing from the superficial part of each node, and generally completely surrounding the culm, appears a leaf, the *sheath* or lower part of which is generally *convolute* or wrapped around the culm. The leaves are *two-ranked* or *distichous*, and are so placed that each leaf is a little above or below any special one selected and exactly half way around the stem, where the blade spreads away from the stem. Usually there is one leaf at a node, but in *Cynodon Dactylon* (Bermuda grass), *Sporobolus arenarius*, and a few others, there are apparently two or three at a node distichously placed above each other.

The Sheaths of the leaves are usually spoken of as split on the side opposite the blade, though exceptional cases are cited where the sheaths are closed, as in *Bromus* (chess), *Melica* (melic grass), and some others. The sheaths of the upper leaves of most grasses are split down to the node, but those of the lower leaves in very many species are closed. In some cases the sheaths are closed at an early stage of development, but later they are split open part of the way down by the enlargement of the growing culm and the young leaves as they push upwards. This is illustrated in Fig. 7.

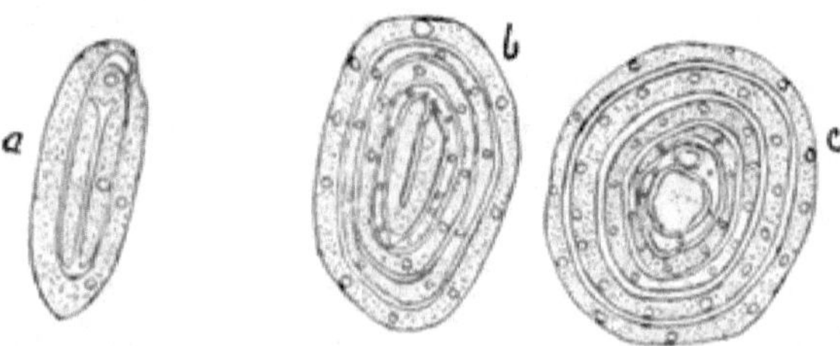

FIG. 7. *a*, A thin cross section of a young leaf of *Poa pratensis* with the blade conduplicate, and the surrounding sheath closed; *b*, a section still lower down, showing three closed sheaths; *c*, still lower down near a node, where five leaves have the sheaths closed. × 10.—(Sudworth.)

At the upper end of the sheath there is often a membranous scale, tongue, or fringe, called the *ligule*. The reader will consult Fig. 51, and observe the ligule of a leaf of June grass. That part of the leaf, which spreads away from the culm, is known as the

Blade or lamina, and is usually sessile and slender, tapering to a point.

To the ordinary observer the blades of grasses seem to be very nearly alike. Even Linnæus thought so, but to the botanist of to-day they present very marked differences.

The abortive leaves on root-stocks generally consist mainly of rudimentary sheaths. Commonly, all the leaves on a stem are much alike, but in some cases the lower leaves are quite unlike those above. The lower leaves of some species of *Bambusa* (bamboo), *Oryzopsis asperifolia* (mountain rice), *Panicum dichotomum* and others, have well developed sheaths, but the blades are rudimentary.

The blades of some leaves, like those of *Leersia* (rice cut-grass) and *Zizania* (wild rice), are not quite symmetrical, or in other words, the midrib is not quite in the middle of the blade.

The blades of many grasses after getting something of a start, may continue to elongate or they may cease to grow. In case of *Poa pratensis* (June grass), *Dactylis glomerata* (orchard grass), and many more, there seems to be scarcely any limit to the length they may attain. In a damp season, when the leaves were sheltered by a hedge, the writer selected a leaf of June grass, still green and vigorous to the end, in which the blade was five feet and four inches long. The place of growth for such leaves is a rather light green semi-circle near the ligule. The tip of such a leaf-blade is the oldest portion. The lower portion may continue to grow as the end is cropped by cattle.

The blade always has upper and lower surfaces unlike each other. Some leaves are *convolute* (rolled into a cylinder), while some are *conduplicate* (or folded), like the two halves of a book, shutting against each other.

When very dry, conduplicate leaves may become convolute, and between conduplicate and convolute vernation, we have all possible gradations passing insensibly into each other. Some leaves, as those of *Lolium rigidum*, are conduplicate towards the apex, and convolute towards the base.

Leaves of many exogenous plants, like most of our trees and shrubs, drop by separating from the stem at a natural joint, but the leaves of most grasses may die, become brown and dry, and still remain attached to the culm. The leaves of a few grasses, as the bamboos and *Spartina* (cord grass), have blades with an articulation or joint at the base; and some leaves have petioles, as *Pharus*, *Pariana*, and *Leptaspis*.

Some leaf-blades, as those of *Panicum Crus-galli* (barn-yard grass), *P. plicatum*, many species of *Sorghum*, taper each way, and are linear lanceolate, but they have as many bundles at the base as in the middle. They are like Fig. 8, only in disguise. Transverse veins are visible to the eye in *Panicum Crus-galli*, *Chloris*, *Bambusa* (bamboo), and in most others they are found to a greater or less extent, but they are not often conspicuous.

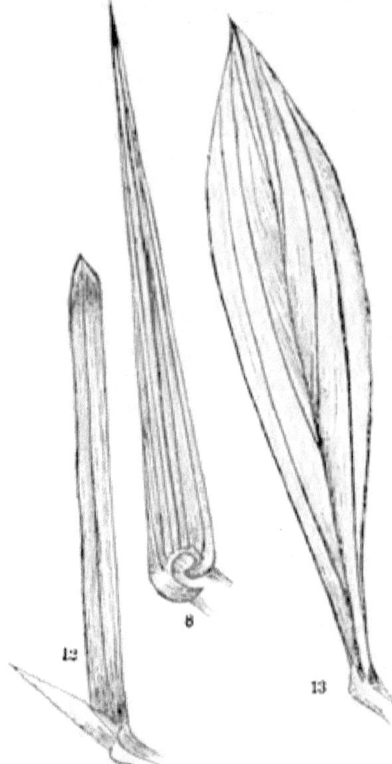

FIG. 8. Leaf-blade of *Arundo donax*, in which the fibro-vascular bundles, one after another, leave the mid-rib for the blade, and those along the margin terminate before reaching the apex. (Duval-Jouve.)

FIG. 12. Parallel-veined leaf of *Poa trivialis*. (Duval-Jouve.)

FIG. 13. Leaf-blade of *Panicum Crus-galli* (barn-yard grass), tapering each way from the middle. (Duval-Jouve.)

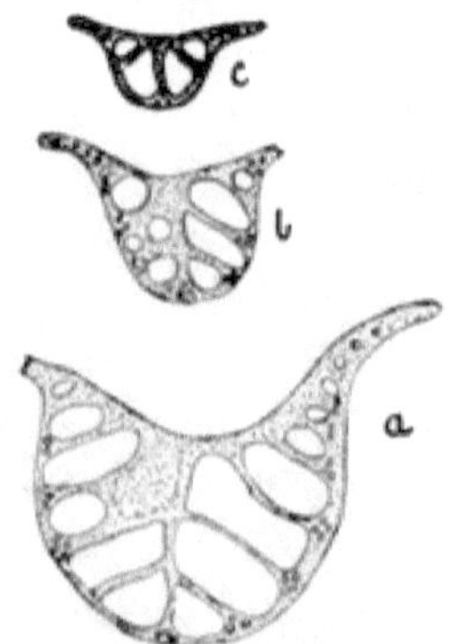

FIG. 9.—Cross sections of a large mid-rib of the leaf of *Zizania aquatica*; *a*, near the base; *b*, farther up near the middle; *c*, still nearer the apex, where most of the bundles have passed into the blade. 1×6. (Sudworth).

Some leaves are quite firm and remain green all winter, even with considerable cold and exposure, while others with a little protection, will remain green for a whole year. Most annual grasses and some perennials are very sensitive and quickly perish and fade on the approach of a frost. Some grasses will make growth at a low temperature, and start early in spring; others need more heat and start slowly. To a limited extent, the less moisture plants contain, the more cold they will endure without injury. When green leaves are exposed to severe cold, if the thawing be gradual, in many cases they will not be injured, but some plants quickly perish with frost, no matter how slowly it is removed.

Minute Structure of the Leaf.—The blade is traversed longitudinally by *fibro-vascular bundles*, which may be distinguished as *primaries*, those the most complete, and those less complete as *secondaries* and *tertiaries*. The bundle is reënforced by a nerve on the upper side of the leaf. That in the middle of the leaf is usually the largest, and is called the mid-vein, mid-rib, or keel.

At the base of a broad leaf, such as that of Indian corn, there is a large concave mid-rib, which contains many fibro-vascular bundles. Following the mid-rib towards the apex of the leaf, we shall see that one after another of these fibro-vascular bundles leaves the mid-rib and passes into the blade. The outer bundles in the lower part disappear in the margin of the leaf, the central ones only extending to the apex.

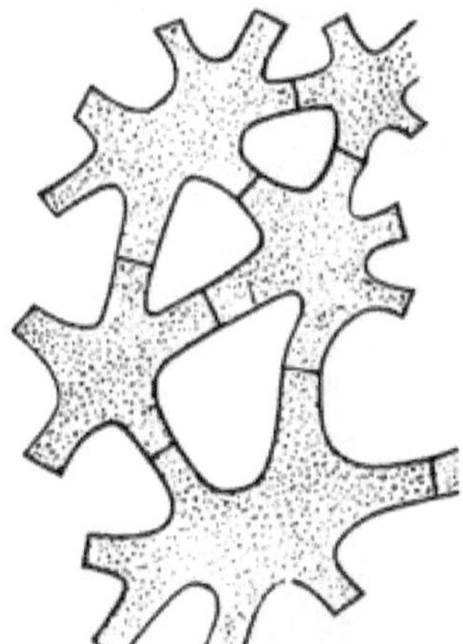

FIG. 10. Stellate cells from the mid-rib of some portions of the leaf of *Zizania aquatica*. 1 × 175.—(Sudworth.)

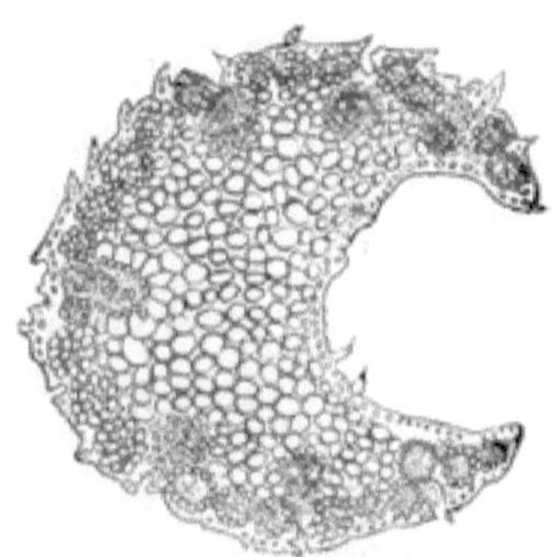

FIG. 11. Section of a leaf of *Andropogon* [illegible], where the whole blade is reduced to what answers to the channelled mid-rib of *Zea mays*. 1 × [illegible]. (Duval-Jouve.)

The blade of a leaf of *Poa pratensis* (June grass) and others, have veins which are exactly parallel, excepting very near the tip where there is an abrupt boat-shaped point

On viewing a thin, magnified transverse section of a mature leaf of *Sesleria cœrulea*, we see: an outer envelope of cells called the *epidermis*, *e*; *fibro-vascular bundles*, more or less developed: *b*, the median bundle, *h*, *h*, lateral bundles; groups of long, thick-walled cells in certain places beneath and next to the epidermis of the upper and lower sides, called the *hypodermal fibers*; *a*, the lower *median fiber*, *d*, the upper *median fiber*, *c*, *d*, lateral groups of *hypodermal fibers*.

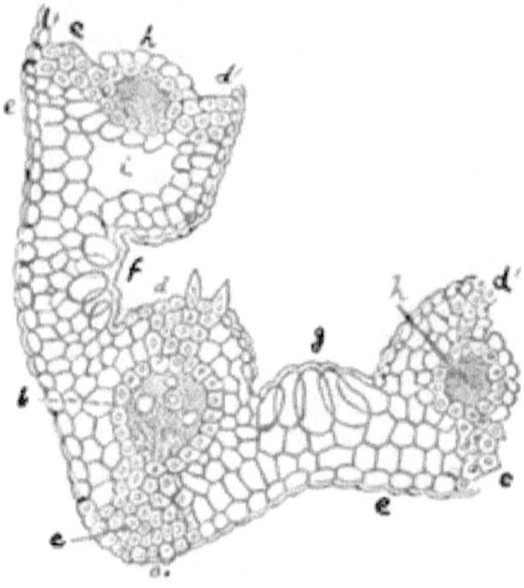

FIG. 11.—Part of a transverse section of a leaf of *Sesleria cœrulea* including the middle; *a*, middle hypodermal fibre; *b*, middle *fibro-vascular bundle*; *c*, *d*, lateral groups of *hypodermal fibers*; *e*, epidermis; *f*, *bulliform* cells, where the blade is closed; *g*, the same where the blade is spread open; *h*, *h*, lateral *fibro-vascular bundles*; *i*, air canal, *lacuna*. 1×120.—(Duval-Jouve.)

The other cells are parenchyma, most of which contain granules of chlorophyll. The vacancy is an air-chamber or canal, *lacuna*, *i*. In aquatic grasses these air-chambers are much larger.

The Epidermal System consists of:

a. Epidermis proper.

b. Bulliform (blister) cells.

c. Stomata.

d. Trichomes.

The Epidermis proper consists of a single layer of cells, the length of which seldom very much exceeds three or four times the width or thickness. The two latter dimensions usually are not very dissimilar.

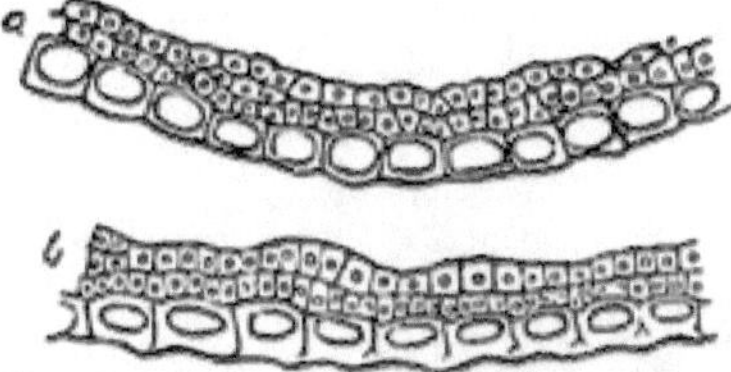

FIG. 15.—Sections of a leaf of *Festuca ovina* var. *glauca* (sheep's fescue); *a*, from a plant grown in the shade with plenty of moisture; *b*, from a plant grown in greater heat, with much light and little moisture. 1 × 180.—(E. Hackel).

The cells of the epidermis adapted to dry, hot climates are very thick, and the cells of those adapted only to moist air are thin, while the cells of the same species may vary much in thickness, depending on a greater or less exposure to light, heat, and moisture.

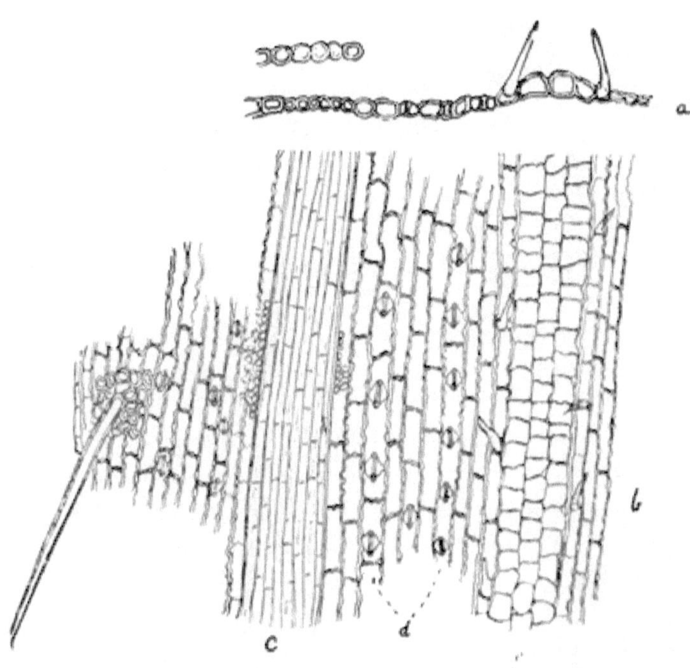

FIG. 16.—This gives some notion of the appearance of the epidermis of *Poa pratensis* (June grass); *a*, cross section ; *b*, seen from the upper side; *c*, over the hypodermal fibers; *d*, rows of stomata; *e*, bands of cells over parenchyma. 1 × 150. (Sudworth).

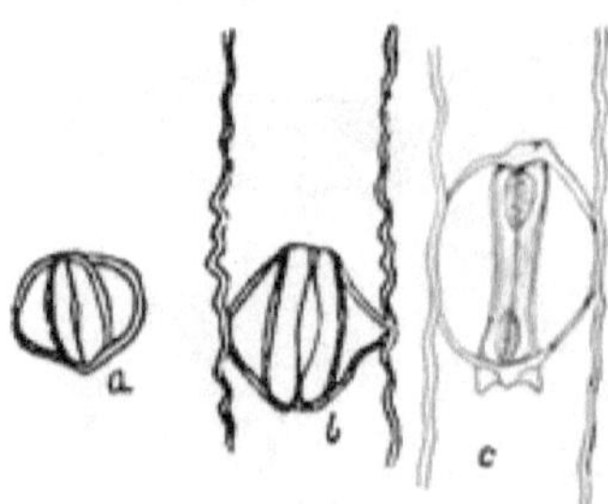

FIG. 17.—*a*, Young stoma of a leaf of Indian corn; *b*, older stoma; *c*, mature stoma. 1×350.—(Sudworth).

The Bulliform Cells are in longitudinal, parallel lines; they are larger and extend further into the leaf than ordinary epidermal cells.

Where the epidermis covers the *hypodermal fibers*, it consists of long, thick-walled cells, which are usually more abundant on the lower than on the upper surface. Sometimes they are reduced to two small rows, or rarely disappear entirely. Sometimes the *hypodermal fibers* cover all the lower side of a leaf, as in many species of *Festuca*. In such cases there may be none, a few, or many on the upper side, or it may be entirely covered excepting a few lines on the sides of the *veins* where the stomata are found.

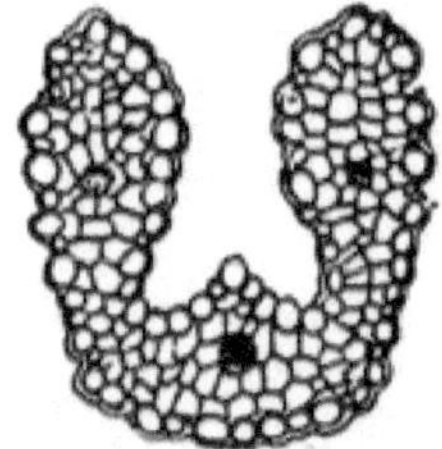

FIG. 18.—Showing a transverse section of a very simple leaf, *Chamagrostis minima* 1×50. (Duval-Jouve.)

The cells of the bands covering the *parenchyma* are larger than those which are over the veins or hypodermal fibres. On the upper surface of the leaf, these bands are often cut in two by bulliform cells.

The Stomata (small mouths) are in regular rows, placed longitudinally on certain parts of the leaf, and are always developed over a small cavity. The plan seems to be the same for all grasses. In some species the stomata are all above, in others all below, while some have them on both sides.

Trichomes.—Some single cells or groups of cells of the epidermis, extend and become *trichomes*, which are straight or curved, stout or feeble. They are real epidermal cells, and are not prolongations from the outer part of a cell, as is the case with a root-hair. They usually point to the apex of a leaf or stem, but in *Leersia* (rice cut-grass), they point downward, and become stout supporting hooks.

Tragus racemosus, *Amphicarpum Purshii*, *Panicum capillare* (hair grass) and others, have stout hairs on the margins of the leaf. On some smooth leaves, the hairs, when young, drop off and leave scars which alternate with the larger cells of the epidermis.

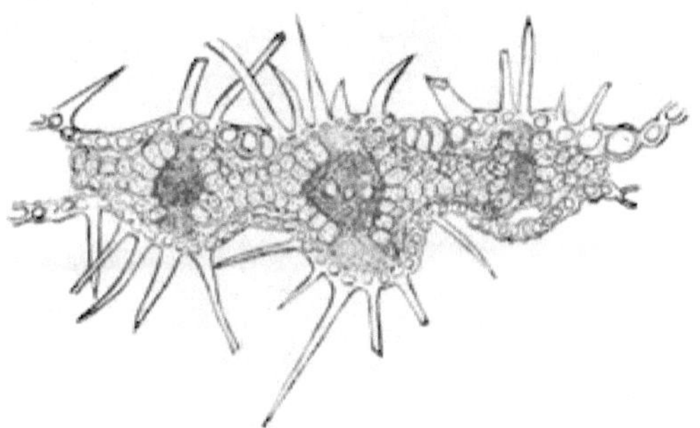

FIG. 12.—Part of a cross section of *Melica stricta*, showing many stiff hairs. × 34.—(Sudworth.)

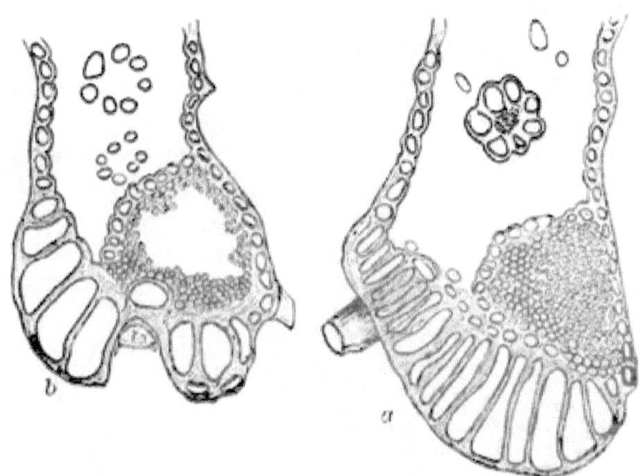

Bulliform Cells.—We will now consider more in detail the bands of bulliform (blister) cells which are larger than other cells of the epidermis, and have thinner walls. They have also

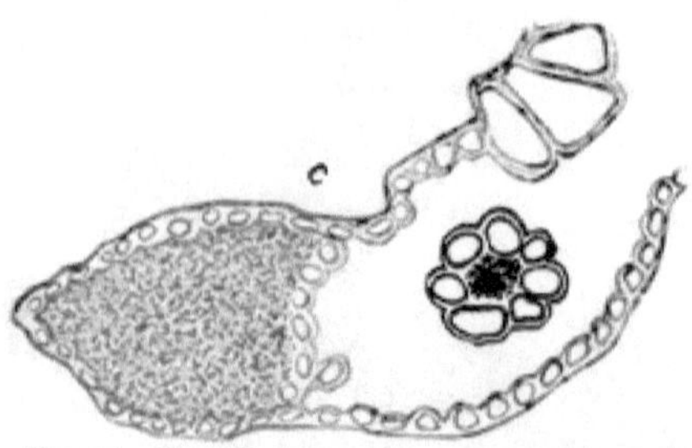

FIG. 20.—This illustrates the cross section at the margin of a leaf of *Amphicarpum Purshii*, shown in three places, at *a*, there is a growth of peculiar cells surrounding the base of a hair, at *b*, we have another view, and at *c*, where no hair is seen, the large group of hypodermal fibers is covered by an ordinary epidermis. 1 × 40. (Sudworth.)

been called *hygroscopic* cells. They are usually more or less wedge-shaped, with the point of the wedge towards the outside of the leaf. In *Zea mays* (Indian corn) these cells are raised above the other cells and puff out like a blister.

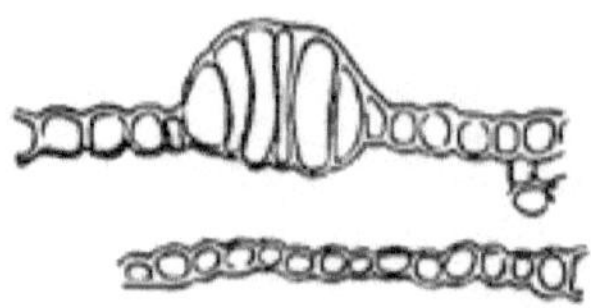

FIG. 21. A portion of a cross-section of a leaf of *Zea mays*, showing one band of bulliform cells raised above the surface. 1 × 17.—(Sudworth.)

When viewed on the surface of the leaf, the bulliform cells are usually seen to have the proportions of length and width much like those next to them. In some cases these cells are as long as wide, with outlines somewhat wavy.

The number of rows in a species is always the same, but varies with the species from 3 to 12 in a band. If there are many rows, the cells are shallow; if few rows, the cells are deep; if three only, those at the side are small, and the middle one is very large. The arrangement of these cells is invariable in a species, but in a genus they vary much. The following examples are given:

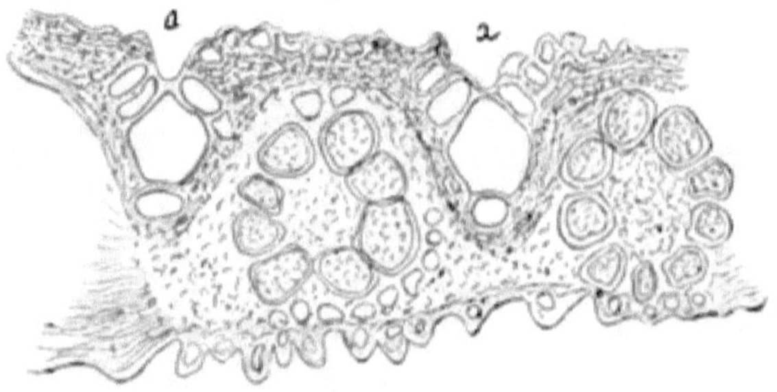

FIG. 22. Cross-section of a leaf of *Cynodon Dactylon*, showing a very large bulliform cell, with one or two small ones on each side of it. 1 × 130.—(Sudworth.)

1°. The leaves of *Dactylis glomerata* (orchard grass or cock's foot) have one median band of bulliform cells.

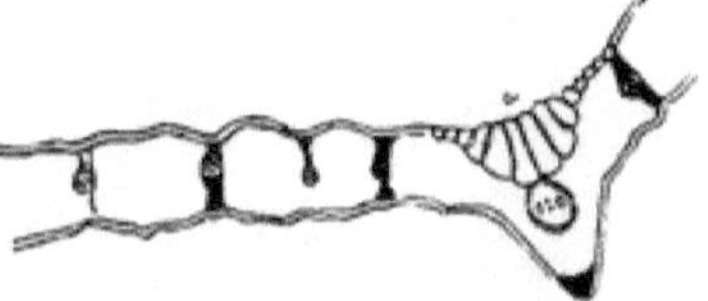

FIG. 23.—Cross-section of part of a leaf of *Dactylis glomerata* showing one band of bulliform cells on the upper side of the middle. 1 × 38.—(Sudworth).

2°. In *Chloris petræa* and others there is one middle band, and one or more on each side.

3°. The leaves of *Poa pratensis* (June grass) and some others have two bands, one each side of the middle.

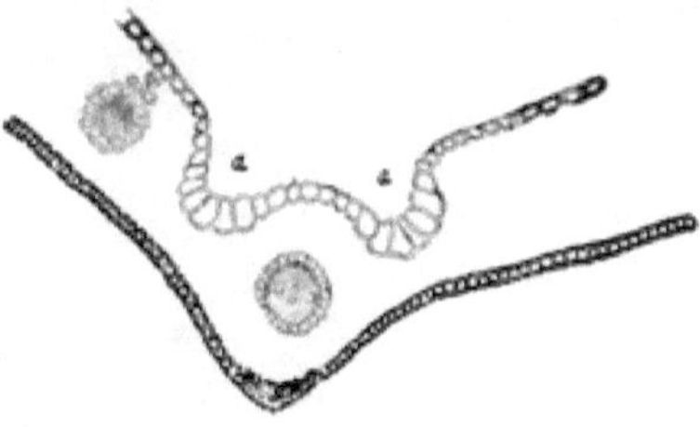

FIG. 24.—A cross-section of part of a leaf of *Poa pratensis* (June grass) showing one band of bulliform cells each side of the middle. 1 × 75.—(Sudworth).

4°. In case of *Andropogon squarrosum* and others there is one band each side of the middle and a small one at each edge.

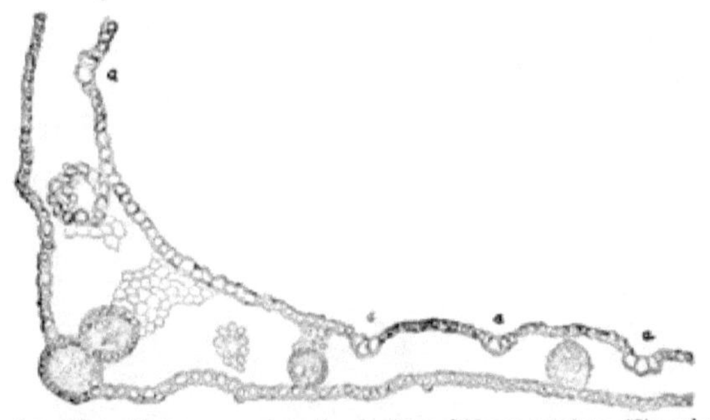

FIG. 25.—A portion of a cross section of a leaf of *Phleum pratense* (Timothy), showing bands of bulliform cells on each side the middle, and others between the veins. 1 × 20.—(Sudworth.)

5°. The leaves of *Phleum pratense* (Timothy), and many others have one band of several shallow cells each side the middle, and others between the veins.

6°. The leaves of *Zea mays* (Indian corn), and many others, have a band between each two primary bundles, and above each tertiary bundle.

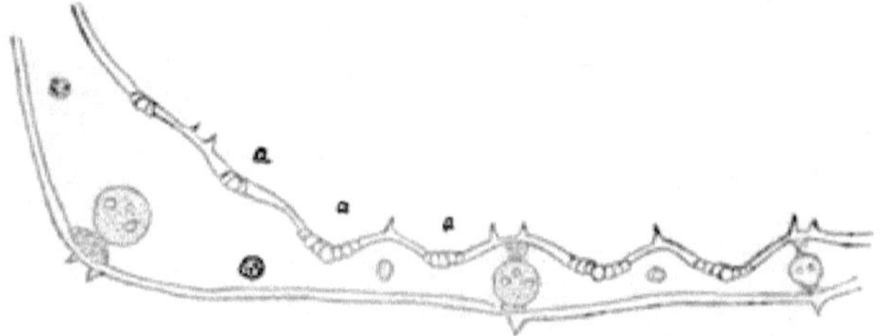

FIG. 26.—A section of *Festuca gigantea*, similar to the previous figure. 1 × 30.—(Hackel).

7°. The leaves of *Leersia oryzoides* (rice cut-grass), have numerous bands of bulliform cells on the upper surface, each side of the middle, and one band each side of the keel on the lower side.

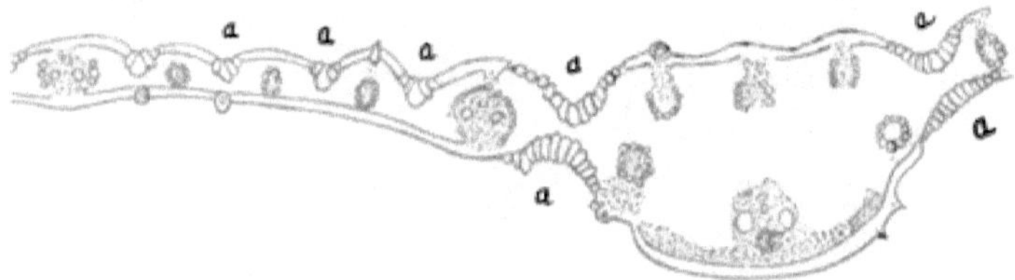

FIG. 27.—Transverse section of a leaf of *Leersia oryzoides* (rice cut-grass), showing lateral bands of bulliform cells on the upper side, and one lateral band below on each side of the keel. 1 × 350.—(Duval-Jouve).

8°. The leaves of *Amphicarpum Purshii* and others, have opposite bands of bulliform cells on both surfaces of the leaf, though those above are the most prominent.

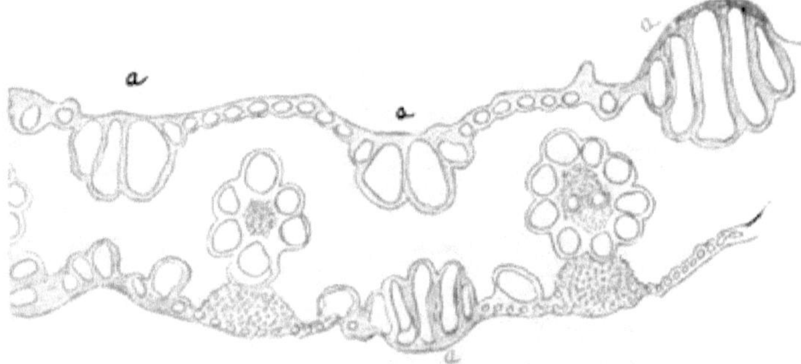

FIG. 28. Transverse section of a leaf of *Amphicarpum Purshii*, showing opposite bands of bulliform cells on both surfaces. 1 × 25. (Southworth).

9°. In the leaves of *Panicum plicatum*, the bands of bulliform cells are first on the upper side and then on the lower, and are found in the grooves.

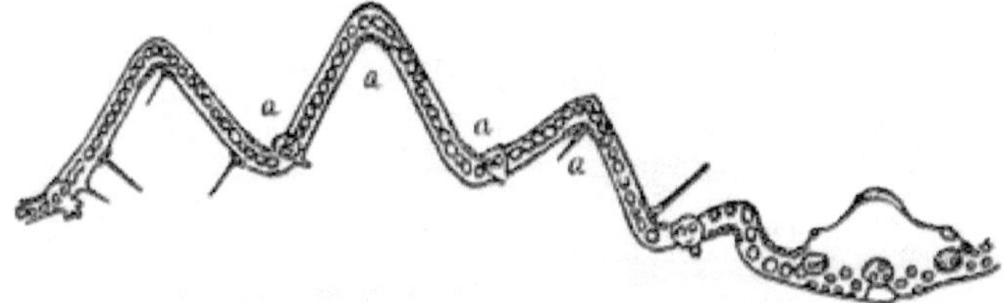

FIG. 29.—Transverse section of a blade of *Panicum plicatum*, in which the bulliform cells are alternately above and below. × 10.—(Duval-Jouve).

FIG. 30.—Section of a leaf of *Andropogon prinoides*, where the bulliform cells are evenly distributed, excepting over the veins. × 50.—(Duval-Jouve).

10°. In case of *Andropogon prinoides*, and other species, these cells are of nearly uniform size, and distributed all along the upper surface, excepting over the veins.

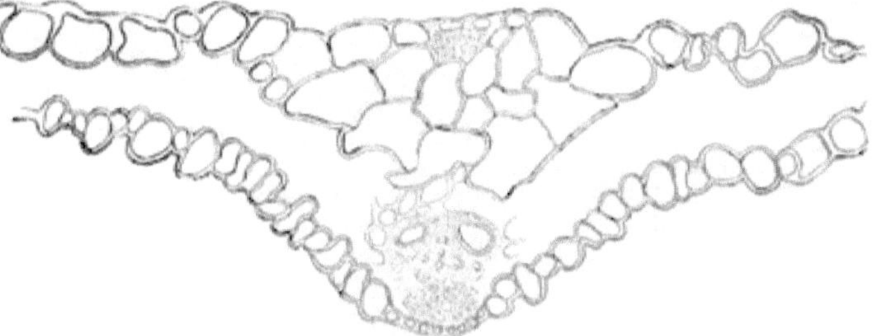

FIG. 31.—Section, through the middle, of a leaf of *Paspalum plicatum*, showing irregular epidermis and many bulliform cells. × 50.—(Sudworth.)

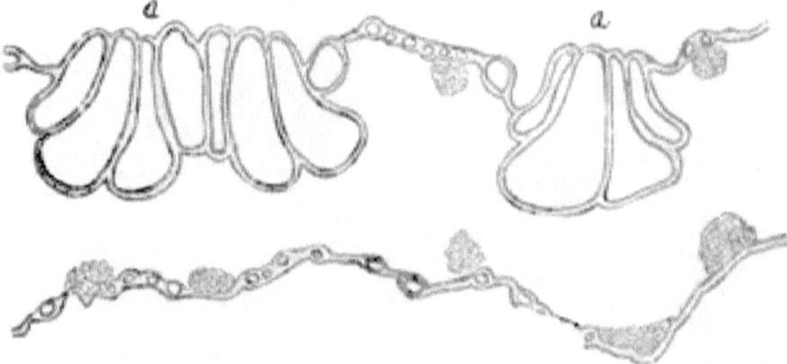

FIG. 32.—Part of a transverse section of a leaf of *Trachypogon polymorphus*, showing small epidermal and very large bulliform cells. × 50.—(Sudworth.)

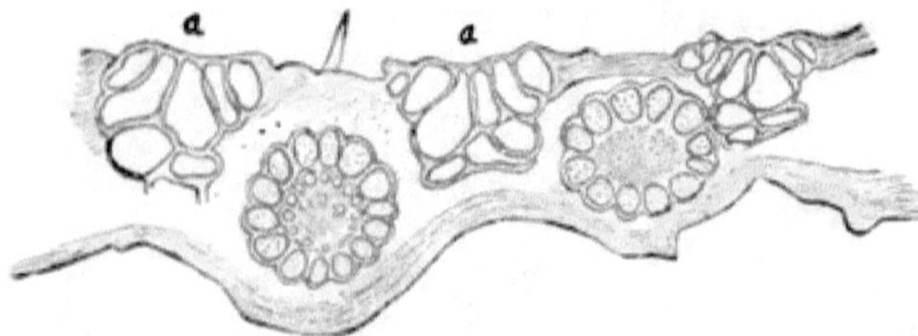

FIG. 33. Part of a section of the leaf of *Munroa squarrosa*, showing three groups of large bulliform cells, extending far into the blade. 1×50.—(Sudworth.)

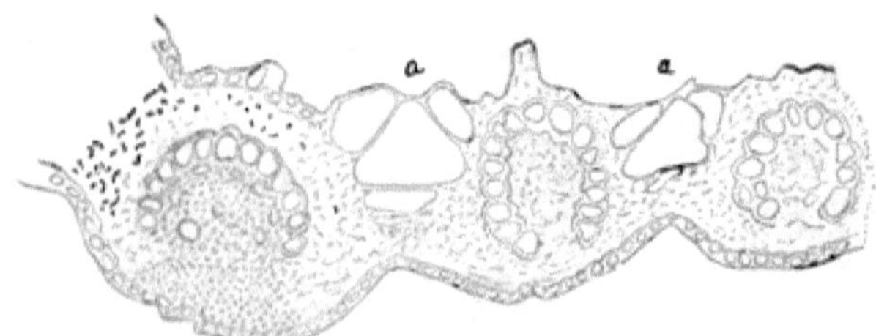

FIG. 34.—Part of a section, including mid-rib, of a leaf of *Cathestechum erectum*, showing two groups of bulliform cells extending two-thirds of the way across the leaf. 1×280.—(Sudworth.)

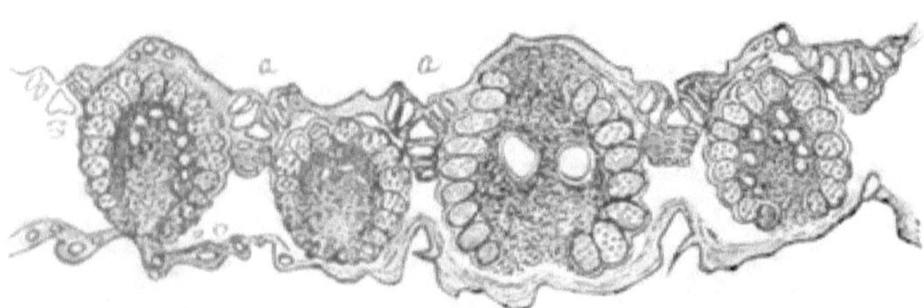

FIG. 35.—Section of part of a leaf of *Epicampes ligulata*, showing five groups of bulliform cells. 1×50.—(Sudworth.)

In vernation (while the leaf is very young in the bud) the leaves take the same positions as when full grown and dried, though the bulliform cells at that time, are very small, as we should expect. The very young leaves of *Dactylis glomerata* (orchard grass) and *Poa pratensis* (June grass) are conduplicate, while those of *Phleum pratense* (Timothy), are convolute. Some young leaves combine these two modes more or less, and may be conduplicate in the middle and convolute on the margins.

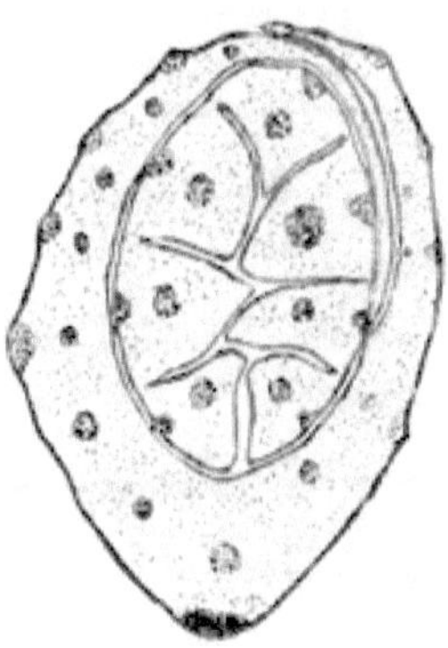

FIG. 36.—A cross section of young leaf of *Aira cæspitosa*, showing its mode of vernation within the sheath. 1×32.

The Movements of the leaves of some Leguminosæ are very quick, in most grasses they are quite slow, depending on the light and change of moisture. In the blades of grass the motion when drying, consists in the approach of the sides when conduplicate, or in rolling or unrolling when convolute, and in torsion when turning the lower surface to the sun.

In 1858, Mr. Duval-Jouve was surprised to see the leaves of *Leersia oryzoides* (rice cut-grass), move quite quickly, as he brushed them. The motion begins at the apex and extends down to the base, and is convolute when closed. Other species of this genus behave in like manner, as also does *Sesleria cœrulea*.

A drop of water on the section of the leaf of the latter causes it to expand instantly. The same is true of a leaf of *Poa pratensis* or of *Dactylis*. Some leaves open very slowly and then only when quite moist, as in case of *Lygeum spartenm*, or *Nardus stricta*.

The annual species of *Aira* and *Chamagrostis minima* are absolutely destitute of motion.

Many remain rolled up when dry, and unroll at night when

the dew is on, while others rarely ever open at all, but remain closed.

The leaves of *Leersia* (rice cut-grass) are most instructive with their bulliform cells above and below. These penetrate the blade deeply and make it very sensitive. In a warm day a brisk rub, or more than one between thumb and finger, causes it to close in a few seconds. After a short interval the leaf opens again, when it will be ready to respond to the same experiment.

The leaves of *Panicum plicatum*, when dry, close in a zigzag manner like a fan.

The bulliform cells of the leaves of *Phleum pratense* (Timothy) and *Alopecurus pratensis* (meadow foxtail) are not very large, and do not penetrate deeply. Such leaves are not good "rollers."

In case of leaves like *Sporobolus* and others, the bulliform cells are large, the groups numerous, and penetrate deeply. These leaves are likely to remain rolled up for a good portion of the time, unless the weather is very moist.

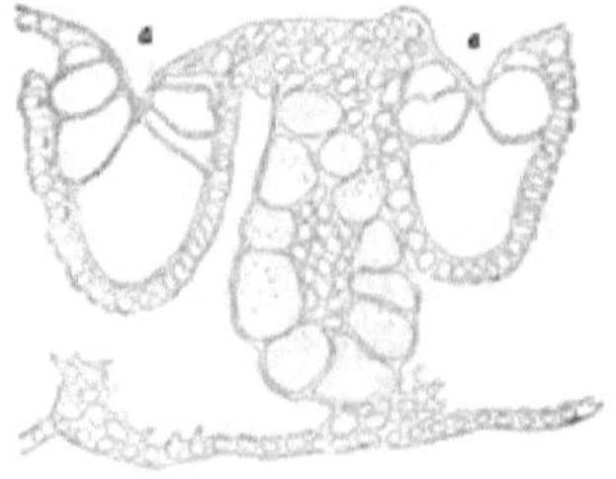

FIG. 37. Transverse section of a small portion of a blade of *Sporobolus cryptandrus* showing bulliform cells, in which there is a single large one, deeply penetrating and some smaller cells at the side. 1×175.—(Sudworth.)

FIG. 38.—Section of a leaf-blade of *Hierochloe alpina*. 1×24. (Sudworth.)

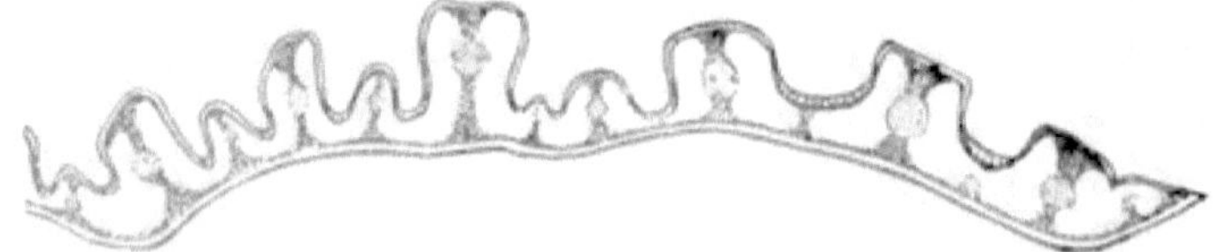

FIG. 39.—Section of a leaf-blade of *Stipa spartea* well adapted for closing in dry weather. 1×54.—Sudworth.

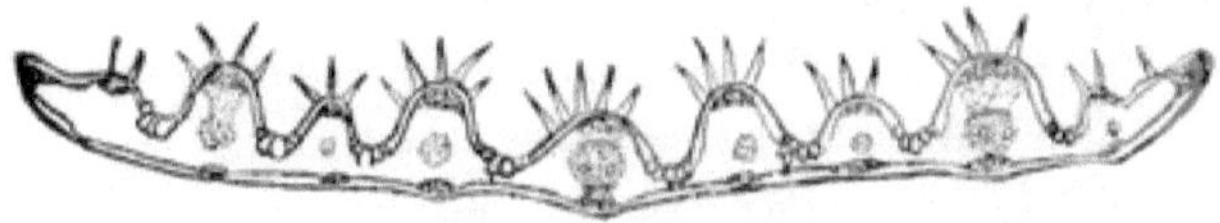

FIG. 40.—Section of a blade of a leaf of *Festuca rubra*. 1×30.—(Hackel.)

The object accomplished by the closing or rolling of the leaves is to cover one surface and assist in preventing excessive evaporation in dry weather.

The bulliform cells in their size, number, and arrangement may be used for critical specific characters.

Sedges, *Cyperaceæ*, often have one band of very large bulliform cells in the median line, and uniformly on the upper side.

These modes of arrangement of the bulliform cells is especially important in a physiological point of view, as they produce various motions of the leaves.

Fibro-vascular Bundles.—In all grasses the structure of these is much the same. There are two, rarely four, large pitted vessels, placed side by side near the middle of a bundle, at equal distances from the lower epidermis. The reader will here find it profitable to consult figure 5 for tracing out details.

Between these is a group of small reticulated cells, as many as fifty in *Festuca arundinacea*, or only two or three in *Panicum Crus-galli* (barn yard grass) and *Leersia oryzoides* (rice cut-grass). Above this group, towards the upper side of the leaf, and in a median line of the bundle is one or more annular or spiral vessels, situated near an air cavity, made by a breaking away of the cells.

On the opposite side, always on a median line, is a group of latticed cells or soft bast.

Surrounding all of the above is the *bundle sheath* formed of long, thick walled cells; and about the whole bundle is the thin-walled parenchyma of the fundamental tissue.

The bundles are not all developed to the same extent. The

primaries are the most complete and have all the elements: the *secondaries* have no annular vessels in the lacuna, and have the other elements much less pronounced: the *tertiaries* lack the lateral vessels, and are reduced to a slender cord of small dotted vessels and latticed cells, or only the latticed cells.

Not unfrequently there are very small transverse bundles running obliquely from one bundle to another. To see them entire, a longitudinal section must be made parallel to the epidermis.

Hypodermal Fibrous Tissue.—Usually this tissue is found in isolated groups just beneath the epidermis, and consists of very long, thick-walled cells, with overlapping, tapering extremities. There are no intercellular spaces. Sometimes these fibers are found at the margins of the leaf only; often opposite the fibro-vascular bundles and in contact with them on the lower side, but separated from them on the upper side by parenchyma.

They protect and strengthen the blade. In some cases they come together and make a continuous band on the lower side of the leaf, but never on the upper side.

In each triangular portion of a leaf of *Deschampsia cæspitosa* we find three fibro-vascular bundles, a large median one, and two small lateral bundles. Below each is a group of hypodermal fibers.

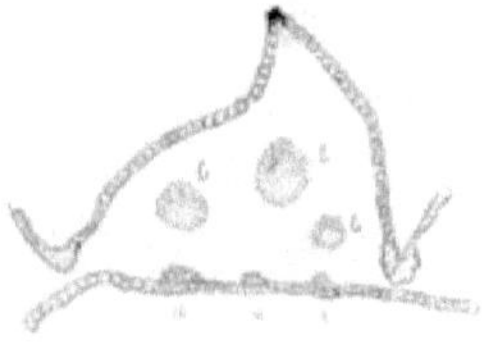

FIG. 41.—A transverse section of about one-seventh of a blade of a leaf of *Deschampsia cæspitosa*, showing one large and two small fibro-vascular bundles, with hypodermal fibers below each bundle. (× 50.—Sudworth.)

In *Stipa tenacissima* there are five fibro-vascular bundles in one nerve.

As examples of hypodermal fibers, we have:

1°. A mere trace in the median line of the blade;

2°. A group at the keel of the blade and one at each margin;

3°. Groups, as in the latter case, with others in certain places on the lower side, or with a continuous layer on the lower side;

4°. Groups above and below the primary bundles only;

5°. Groups above and below each bundle, but not continuous;

6°. Groups above and below each bundle, and contiguous;

7°. Groups covering the mesophyll, except some cells bearing chlorophyll on the sides of the nerves.

The first three of the above are conduplicate in vernation, and the fourth includes all of the species of *Andropogon* and *Panicum*, except *P. plicatum*. So far as the development of hypodermal fibers are concerned, *Chamagrostis minima* and *Stipa tenacissima* are extremes. The former is illustrated by figure 18, and figure 14 will answer as a substitute for the latter.

In aquatic and in annual grasses these fibers are feebly developed, while those grown in extreme dry, hot countries are remarkable for the development of this tissue. Upland grasses grown in the shade, with an ample supply of moisture, have their woody fibers feebly developed.

When this tissue is well developed it helps prevent the free evaporation of moisture. The closing of the stomata also helps to retain the moisture.

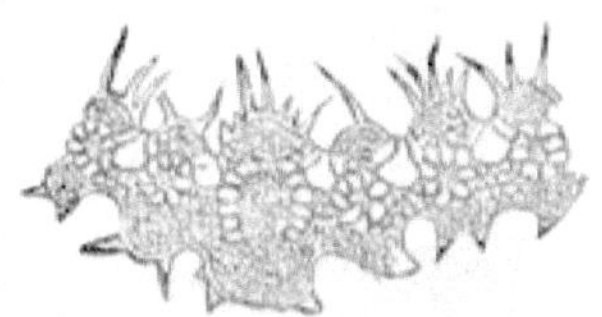

FIG. 42. Section of a leaf of *Pappophorum scabrum*, with well developed hypodermal tissue. 1×50.—(Duval-Jouve.)

FIG. 43.—Section of a blade of a leaf of *Festuca ovina* var. *laevis*, with a group of hypodermal fibers below the mid-vein, and one at each margin of the leaf. 1×30.—(Hackel.)

FIG. 44. Section of the blade of *Festuca ovina*, with hypodermal fibers extending over the lower side. 1×30.—(Hackel.)

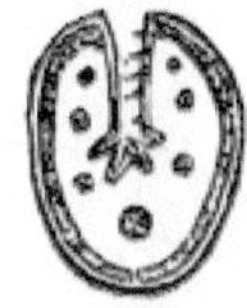

FIG. 45. Section of a leaf of *Festuca ovina* var. *duriuscula* (hard fescue), with hypodermal fibers extending over the lower side. 1×30.—(Hackel.)

In the last three the bulliform cells are wanting or only feebly developed, and the blades remain closed or nearly closed even when mature.

Parenchyma of the Leaf.—This is a name applied to all the rest of the leaf-blade after taking out the epidermis, the fibro-vascular bundles, and the hypodermal fibers. It presents three forms, which are quite distinct:

a. Cells containing chlorophyll and found in the leaves of all grasses without exception.

b. Cells without color inside, found in certain species only.

c. The star shaped and branching cells found in the air canals of species (Fig. 10.) more or less aquatic.

The chlorophyll-bearing parenchyma is of two sorts:

a. Where the grains are rather large and compact.

b. Where some of the chlorophyll is in the form of grains, and some of it is diffused more or less like jelly.

Where a part of the chlorophyll is more or less diffused, the rest is in cells which form concentric cylinders, or the cylinders may be open in one or two places.

FIG. 46. Section of a blade of *Bouteloua Havardii*, showing some closed and some open cylinders of cells containing grains of chlorophyll.—(Sudworth.)

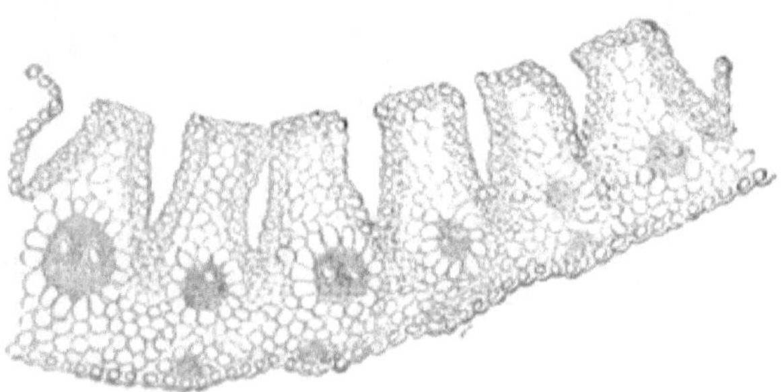

FIG 47.—Section of part of a blade of *Spartina stricta* var., showing large cells of parenchyma destitute of chlorophyll: these are situated above the fibro-vascular bundles, and in the middle of the lobes which extend upward. 1×34.—(Sudworth.)

The cells of parenchyma, which contain chlorophyll, reach their maximum in species which grow in cool, shady places.

FIG. 48.—Cross-section of a blade of *Spartina juncea*, in which the upper surface is deeply furrowed. 1×34.—(Sudworth.)

The Torsion of Leaves.—The leaves of most flowering plants quite uniformly turn the upper surface to the light and keep the lower surface in the shade. This rule does not hold good with the grasses nor with quite a number of others, such as *Typha* (cat-tail flag) and *Gladiolus* among endogens; and some species of *Liatris* (blazing star), and others among exogens.

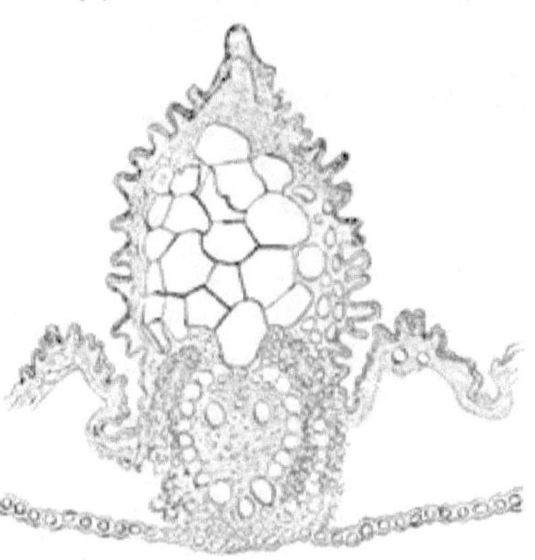

FIG. 49.—Cross-section of the central part of a blade of *Spartina juncea*, showing above the midrib, a remarkable enlargement, which is occupied by large cells of parenchyma, destitute of chlorophyll. 1×100. —(Sudworth.)

In half or more of the grasses examined, the whole or a majority of the leaves, by a twist of the lower portion of the blade, turn "wrong side up," and expose the "lower side" to the sunlight. In most other cases, we have seen that during the warmest and dryest weather, when the sun's rays are the most trying to the life of the plant, the leaves shut up or roll up, leaving the under surface alone exposed. Whether right side up or wrong side up, the surface most exposed generally possesses the firmer epidermis.

Young leaves of *Phleum pratense* (Timothy), several species of *Bromus* (chess), *Triticum* (wheat), and *Agropyron* (quack grass) *Secale cereale* (rye), and others, twist once or more with the sun, or in the direction which they would twist were the sun the cause of torsion.

Young leaves of *Avena sativa* (oats) and *Setaria glauca* (pigeon grass) quite uniformly twist against the sun, while those of *Poa pratensis* (June grass) and *Panicum capillare* seem indifferent as to the direction in which they twist. The sun does not seem to dictate the direction of the twisting.

The leaves of grasses generally twist best and with greatest uniformity when young, even though they are much shaded from the rays of the sun.

Many leaves twist most towards the apex, while others twist most, or entirely, at or near the base of the blade.

The margins of many leaves grow a little longer than the central portion, and if the mid-rib is not very prominent, this will produce torsion of the blade. In cases of Indian corn, the margins of the older leaves are often longer than the mid-rib, but there is no torsion. The margins are undulating. With a light mid-rib and stouter margins, the leaves of this plant would show torsion. Probably one reason why most of the torsion is towards the apex of many leaves is because the mid-rib is not very strong at that part of the leaf.

When young and quite erect, the lower side of many leaves seems to grow a trifle faster than the upper side, and this perhaps tips the leaf over "bottom side" up.

Duval-Jouve believes that torsion of the blades of grasses depends on the distribution of the fibrous tissue. In dry weather this tissue contracts least, so the blade twists.

In some the air canals, *lacunæ*, let in dry air, which contracts the delicate cells of parenchyma. The writer has not yet been able to find the reason for a uniformity in the direction for the torsion of the leaves of any species of grass.

Generic and Specific Characters in the Leaf.— E. Hackel, in his *Monographia Festucarum Europæarum*, says: "The histological characters of the leaf-blade unquestionably include those most important for the discrimination of the forms of Festuca, but the degree of constancy or value of each character must first be determined."

By experimenting he claims to have found a solid foundation for the estimation of these characters.

He finds the mesophyll and fibro-vascular bundles quite uniform with all sorts of treatment of the plants, but the epidermis offers remarkable differences, especially that on the lower side of the leaf. This difference is apparent in the thickness of the outer walls, the size of the cavities, and the existence or absence of projections on the partition walls. The dry, cultivated plants had their epidermis strongly thickened toward the outside, the cavities diminished, and over the partition wall had developed cuticular projections. The moist cultivated plants produced slightly thickened epidermis cells, broad cavities, and no trace of cuticular projections.

The sclerenchyma or bast, or hypodermal fibers, varies much with different soils and amount of moisture. Species of moist, shady habitats, show in their leaves a clear preponderance of the assimilating over the mechanical system.

In very many respects, it will be seen, that a critical study and close comparison of the leaves of grasses will reveal a wonderful variety in their structure and cannot fail to excite the admiration of every student. In certain portions of the preceding account of the leaf, the writer has followed Duval-Jouve.

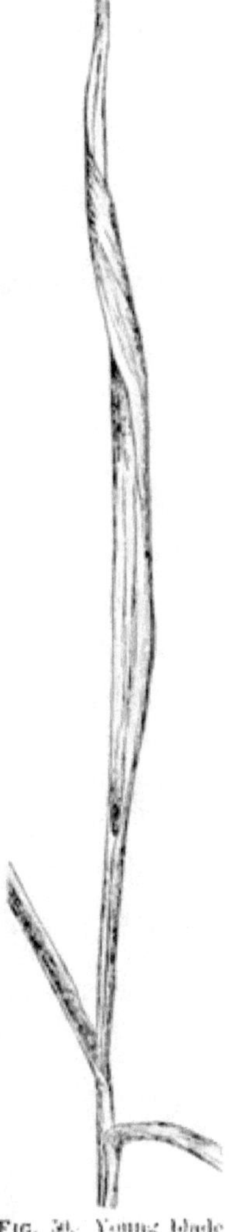

FIG. 30. Young blade of *Triticum vulgare* (wheat) twisting with the course of the sun. Reduced ⅛. (Sudworth.)

Fig. 51.

The Bracts and Flowers.—The grasses form a natural order which is one of the easiest to learn to recognize, but for this very reason it is generally difficult to distinguish the several species.

The best characters for describing grasses are found in connection with the bracts, flowers and the ripened ovary.

A great diversity of views has been entertained by leading botanists in relation to the morphology of the flower and the names to be given to each part.

According to our best modern authorities, including Bentham, Hooker, Gray, Sachs, Munro, and Döll, the three outer scales constitute no part of the flower, but answer to bracts.

Morphology of the Bracts and Flowers.—The following is a full abstract of an able essay on this subject, by the late Geo. Bentham, and is taken from the Transactions of the Linnean Society:

The terminology adopted by botanists has been very unsettled and repeatedly modified. The absence of all homology between the so-called sepals in grasses and those of perfect flowers has been repeatedly demonstrated. "Some years ago, when preparing my *Handbook of the British Flora,* I purposed following Kunth, but I was soon brought to a standstill by the anomaly of the spikelet of *Milium* being described as having two flowers

---

Fig. 51.—A plant of *Poa pratensis*, L. (June grass). At 1, a small plant, with roots, root-stocks, leaves, culm and flowers; *c*, part of a sheath of a leaf with a white *ligule*, above which is part of a blade; *a*, *spikelet*, closed, containing four *florets*; *b*, spikelet spread open, containing five florets, as seen when in flower; the lower scales as seen in *a* and *b*, are the *empty glumes*, *c*, a *floret*, with *floral glume* at the right, *palea* at the left, including three stamens; *f*, cross-section of the *floral glume* which is 5-ribbed, and keeled; *d*, a pistil with the ovary below bearing two short *styles*, each terminated by a feathery *stigma*; at the base on each side is a *lodicule*. (Scribner.)

According to Robert Brown, the two lower scales of *a* and *b* are the *glumæ*, and constitute an *involucre*. They are the *empty glumes* or *basal glumes* of many authors; *gluma* of Dumortier; *tegmen* of Palisot de Beauvois. According to Robert Brown and Jussieu, the two scales at *c*, are the *paleæ*, and represent the sepals; *glumellæ* of Dumortier; *stragula* of Palisot de Beauvois; *perianthium* of authors. According to R. Brown and Jussieu, the right hand scale in *c* is the *lower or outer palea*, *gluma fertile* of Germain de St. Pierre; flowering glume of Bentham, Hooker, Döll.

According to R. Brown and Jussieu, the blunt scale at the left in *c* is the *interior palea*, *paleola interior* of Dumortier; *spathella* of Döll. According to R. Brown and Kunth, the small scales at *d* are the *squamulæ*; *lodicules* of Bentham and others; *nectaria* of Schreber; *glumellulæ* of some authors. By many, these scales were thought to represent petals.

and one glume, when I could not expect any of my readers to see more than one flower with three glumes.

"After carefully examining a great variety of genera, and comparing them with the nearest allied orders, it appeared to me that no distinct and universally applicable definition of the term glume could be given unless it were applied, as in *Cyperaceæ*, to the whole of the primary scales attached to the main axis of the spikelet. After printing, I ascertained that similar views had been independently propounded by Hugo, Mohl, Döll and others in Germany, and by Germain de St. Pierre, in France."

In several of our large genera of grasses, the only difference between the one or two outer empty glumes and the flowering ones is that they are rather smaller or rather larger, and there is often more difference between the first and second empty glumes than between the upper empty glume and the first flowering one. In couch grass the empty and flowering glumes are precisely similar, very gradually diminishing in size from the outer empty to the uppermost flowering glume. An empty glume in one spikelet may correspond to a flowering one in another spikelet of the same plant. In rye-grass the spikelets are alternately placed in one plane, right and left, the single empty glume of each spikelet being the lowest and outer one, whilst the second glume next the axis of inflorescence, is the lowest flowering one. In the uppermost spikelet there are two empty glumes, and this is not owing to the development of an additional outer glume, for the lower of the two empty ones is on the side it ought to be in the regular alternation with the lower spikelets, but the second glume, which in the lower spikelets encloses a flower, is in this subterminal one empty. So in several Paniceæ, the second or third glume, according to the genus or species, has been observed sometimes, to enclose a rudimentary or male, or even a perfect flower, and at other times to be quite empty, without any change in its appearance.

In *Panicum*, according to the Kunthean terminology, the first minute scale is a glume, the second, many times larger, is also a glume, the third, often precisely similar to the second, is not a glume, but a flower, and the fourth, whether similar or more or less dissimilar, is a part of a flower. In some gramineæ there are additional empty glumes, usually small and often different in form, either immediately below the flowering ones, as in *Anthoxanthum* and *Phalaris*, or at the end of the spikelet, as in *Melica*. These have no pretensions to be flowers at all. In some genera, as in *Uniola*, from three to six of the lower glumes are empty, and precisely similar to each other, and yet we are only allowed to call the two lowest ones glumes, the others are termed flowers. We are not even allowed to define glumes as the two lowest scales of the spikelet; for that of *Leersia*, which has two glumes, one empty, the other flowering, is described as having no glumes but two flowers. In *Kyllinga* and *Courtoisia*, in Cyperaceæ, where the fruit is similarly enclosed in two glumes, they are correctly described as such, one empty, the other a flowering one.

The so-called upper palea is neither homologous nor similar to the so-called lower palea or flowering glume. It is inserted on the axis of the flower, and not on that of the spikelet, as may be seen in cultivated wheat. It is differently shaped, and having instead of one central rib or keel two prominent nerves, it is generally supposed to be a double organ composed of the union of two scales. These two scales are probably the homologues of the two bracteoles of *Hypolytrum* and *Platylepis*. It is convenient to designate them by a special name, for which the generally received term palea is not inappropriate, and commits one to no special theory in regard to it. "It appears to me that flowering glume and palea is not more cumbrous than the deceptive one lower palea and upper palea."

The two or rarely three small scales above the palea and

alternating with the stamens in most grasses, have been supposed to represent a reduced perianth; but their homology is not satisfactorily demonstrated.

"To sum up, therefore, the spikelets of *Gramineæ* may be described as composed of a series of alternate *glumes*, distichously imbricated along the axis. To be really useful, descriptions should be clear and intelligible, and enable the reader to identify the plant. He should describe only what he actually *sees*, not what it may be theoretically imagined he ought to see."

The empty glumes are often more or less boat-shaped, and with the one to many flowers which they include, constitute a *spikelet, spicula* or *locusta*. One or both empty glumes may be absent in certain cases. The spikelets are arranged in panicles, racemes, spikes or heads.

The floral glume usually resembles the two empty glumes in having a midrib with an equal number of ribs on each side, while the *palea* often has two ribs, with a thin membrane between which is often notched at the apex.

It is of much importance in describing grasses to observe the relative lengths, sizes, shape, number of ribs, the nature of the awn, and the texture of the glumes and palea.

The midrib of one or more glumes often extends upwards from the apex into an awn, and in case of the floral glume, the awn sometimes starts from a notch at the top; sometimes from the back below the apex, and is then said to be *dorsal*.

The lower part of the awn is often twisted when dry, but straightens when moist. If the lower part twists, the upper part inclines at an angle.

The glumes and the palea probably represent the sheaths of leaves, and where an awn exists it sometimes represents the blade of a leaf. This is quite well shown in *proliferous* flowers of grasses, as seen in Figure 52, a proliferous floral glume of *Phleum pratense* (Timothy).

We say flowers are *proliferous* when either the glumes, palea, stamens or ovary, or all of these develop into small leaves in place of flowers. This is not uncommon in Indian corn and Juncus. The bulblets of onions or "onion sets" are familiar examples.

Fig. 52. Proliferous floral glume of *Phleum pratense* (Timothy), with a portion representing the sheath and a portion representing the blade of a leaf, slightly enlarged. —(Sudworth.)

In this connection a reference to figure 53 will show several forms of *ovaries and styles*, and impress the reader with the importance of examining these minute and delicate organs for generic and specific characters.

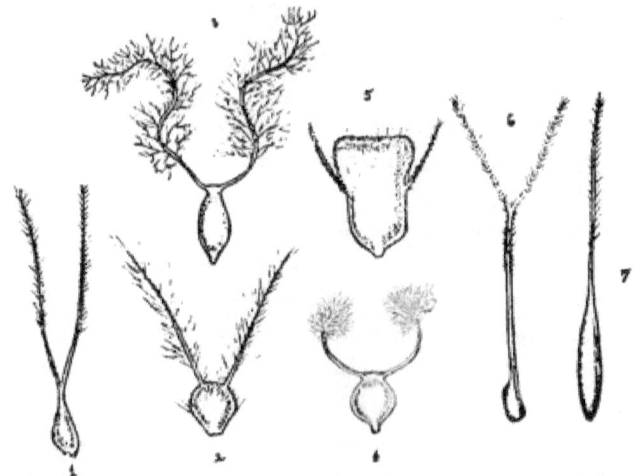

Fig. 53. *1*, Pistil of *Mibora minima*; *2*, Pistil of *Arrhenatherum avenaceum*; *3*, Pistil of *Glyceria aquatica*; *4*, Pistil of *Melica uniflora*; *6*, Pistil of *Bromus mollis*; *6*, Pistil of *Alopecurus pratensis*, meadow fox tail; *7*, Pistil of *Nardus stricta*. All magnified.—(From *Agrostographia synoptica* by Kunth.)

Fertilization of the Flowers.—When the flowers arrive at a certain stage of growth, the stigmas are ready to receive the pollen, which sends a miniature thread down the style to the ovule. The pollen of grasses is in the form of round, smooth cells, and escapes readily. The flowers of grasses, except where close fertilized, are usually *anemophilous*, fertilized by the aid of the wind. In a few cases, insects visit the flowers quite regularly for pollen, and most likely render aid in the fertilization. The writer has several times seen large numbers of honey bees, early in the day, gathering the pollen of *Festuca arundinacea*.

Buchloë (buffalo grass) is an example of those which are diœcious, and of course the flowers are all cross-fertilized. Indian corn, *Zizania* (wild rice) and *Tripsacum* (gama-grass) are monœcious and are very likely to be crossed. In some cases of Indian corn, and probably it is so with some other species, the flowers are *protogynous*, i. e., the pistils come forth a day or more in advance of the anthers. In other cases, as for example, sweet vernal grass and meadow fox tail, the flowers are *proterandrous*, i. e., the anthers mature in advance of the pistils. In either plan, cross fertilization is secured.

The spikelets of *Arrhenatherum avenaceum* (tall oat-grass), and others, contain a staminate or sterile flower to every perfect one, and the flowers of *Hierochloa borealis* (vanilla grass), are two of them staminate to one that is perfect. The use of these staminate flowers can only be for crossing. Some cultivated plants of *A. avenaceum* bear only staminate flowers.

In many cases where the flowers are perfect, the stamens shed their pollen before the stigmas are ready, or the reverse is the case.

In some instances the stamens and pistils appear to mature at the same time, as in most, if not all, sorts of cultivated wheat, barley, oats and rye. In the three former, the glumes and paleæ usually closely cover up the stigmas till they are fertilized or covered with pollen. The glumes of rye spread so that cross-fertilization may take place. *Amphicarpum*, *Oryza clandestina*, some species of *Hordeum* and *Cryptostachys*, and most likely others, produce fertile flowers below ground, and are called *cleistogamic*.

There appears to be no fixed rule with regard to the fertilization of the flowers of a genus.

There are instances among plants in which the flowers of the same species are fertilized in a different manner in different seasons and in different countries, and certain specimens of a species are fertilized in an exceptional manner during the same season or in the same neighborhood.

As a rule, a certain specified flower of a grass remains open only for a very short time, but different flowers of a plant may appear at successive periods, extending over eight days, more or less, in Indian corn; seven days, more or less, in Timothy, several days in oats and wheat, and for a much longer period in branching grasses like *Eragrostis* and *Muhlenbergia*.

As an example of the fertilization of grasses, we find the following, by A. S. Wilson, in an admirable, illustrated paper on "Fertilization of Cereals," in the Gardeners' Chronicle, for March 1874, and February, 1875:

"From the time at which the ears, or part of the ears, of the four European cereals, wheat, rye, barley and oats, appear above the sheath, till the time of flowering, the styles and the anthers remain in nearly the same position. During this time the filaments are of such length as to place the lower ends of the anthers in contact with the upper part of the ovary, while the styles lie embraced by the anthers, the whole being straight and running in the same direction as the axis of the closed pales. If a floret of two-rowed barley is held up between the eye and the light before fertilization has taken place, the anthers will be seen

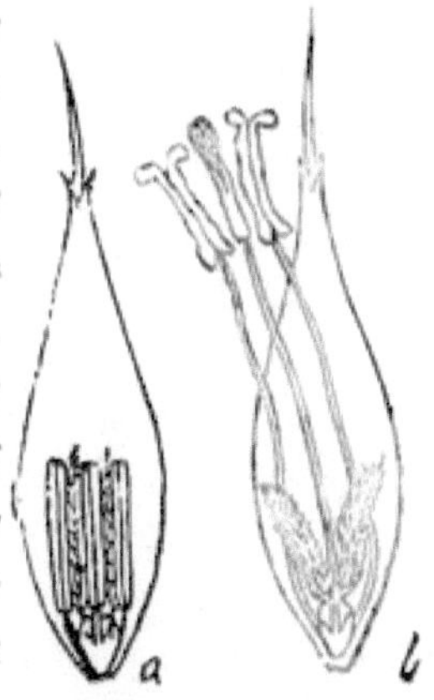

Fig. 54.—Flowers of wheat, *a*, young; *b*, older. Enlarged. —(Gardeners' Chronicle.)

through the pales lying in their original position, and if the flower is then opened and inspected, it will be found that the anthers are still unopened, and still retain their bright yellow color. But if, on looking through the semi-transparent pales, the anthers are seen in the upper part of the cup, fertilization has taken place, and if the floret is opened the anthers will be found open, with the pollen scattered about on the feathers and inner surfaces of the pales, and the bright color of the anthers passing away. The inner pale in this form of barley is so tightly embraced by the overlapping edges of the outer pale as prevents further opening.

"The different varieties of wheat, so far as known to the writer, observe conditions of opening the flower similar to those of the barleys. Many wheat florets never open so far as to give room for the egress of the anthers. Some open so far as to allow one or more anthers to get half out, in which position they are caught and held by the reclosing of the pales. In many the anthers are wholly retain d, but the general rule is for the floret to open so far as to throw out the anthers.

"Opening of the cereal flowers takes place at all hours of the day. I have obs rved that it also takes place in all kinds of weather, wet or dry. I have observed that spelt flowers open in the morning before the sun touched them. I have also seen them open in a dead calm after sunset, many of them had opened and closed within an hour previously. I have likewise seen wheat and spelt flowers open during heavy rain, and in dull, cloudy weather. Fertilization seems to take place when the flower is ripe, independently of any particular state of the weather. In respect of all florets which do not open so far as to eject their anthers, the falling of rain or the blowing of the strongest wind is perhaps a matter of indifference. The opening of the flowers may be induced by handling the ear in a gentle way when the natural time of flowering has nearly arrived. I have seen eleven rye florets throw out their bright yellow anthers at the same time on one spike, by simply drawing them through the hand.

"Break off a barley floret from an ear which is just coming into blossom: open the pales gently, and put it under a low magnifying power. Presently a slight tremor takes place. The anthers begin to move upward. The filaments are visibly growing before the eye at the real rate of six miles an hour. The anthers get more and more distended. They are now half way up the unpretending green chalice. Observe the little slit commencing near the apex of the most advanced. Out darts a little spurt of tiny bullets. Presently the next and the next opens. Instantly another and another spurt of tiny bullets are sent dancing from each half-open suture over the enclosing sides of the pales, or down upon the spreading feathers. Now and then a solitary ball bounds out of the opening cavity over the plain in front of it. In various wheats and spelts the points of the feathers are frequently thrown outside, where they are sometimes fixed permanently by the reclosing of the valves. But the rule in wheat, oats and barley is, not to expose the feathers. These are fertilized before the anthers are visible outside. By estimate, a single anther of rye contains 20,000 cells of pollen, and an acre of rye produces 200 lbs. of pollen."

The Caryopsis or Grain, as will be seen, is the ripened ovary which is closely filled by the seed.

Here, also, the reader should consult figures 55 and 56 to notice the structure of a *caryopsis* or grain and its germination.

The Seed is a miniature plant in its simplest form, as Prof. W. W. Tracy says, "packed ready for transportation," and supplied with concentrated food destined to nourish the young plant till it form roots and leaves.

As the young chicks feed upon the yolk of the egg, so the young grass-plant subsists on the starch stored up in the endosperm. The starch in the plant takes the place of milk to the colt, calf, or pig. The milk is secreted by the mother animal; the starch was formed in the leaves of the mother plant and deposited in the seed for future use of the seedling.

As the water ram needs some water to move it, to enable it to send some of the water higher, so the young grass-plant throws away, if we may use the expression, some of its substance to enable it to organize the remainder into roots and terminal bud. During the growth of seeds and bulbs in the dark, the actual dry weight is diminished, although the size may increase.

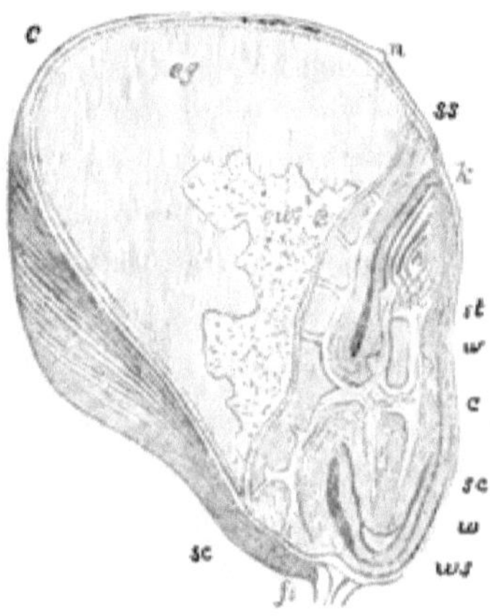

Fig. 55.—Longitudinal section of the grain of Indian corn; *c*, thin wall of the ovary; *n*, remains of the lower part of the style, known as the "silk;" *fs*, base of the grain; *eg*, *ew*, endosperm, which feeds the young plant as it germinates; *sc*, *ss*, scutellum or cotyledon of embryo; *e*, its epidermis; *k*, plumule or terminal bud; *w*, (below), the main or primary root; *ws*, the root sheath; *w*, (above) adventitious or secondary roots springing from the first internode of the stem; *st*, the stem. Enlarged about six times. (Sachs).

Fig. 56.—Germination of Indian corn. *a* and *b*, front and side views of the embryo removed from the kernel; *w*, the primary root; *ws*, its root sheath. (Sachs).

Fig. 57.—*P*, The plumule; *l*, fragment of wall of ovary; *w*, root with root-hairs above and naked below. (Sachs).

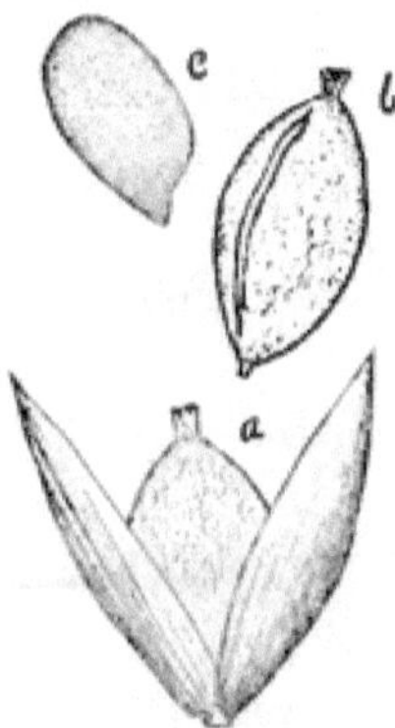

Fig. 59.—*a*, The caryopsis of *Sporobolus cryptandrus* within the glumes; *b*, the empty ovary split open; *c*, the seed which has escaped from the ovary. (Sudworth).

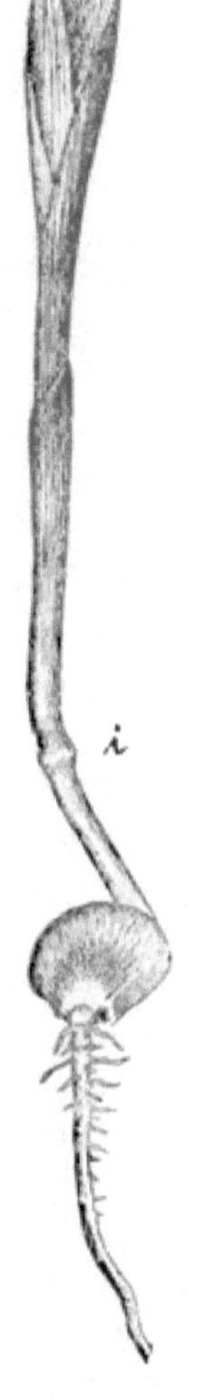

In case of most grasses, *the caryopsis* consists of the seed permanently inclosed in the adherent walls of the ovary. The seeds of *Sporobolus*, when mature, freely escape from the delicate ovary as shown in Fig. 59.

Fig. 58.—Young plant with remains of the kernel, part of root with lateral roots starting, apex of main root removed. (Sudworth).

CHAPTER II.

THE POWER OF MOTION IN PLANTS.

It would be foreign to the object of this work to say much on this interesting subject. Reference has previously been made to the motion of protoplasm in living cells, to the closing and opening of the leaves when dried or moistened, and the growth of nodes on the lower side to aid in straightening up a culm which has fallen down. The following will serve to illustrate what may not be new to all of the readers of these pages:

A thrifty hop-vine went winding up nine or ten feet to the top of a stake, and then four feet and two inches above any support, when it tipped over in the direction of the prevailing wind. It swung slowly around, sometimes making a revolution in from one to two hours. If another stick be within reach of the revolving top, it will seize the support and go on climbing as before.

Every one knows that asparagus or celery, placed on the side, will soon show the tips bending upward, and that the stems and leaves of a geranium set in the window will soon bend towards the light. These are familiar, and on that account may not awaken much curiosity, but it must seem wonderful to learn, for the first time, that the power of moving in circles or ellipses, or zigzag lines, is universal, so far as we know, to all young growing stems and all their branches. The same is true of the young leaves and all of the young roots. "Every growing part of every plant is continually circumnutating or bowing around." (Darwin). This motion is produced by the increased turgescence of the cells, together with the extensibility of their cell-walls on the convex side.

As Darwin says, "It would appear as if the changes in the cells required periods of rest. A young root may be compared with a burrowing animal, such as a mole, which wishes to penetrate perpendicularly down into the ground. By continually moving from side to side, he will find the easiest place for descending. If the earth is damper on one than on the other side, he will turn thither as to better hunting ground. The root, doubtless, can only distinguish water which touches it, having no power to 'scent' moisture in the distance. It hunts like a blind worm, by feeling, rather than as a hound by scent or vision. The tip alone of the root is sensitive, and when excited causes the adjoining parts to bend. It acts like the brain of one of the lower animals; the brain being seated within the anterior end of the body. And yet the tip of the root of Indian corn, unless held in place, has not power enough to penetrate or indent the thinnest tinfoil. It does not act like a nail when hammered into a board, but more like a wedge of wood, which, whilst slowly driven into a crevice, continually expands at the same time by the absorption of water."

A young stem of corn, the plumule of the seed, bends here and there in every direction, finding the easiest place out of the soil, and after reaching the surface and growing several inches above, it swings about, making the opening at the surface of the ground slightly funnel-shaped. Some plants are sensitive to jarring or friction. Previous mention of this has been made in the chapter on leaves.

Nature not only sows and distributes the seeds of grasses, but often buries them in the soil. Seeds are generally produced in profuse abundance, enough for perpetuating the species and enough to spare for the food of small animals, and enough to provide against numerous accidents and failures. After the seeds are scattered they are often shaded by other plants which aid in keeping them moist till they germinate. Freezing and thawing, rains and melting snows cover a portion of the seeds; the wind drifts soil or leaves or other small particles over others.

In countries subject to drought, where the soil is sandy and light, the awns of *Stipa, Danthonia, Avena, Heteropogon, Anthistiria, Aira*, and some others, assist the seeds in thrusting themselves beneath the dry surface to a place of moisture, where they may germinate. Some of these literally bore their way into the soil.

When dry, the lower part of the awns of these grasses twists about, the upper portion bending off at an angle; when wet, the awn untwists and finally becomes straight. The lower part of the chaff which envelopes some grass seeds is furnished with a sharp or oblique beak, provided with stiff hairs, which act as beards. By dropping such seeds on the surface, and alternately drying and wetting them, an experimenter will find that they penetrate the sand, even in some instances extending down six inches below the surface. It seems to make no difference whether the grains are dropped among sticks or stubble, or on smooth sand, they alike penetrate the soil. Even in clay soil the seeds work themselves into the cracks where the sun has dried it: on the return of rains the cracks close, or soil covers the seeds.

If the stubble, straw or any other objects prevent the awn from turning around, the seed will revolve on its axis. Besides, if the awn be wet and held down by any object, as it tries to straighten itself it will help push the seed, like a brad-awl, into the ground. On wetting the grain and awn of *Stipa pennata*, Francis Darwin (Transactions of the Linnean Society, p. 149, 1876), found that the rate increases up to the fifth revolution, and then diminished quickly. This is shown in the table:

Turn.	Completed in		Turn.	Completed in	
	M.	S.		M.	S.
No. 1	2	30	No. 6	1	30
No. 2	2	00	No. 7	1	45
No. 3	1	45	No. 8	2	10
No. 4	1	35	No. 9	3	20
No. 5	1	25	No. 10	..	..

In three wettings and three dryings, a little over an inch was buried in dry sand. A rise of temperature affects the awns in the same way as increased moisture; a fall of temperature acts like dryness. Mr. Darwin found that minute strips of the awn, consisting even of two long cells, twisted just as well as the entire awn. He thinks the torsion is produced by the striation or stratification of the cell walls. There are series of parallel lines, alternately light and dark, traversing the surface of the cell. Very frequently the two systems wind spirally round the axis in opposite directions. When the tissue expands during the absorption of water, it is due mainly to the swelling of the less dense striæ. This is thought to be the cause of torsion in cotton wool. Soon after being buried, where the soil is moist, the awn breaks off at a joint from the apex of the grain. The seeds of some of these species, such as those of *Stipa spartea*, are very annoying to sheep and other animals, such as are covered with thick hair. They sometimes even cause death. [Dr. M. Stalker in Am. Nat., p. 929, 1884]. Where plenty, they penetrate clothing about the ankles of people, and produce considerable trouble. Those, like sweet vernal, which are provided with feeble awns, work their way under leaves, sticks, rubbish, and find every little hole and crack in the dried earth, when the first rain covers them with soil.

Fig. 60.—Ovary and awn of *Stipa avenacea* 1×1.—(Sudworth).

Fig. 61.—Two long cells of an awn isolated and twisting when dry. Much enlarged.—(Trans. Lin. Soc).

The fertile flowers of *Amphicarpum* are not those on top of the culms, but those out of sight and among the roots under ground.

Moles, ants, and other small animals move earth and cover seeds.

CHAPTER III.

PLANT GROWTH.

Germination of Seeds.—Figures 56, 57, 58, on a former page, illustrate the parts of a kernel of corn and its mode of growth. For the account of the structure of the seed consult a former paragraph on this subject. It will be seen that the grain of corn, as is true of all the grasses, remains stationary where planted, at the base of the ascending axis.

"For germination to take place, moisture, oxygen and a suitable temperature are necessary. Under these conditions the seed swells, oxygen is absorbed, a part of the carbonaceous ingredients is oxidized, heat is developed, and carbon dioxide is evolved. During these changes the solid ingredients of the seed gradually become soluble; the starch and fat are converted into sugar; the albuminoids are converted into amides."—(*Warington's Chemistry of the Farm.*)

In the Temperate Zones, the seeds of grasses germinate quickly at a rather low temperature, though there is considerable difference in this particular. Some germinate a little above the freezing point, while every farmer knows that Indian corn, sorghum and millet start slowly, unless the weather be quite warm, and that the seeds will decay if kept wet and cold.

Old or light seeds often sprout quickly, but produce weak plants.

Though the seeds of grasses be secured when quite young, in the milk, a short time after flowering, when the endosperm is very small and the seeds shrivel as they dry, yet, if the embryo be formed and the seed well cured in a dry place, it will germinate. Dry seeds will endure much cold; wet or green seeds will endure but little.

Grass seeds may be covered deeper in a sandy soil than in clay, deeper in a rather dry soil than in a wet one.

The seeds of grasses are quite small, and should seldom be covered more than the eighth or fourth of an inch deep. The only need of a covering is to keep the seeds moist, and in some cases to prevent birds or other small animals from eating them. If planted deep, the supply of oxygen is liable to be insufficient, or if it be present, the seed is likely to become exhausted or much weakened in thrusting its young stem and leaves to the surface, where it may reach the light and begin to make a permanent growth of green leaves. All experiments are so much modified by the weather and condition of the soil, that to give definite rules is difficult.

The Function of Green Leaves.—Leaves not unfrequently absorb water in a liquid state as well as in the form of vapor, yet the roots absorb most of the water for plant growth. That leaves sometimes absorb water is most easily tested by observing the revival of cut flowers or plants when placed in a moist, tin box.

"The paramount function of the leaf is the absorption and assimilation of carbon, as such does not exist in the atmosphere, unless, indeed, as an impurity in air of towns, and a very prejudicial one to plants." (*Masters*). The carbon of plants comes from carbon dioxide, and is decomposed through the agency of chlorophyll under the influence of light.

Plants can endure darkness for a short time, but if long continued the chlorophyll disappears and the leaves fade, and finally perish, as may be seen in warm weather where a board is placed on the grass for some time.

Unless the air be saturated with moisture, which is not commonly the case for long periods, the leaves evaporate large quantities of water. The surplus passes off, leaving the condensed assimilated matter for building up the plant. In a growing season, while everything is thrifty, a grass-plant contains 70 to 80 per cent or more of water.

The Plant is a Factory.—"All the labour of the plant by which out of air, water, and a pinch of divers salts scattered in the soil, it builds up leaf and stem and roots, and puts together material for seed or bud or bulb, is wrought and wrought only by the green cells, which give greenness to leaf and branch or stem. We may say of the plant, that the green cells of the green leaves are the blood thereof. As the food which an animal takes remains a mere burden until it is transmuted into blood, so the material which the roots bring to the plant is mere dead food till the cunning toil of a chlorophyll-holding cell has passed into it the quickening sunbeam. Take away from a plant even so much as a single green leaf, and you rob it of so much of its very life blood." (*Masters*, quoted from *Gardener's Chronicle*).

A living plant is a machine or a factory, which, under the influence of light and heat, transforms raw materials into organic matter, suitable for enlarging the plant or enabling it to grow. In nearly all cases, some portions of a plant are dying while others are growing, and to some extent, one part is independent of other portions. This enables a plant to change its place of growth, to feed on its own stock of nourishment, or to recuperate when injured. The formation and enlargement of new cells constitute growth. To be ready for absorption by plants, matter must be in a liquid or gaseous condition. To a great extent a plant takes what it likes best, or is capable of controlling the quantity of any substance absorbed.

Of the materials assimilated, a part goes at once to form cell walls, cork, mucilage, etc., and can never be changed by the plant into matter for constructing other parts of the plant, while other portions of assimilated material take the form of starch, oil, inulin, and are likely to be again changed and transferred once or more times to other portions of the plant.

Only a very small part of the most fertile soil is in condition to be used for plant food. Some soils may contain a large amount of materials which the plants cannot take, or do not need. A fertile soil is capable of retaining plant food, while sandy soils, owing to their excellent natural drainage, are not fertile unless frequently supplied with manure.

Two different kinds of plants growing in the same field will usually be found to contain certain substances in different proportion. Some are essential, others not; some in large quantity, others in small quantity, yet, strange as it may seem, by the chemical composition of a plant we cannot always tell what manures will benefit it most.

Composition of Plants.—The combustible part of plants is made up of five chemical elements—carbon, oxygen, hydrogen, nitrogen and sulphur; without these no plant is ever produced. The carbon, hydrogen and oxygen form the cellulose, lignose, pectin, starch, sugar, fat and vegetable acids. The same elements united with nitrogen form the amides and alkaloids; and further united with sulphur the still more important albuminoids.

The incombustible ash always contains five elements—potassium, magnesium, calcium, iron and phosphorus, besides sulphur. Iron is present in only very small quantity. Besides these, an ash will generally contain sodium, silicon and chlorine, sometimes manganese, and perhaps minute quantities of other elements.

The earlier chemists spoke of the combustible portions of plants as "organic," and the incombustible portions as "inorganic." This distinction is no longer considered accurate.

Excepting oxygen, these elements are taken from compounds, such as water, carbon dioxide and the substances combined as shown in the following:

Nitrates. Sulphates. Carbonates. Phosphates. Silicates. Chlorides.	..............OF..............	Ammonia. Potash. Lime. Iron. Soda. Magnesia.

COMPOSITION OF A CROP OF MEADOW GRASS.

Water			8,378 lbs.
Carbon	1315	Combustible matter	2,613 lbs.
Hydrogen	144		
Nitrogen	49		
Oxygen and Sulphur	1105		
Potash	56.3	Ash	209 lbs.
Soda	11.9		
Lime	28.1		
Magnesia	10.1		
Oxide of Iron	.9		
Phosphoric Acid	12.7		
Sulphuric Acid	10.8		
Chlorine	16.2		
Silica	57.5		
Sand, &c	4.5		
Total crop			11,200 lbs.

From the soil plants obtain, by means of their roots, all their ash constituents, all their sulphur, and nearly the whole of their nitrogen and water. From the atmosphere they obtain, through their leaves, the whole, or nearly the whole, of their carbon, with probably small quantities of nitrogen and water. The amount and composition of the ash of succulent plants, as meadow grass and clover, is greatly influenced by the character of the soil and the manure applied.

For most of the above paragraph the writer is indebted to the *Chemistry of the Farm*, by R. Warington.

Meadow hay contains a much larger proportion of potash and lime than is found in the ripened grain of the cereals.

The Chemical Composition of American Grasses.* "In submitting grasses to chemical analysis, with a view of judging of their nutritive value, it is usual to determine the amount present of water, ash, fat or oil, fiber and nitrogen. From the latter the amount of albuminoids to which it is equivalent is readily calculated by multiplying by a factor which represents the per

* Taken by permission from the Agricultural Grasses, by the United State Department of Agriculture, 1884, Clifford Richardson, Assistant Chemist.

cent of nitrogen present in the average albuminoid, and by subtracting the sum of all these constituents from one hundred, the percentage of undetermined matter is obtain.d, and as it of course contains no nitrogen, and consists of the extractive principles of the plant, it is described as 'Nitrogen free extract.' It includes all the carbo-hydrates, such a sugar, starch, and gum, together with certain other allied substances, with which we are less intimately acquainted, but which have a certain nutritive value.

"Although it has been customary to state as albuminoids the equivalent of the nitrogen found, this is rarely entirely correct, as a portion is generally present in a less highly elaborated form of a smaller nutritive value. This portion is described as non-albuminoid nitrogen, and in analysis of the present day the amount is always given as an additional source of information, although our knowledge of its exact value to the animal is rather uncertain.

"The wide variations in fiber and albuminoids must be regarded as being entirely due to physiological causes, which are difficult to explain. *Digitaria sanguinale*, for instance, which in one specimen contains the extreme amount of albuminoids and a small amount of fiber has in another only half as much albumen and one and three-quarter times as much fiber. We learn then, that species are not in themselves at all fixed in their composition, there being as large variations among specimens of the same as between specimens of different species.

Analysis of Phleum pratense (Timothy) from various localities.

FULL BLOOM.

LOCALITY.	Ash.	Fat.	Nitrogen Free Extract.	Crude Fiber.	Albuminoids.	Total Nitrogen.	Non-Albuminoids Nitrogen.	Per Cent. of Total Nitrogen as Non-Albuminoids.
Department Garden, 1881	7.16	1.17	50.03	27.35	10.99	1.75	.51	29.1
Department Garden, 1880	5.06	3.58	58.93	21.93	9.90	1.58	.38	24.0
Maryland	4.93	1.22	52.83	30.43	7.09	1.25	.15	12.2
New Hampshire	4.57	4.20	57.16	28.28	5.79	.93	.10	10.8
Indiana	7.05	2.18	52.99	32.24	5.52	.88	.00	.0

Analyses of Dactylis glomerata (orchard grass) from various localities.

FULL BLOOM.

LOCALITY.	Ash.	Fat.	Nitrogen Free Extract.	Crude Fiber.	Albuminoids.	Total Nitrogen.	Non-Albuminoids Nitrogen.	Per Cent. of Total Nitrogen as Non-Albuminoids.
North Carolina	7.42	3.56	56.03	23.08	9.91	1.58	.30	19.0
District of Columbia	8.07	3.24	53.76	25.40	9.53	1.53	.16	10.5
Maine	8.02	3.33	51.80	26.65	8.74	1.40	.36	25.7
District of Columbia	6.00	3.62	57.34	24.42	8.62	1.38	.42	30.4
Pennsylvania	6.23	2.93	54.60	27.51	8.56	1.37	.51	37.2
New Hampshire	8.44	3.49	51.75	27.91	8.41	1.35	.42	30.9

"The different sections furnish very different qualities of grasses, and for the reason that those from the north were almost entirely from cultivated soil, while those from the other sections were many or most of them wild species from old sod. The improvement brought about by cultivation is marked, and the difference between a ton of wild western and eastern cultivated hay is apparent.

"In comparison with German grasses our best do not equal in amount of albuminoids, those classed by Wolff as *fair*, but they are far superior in having a much smaller percentage of fiber, and consequently a large amount of digestible carbohydrates. In the grasses of both countries the fiber increases with regularity as the nitrogenous constituents decrease, and of the latter the non-albuminoid portion is relatively greatly the poorer the quality of the grass.

"Analyses have been made of series illustrating the changes in composition of several species from the appearance of the blade to the maturity of the seed.

"With a few exceptions the specimens were personally collected in the grounds of the Department. They all grew in the summer of 1880 except the few series illustrative of the first year's growth of certain species. The specimens were cut close to the roots, weighed and dried rapidly in a current of air at 60° C.

DESCRIPTION.	When Cut.	Height in Centimeters.	Dry Substance. Ash.	Fat.	Nitrogen Free Extract.	Crude Fiber.	Albuminoids.	Total Nitrogen.	Non-albuminoid Nitrogen.	Per Cent. of Nitrogen Non-Albuminoid.	Fresh Substance. Water.	Ash.	Fat.	Nitrogen Free Extract.	Crude Fiber.	Albuminoids.
I.—AGROSTIS VULGARIS.																
DEPARTMENT GROUNDS.																
Good soil:																
Panicle not out	June 1	42	8.19	3.77	53.88	20.97	13.19	2.11	.82	38.9	67.8	2.64	1.21	17.35	6.75	4.25
Panicle out; closed	June 1	58	7.34	4.05	54.13	20.87	13.61	2.18	.80	36.7	68.1	2.34	1.29	17.27	6.66	4.34
In early bloom	June 19	48	7.55	3.62	54.46	21.64	12.73	2.04	.54	26.4	70.1	2.26	1.08	16.29	6.47	3.81
In full bloom	June 23	45	7.27	2.87	56.82	22.02	11.02	1.76	.53	30.1	61.4	2.80	1.11	21.93	8.50	4.25
Seed in the milk	July 1	43	6.60	3.54	60.02	19.43	10.14	1.67	.36	21.6	53.3	3.08	1.61	28.03	9.07	4.88
Seed hard	July 1	47	6.74	4.25	58.88	20.66	9.47	1.52	.18	11.8	51.5	3.27	2.06	28.56	10.02	4.59
Seed mature	July 9	55	5.30	2.74	61.32	21.75	8.89	1.42	.09	6.3	57.0	2.28	1.18	26.37	9.35	3.82
Poorer soil:																
Panicle spreading	June 16	43	8.41	3.88	57.41	20.49	9.81	1.57	.28	17.8	68.2	2.67	1.23	18.26	6.52	3.12
Early blooming	June 18	53	5.84	5.30	58.49	20.44	9.95	1.59	.32	20.1	58.8	2.41	2.18	24.10	8.41	4.10
II.—PHLEUM PRATENSE.																
DEPARTMENT GROUNDS.																
Good soil:																
Spike invisible	June 1	42	8.68	4.56	54.31	19.91	12.54	2.01	.70	35.0	70.7	2.54	1.34	15.92	5.83	3.67
Spike visible	June 1	62	6.41	3.40	47.26	21.03	11.90	1.86	.55	29.5	71.9	1.80	.96	16.00	5.91	3.34
Before bloom	June 23	45	9.82	3.63	54.19	22.03	10.33	1.65	.36	21.8	67.5	3.19	1.18	17.61	7.16	3.36
In early bloom	June 21	60	6.01	3.85	57.21	22.70	10.20	1.63	.30	18.4	64.9	2.12	1.35	20.08	7.97	3.58
In full bloom	June 18	58	5.06	3.58	58.93	21.96	9.90	1.58	.38	24.0	67.2	1.66	1.17	19.33	7.19	3.25
Early seed	June 18	52	10.53	3.40	51.07	22.90	12.40	1.93	.51	26.4	77.8	2.34	.76	11.34	5.08	2.68
Poorer soil:																
In bloom	June 4	60	6.56	3.95	57.48	23.53	8.48	1.36	.30	22.0	63.4	2.40	1.45	21.01	8.61	3.10
In full bloom	July 1	70	5.64	2.98	61.08	22.84	7.46	1.19	.36	30.3	71.9	1.58	.84	17.16	6.42	2.10

DESCRIPTION.	When Cut.	Height in Centimeters.	Dry Substance: Ash.	Fat.	Nitrogen Free Extract.	Crude Fiber.	Albuminoids.	Total Nitrogen.	Non-albuminoid Nitrogen.	Per Cent. of Nitrogen Non-Albuminoid.	Fresh Substance: Water.	Ash.	Fat.	Nitrogen Free Extract.	Crude Fiber.	Albuminoids.
II. PHLEUM PRATENSE.																
DEPARTMENT GARDEN.																
First year's growth:																
Head out	June 19	49	8.58	6.10	47.22	23.95	14.15	2.26	.39	17.3	78.30	1.84	1.31	10.42	5.14	3.03
In bloom	June 26	76	7.16	4.47	50.66	27.35	10.90	1.75	.51	29.1	66.75	2.38	1.49	16.64	9.09	3.65
After bloom	July 3	65	6.52	4.69	51.79	28.26	8.74	1.40	.25	17.9	56.63	2.83	2.03	22.46	12.26	3.79
" "	July 10	75	5.63	3.72	55.29	27.08	8.18	1.27	.15	11.3	58.86	2.31	1.53	22.79	11.14	3.37
INDIANA.																
Head not out	June 8	....	7.94	1.97	49.93	29.19	10.97	1.75	.18	10.3	70.60	2.38	.59	14.68	8.56	3.26
Before bloom	June 15	....	7.64	2.47	52.64	29.65	7.80	1.25	.28	22.4	67.50	2.48	.74	17.11	9.61	2.53
In bloom	June 26	....	7.05	2.18	52.99	32.26	5.52	.88	.00	.00	64.50	2.50	.78	18.81	11.45	1.96
After bloom	July 6		6.63	2.55	53.93	31.32	5.57	.89	.03	3.3	56.30	2.91	1.11	23.57	13.69	2.43
Early seed	July 16		5.96	3.74	60.77	24.70	4.84	.78	.00	.00	53.00	2.80	1.76	28.56	11.61	2.27
NEW HAMPSHIRE.																
Spike invisible	....		5.19	4.60	57.09	23.46	7.66	1.55	.30	19.4	....	....	....			....
Spike visible	....	....	4.73	4.22	56.10	25.34	9.61	1.54	.45	29.2	....	....	....	....	....	....
In bloom	....	....	4.57	4.20	57.16	28.28	5.79	.93	.10	10.8	....	....	....	....	....	....
After bloom	....	....	3.88	3.23	58.72	28.92	5.25	.84	.15	17.9	....	....	....	....	....	....
Early seed	....		3.2	2.7	62.5	26.03	5.41	.87	.18	20.7	....	....	....	....	....	....
III. DACTYLIS GLOMERATA.																
DEPARTMENT GARDEN.																
Panicles not out	April 23	35	10.29	4.12	50.86	18.76	15.97	2.49	1.01	40.6	78.8	2.18	.87	10.78	3.98	3.39
Panicles closed	May 4	55	8.26	3.13	55.01	23.18	10.30	1.65	.00	0.0	79.3	1.71	.64	11.40	4.80	2.15
In full bloom	May 13	87	8.07	3.24	53.76	25.40	9.53	1.53	.16	10.5	77.3	1.83	.74	12.20	5.77	2.16
After bloom	June 1	125	9.01	2.83	52.65	27.26	8.25	1.32	.33	25.0	73.5	2.39	.75	13.95	7.22	2.19
Later growth:																
In bloom	June 18	80	8.64	3.08	50.29	24.67	12.51	1.99	.77	38.7	66.9	2.86	1.32	16.62	8.16	4.14
Late bloom	June 23	75	6.00	3.62	57.34	24.42	8.62	1.38	.42	30.4	60.2	2.38	1.44	22.82	9.72	3.43
Seed nearly ripe	July 4	75	6.73	3.34	57.34	25.09	7.30	1.16	.45	38.8	62.3	2.54	1.26	21.03	9.46	2.75

DEPARTMENT GARDEN.																		
First year's growth:																		
Head not out	June 12	28	11.50	6.89	48.06	20.63	12.92	2.07	.15	7.3	79.50	2.36	1.41	9.85	4.23	2.65		
Green	July 15	....	10.52	6.86	46.95	21.61	14.03	2.25	.39	17.3	72.30	2.91	1.30	13.00	6.00	3.80		
Yellow	July 15	....	10.11	5.95	52.37	22.44	9.10	1.46	.18	12.3	74.60	2.58	1.51	13.30	5.70	2.31		
....	Oct. 25	....	10.95	6.50	47.96	21.24	13.33	2.14	.54	25.2	68.70	3.43	2.03	15.02	6.65	4.17		
IV. ALOPECURUS PRATENSIS.																		
Head just appearing	April 19	.....	9.21	4.09	52.16	18.21	15.73	2.52	.06	38.2	77.1	2.11	1.08	11.91	4.17	3.00		
Before bloom	April 19	.....	7.90	4.46	51.06	22.40	13.58	2.17	.53	40.9	76.7	1.84	1.04	12.03	5.22	3.17		
In bloom	May 1	.....	7.75	3.36	54.30	23.78	10.81	1.73	.00	0.0	60.0	3.10	1.34	21.73	9.51	4.32		
After bloom	May 12	.....	8.17	3.50	54.35	25.36	8.62	1.38	.07	5.0	66.6	2.73	1.17	18.15	8.47	2.88		
V. POA PRATENSIS.																		
DEPARTMENT GARDEN.																		
Set No. 1 grown on good soil:																		
Panicle just visible	April 27	30	8.07	4.88	48.74	18.43	19.88	3.18	.48	15.1	76.70	1.88	1.14	11.34	4.30	4.64		
Panicle spreading	May 1	30	5.51	4.07	51.32	22.83	16.21	2.68	.30	11.2	70.80	1.61	1.19	14.99	6.67	4.74		
In full bloom	May 21	30	8.30	3.90	51.63	23.56	12.61	2.01	.02	1.0	71.90	2.39	1.10	14.45	6.68	3.54		
In seed	June 5	30	6.38	4.25	52.54	24.34	12.49	2.00	.37	18.5	55.90	2.81	1.87	23.17	10.74	5.51		
Set No. 2, grown on poor soil:																		
Panicle closed	April 27		6.63	3.92	55.32	21.92	12.23	1.96	.12	6.1	......	......	......	......	......	......		
In full bloom	May 8	65	7.02	2.85	56.85	25.46	7.82	1.28	.10	7.8	60.00	2.18	.88	17.62	7.90	2.42		
Set No. 3 grown on poor soil; wayside:																		
After bloom; brown	June 1	65	7.23	3.92	56.12	23.85	8.88	1.42	.25	17.6	55.40	3.22	1.75	25.03	10.64	3.96		
In full bloom	May 19	78	7.73	3.41	55.32	23.10	10.44	1.67	.14	8.4	66.20	2.62	1.15	19.02	7.81	3.53		
In seed; brown	June 8	75	6.21	3.51	58.58	24.31	7.36	1.18	.15	12.7	51.00	2.81	1.59	26.61	11.05	3.34		
QUINCY, ILL.																		
Set No. 4:																		
Before bloom	May 10		8.42	4.99	45.34	21.87	19.38	3.10	.63	20.3	......	......	......	......	......	......		
In bloom	May 17	....	7.82	3.77	48.29	24.93	15.09	2.41	.51	21.1	......	......	......	......	......	......		
After bloom	May 27	...	9.07	3.50	52.51	22.75	12.37	1.97	.35	17.8	......	......	......	......	......	......		
VI. POA COMPRESSA.																		
Poor soil:																		
Panicle not out	June 1	14	7.75	5.29	58.08	18.19	10.69	1.71	*.10	5.8	67.90	2.49	1.70	18.64	5.84	3.43		
Panicle well out	June 1	28	6.81	4.41	55.18	21.30	12.30	1.97	.52	26.4	68.70	2.13	1.38	17.27	6.67	3.85		
In bloom	June 17	30	6.08	4.32	58.18	18.53	12.69	2.03	.45	22.3	70.70	1.78	1.32	17.16	5.43	3.72		
After bloom	June 23	30	5.14	3.85	63.89	18.15	8.97	1.13	.35	24.5	51.80	2.17	1.86	30.79	8.25	4.33		

8

DESCRIPTION.	When Cut.	Height in Centimeters.	Dry Substance. Ash.	Fat.	Nitrogen Free Extract.	Crude Fiber.	Albuminoids.	Total Nitrogen.	Non-albuminoid Nitrogen.	Per Cent. of Nitrogen Non-Albuminoid.	Fresh Substance. Water.	Ash.	Fat.	Nitrogen Free Extract.	Crude Fiber.	Albuminoids.
VII. Holcus lanatus.																
Very young	April [illegible]		9.58	4.53	54.48	18.64	12.37	1.98	.21	10.6	82.3	1.77	.80	9.64	3.30	2.19
Late bloom	May 25	72	8.23	3.89	55.52	25.01	7.35	1.30	.60	46.2	50.6	4.06	1.92	27.43	12.35	3.63
VIII. Arrhenatherum avenaceum.																
In full bloom	May 25	85	7.93	4.63	54.83	24.33	8.78	1.41	.15	10.6	62.3	2.99	1.72	20.71	9.17	3.31
After bloom	June 4	60	7.88	4.19	51.76	21.51	14.06	2.25	.96	40.9	71.4	2.02	1.07	13.25	5.51	8.75
IX. Anthoxanthum odoratum.																
Very young	May 1	15	6.29	4.27	61.58	17.17	10.59	1.70	.06	3.5	76.9	1.47	.99	14.22	3.97	2.45
In full bloom	May 1	40	7.09	3.36	59.15	21.63	9.47	1.52	.15	9.9	78.8	1.50	.71	12.62	4.57	2.00
After bloom	June 19	45	7.27	4.86	53.40	21.17	13.30	2.13	.51	23.9	69.9	2.20	1.46	16.07	6.37	4.00
After blooming	July 19	55	5.79	4.08	58.02	25.00	7.11	1.14	.35	30.7	53.4	2.70	1.90	27.04	11.65	3.31
X. Lolium perenne.																
Head invisible	May 1	35	8.66	3.58	57.70	18.39	11.67	1.87	.28	15.0	78.6	1.85	.76	12.35	3.94	2.50
" "	May 4	28	9.48	4.34	55.08	18.00	13.10	2.09	.39	18.7	82.4	1.67	.76	9.69	3.17	2.31
Head well out	May 4	30	7.96	3.64	56.75	20.55	11.10	1.78	.33	18.5	74.0	2.07	.94	14.71	5.34	2.89
Before bloom	May 12	55	8.40	3.75	54.97	23.98	8.90	1.43	.09	6.3	56.4	1.98	.89	12.96	5.65	2.12
After bloom	June 1	52	7.50	2.64	56.84	25.42	7.60	1.21	....	....	63.1	2.77	.97	20.97	9.38	2.81
XI. Bromus unioloides.																
Panicle not out	April 23	35	10.65	5.03	48.73	18.54	17.05	2.73	1.06	38.8	80.00	2.05	.97	9.45	3.60	3.31
Panicle closed	May 4	64	8.95	3.44	51.61	22.22	14.33	2.31	.55	23.8	75.40	2.20	.85	12.55	5.47	3.53
In full bloom	May 13	75	9.28	3.96	51.46	22.69	12.63	2.02	.34	16.8	79.00	1.91	.82	10.69	4.67	2.60
After bloom	June 1	78	6.98	2.37	54.77	25.33	10.83	1.74	.35	20.0	67.50	2.17	.77	17.80	8.23	3.52
In seed; brown	June 1	85	8.55	2.10	59.71	19.85	9.79	1.57	.33	21.0	61.70	3.02	.74	21.08	7.01	3.45

"The preceding analyses furnish the data from which is derived the general conclusion that as a grass grows older its contents of water decreases, ash decreases, fat decreases, albuminoids decrease, carbo-hydrates increase, crude fiber increases, non-albuminoids decrease till bloom or slightly after, when they are at their lowest, and then increase again during the formation of the seed.

"There are exceptions to these rules, but for the large majority of species, under ordinary conditions of environment they hold good.

"There are almost no exceptions to the fact, that water decreases in the maturer specimens; that is to say the plant gradually dries up and becomes less succulent. The ash is very dependent on locality and surroundings.

"The albuminoids decrease in amount with great regularity, the few cases where an increase appears being owing to the fact that the specimens were probably grown under varying conditions.

"Although largely a matter of opinion, it would seem from the foregoing results that the time of bloom or very little later is the time for cutting grasses to be cured as hay. The amount of water has diminished relatively, and there is a proportionately larger amount of nutriment, in the material cut, and the weight of the latter will be at its highest point, economically considered. Later on, the amount of fiber becomes too prominent, the stalk grows hard, arid, and indigestable, and the albuminoids decrease, while the dry seeds are readily detached from their glumes and lost with their store of nitrogen.

"For different species, however, different times are undoubtedly suitable, and experience must be added to our chemical knowledge to enable a rational decision to be arrived at.

"This work was inaugurated by Dr. Peter Collier, as chemist to this Department, and the laboratory work for the first year was in the hands of Henry B. Parsons, Mr. Charles Wellington, and myself. The remainder of the work has been under my own supervision, and has been almost entirely carried out by Mr. Miles Fuller and myself."

From the foregoing statements, we conclude that grasses of the better grazing districts, when grown in a dry season, make the best feed, but usually less in quantity. Grasses grown in sunny weather are better than those grown in cloudy weather or in the shade. Woodland pastures are proverbially lacking in "heart" or nourishment. Grasses grown on marshes or wet land are not so nutritious as those grown on dry land. Grasses grown on rich loam or clay, in fine condition, are more nutritious than those grown on poor, thin soil.

Further statements in regard to the chemistry of plant growth will be found in the chapter on red clover.

CHAPTER IV.

CLASSIFYING, NAMING, DESCRIBING, COLLECTING, STUDYING.

Plant Affinity.—In the plant kingdom there are certain genera so closely related to each other that the botanist calls them *families* or *natural orders*. The plants of a family resemble each other in many respects.

"That which really determines affinity is correspondence in structure. It may be said that those plants are most nearly related which correspond in the greatest number of points, and those the most distantly in which we find the fewest points of correspondence. The organs of vegetation are of very different degrees of value in determining resemblance of structure. All constant characters of whatever nature, require to be taken into account in classifying plants according to their natural affinities. Whatever points of structure are variable in the same species, or in species nearly allied to each other, are unessential and should be set aside, or be regarded as of comparative unimportance." (Lindley's Vegetable Kingdom.)

Those who have given little attention to the subject are liable to make mistakes in judging of natural affinity, because they draw conclusions from unimportant circumstances, the chief of which are size, form, color, and minute details.

An artificial classification is founded on some one or a few characters, disregarding all others. For example, it would place all trees by themselves in one group, all shrubs in another: all those which had five stamens together, in distinction from those which had any other number of stamens, while a natural classification aims to consider all structural features while young as well as when mature, placing plants together which resemble each other in numerous particulars, and show real relationship.

Families of Greatest Worth.—All the flowering plants growing in the United States are included in about one hundred and seventy families. In this portion of country, most of the plants which are cultivated to supply man and his domestic animals with food are included within sixteen of these families.

The Cruciferæ (Mustard Family), includes peppergrass, water cress, horse-radish, mustard, sea kale, turnip, ruta-baga, cabbage, kale, broccoli, brussels spruts, cauliflower, coleworts, kohlrabi.

The Rutaceæ (Orangeworts), includes the orange, the lime, the lemon, the shaddock.

The Vitaceæ (Grape Family), gives one species of grape to Europe and eleven to North America, besides the beautiful Virginia creeper.

The Leguminosæ (Pulse family), is second in size to the *Compositæ* and is one of great value. It includes peas, beans, and the clovers, and is noticed in the appendix, which treats of the clovers.

The Rosaceæ (Rose Family), is not a very large one, but is of much importance on account of the great value of many plants which it contains. Here belong the *pomes*, such as apples, pears, quinces, medlars, service berries; and here are the *drupes*, such as peaches, almonds, apricots, nectarines, plums, prunes, and cherries. Here are found strawberries, red raspberries and black raspberries, and blackberries. This may well be called the "fruit" family. There are also many choice flowers, including the rose, potentilla, spiræa, hawthorn, and Japanese quince.

The Saxifragaceæ (Saxifrage Family) affords currants and gooseberries, mock orange, deutzia, hydrangea, and saxifrage.

The Cucurbitaceæ (Gourd Family) contains squashes, pumpkins, melons, musk melons, gourds, and cucumbers.

The Umbelliferæ (Parsley Family) includes the carrot, chervil, celery, turnip-rooted celery, parsley, parsnip, caraway, coriander, fennel, lovage, and sweet cicely.

The Compositæ (Sunflower or Aster Family) is the largest family of flowering plants, and contains about one-eighth of all species in the United States. It affords a large number of weeds, such as thistles, ox-eye daisy, rag-weed, May-weed, yarrow, fire-weed, dandelion, burdock, cocklebur, flea-bane, and many more. It contains a large number which are valuable for ornament, as asters, zinnias, dahlias, feverfews, cinerarias, chrysanthemums, and sunflowers. Considering the enormous size of the family, about 10,000 species, we should expect something profitable in the line of field and garden products. The best it can do is to furnish lettuce, two kinds of artichokes, dandelion, salsify, chicory, endive, and sunflower. There is not a fruit nor a valuable vegetable, properly so-called, nor a good forage plant, so far as we know, in the entire list.

The Ericaceæ (Heath Family) is one of much interest to the florist. It includes the cranberry, blueberry, huckelberry, rhododendron, azalea, laurel, heath, and trailing arbutus, or May-flower.

The Convolvulaceæ (Convolvulus Family) affords the sweet potato, morning glory, and cypress vine.

The Solanaceæ (Night-shade Family) furnishes the potato, tomato, egg-plant, pepper, ground-cherry, tobacco, belladonna, bittersweet, petunia.

The Chenopodiaceæ (Goosefoot Family) affords the beet, spinach, orache, Swiss chard, and several weeds.

The Polygonaceæ (Buckwheat Family) contains buckwheat, rhubarb, knot-grass, sorrel, several docks, and swartweed.

The Scitamineæ (Gingerworts) includes the cinnamons, gingers, bananas, and arrow-roots.

The Liliaceæ (Lily Family) gives us the lily, hyacinth, tulip, asparagus, chives, garlic, leek, onion, shallots.

There are several other families which contribute more or less to the crops of the field, orchard and garden.

The Gramineæ (Grass Family) is by far the most important of any, and is noticed on the succeeding page.

For making clothing, there are two families of much value, viz:

The Malvaceæ (Mallow Family) containing okra, mallow, hollyhock, hibiscus, abutilon, and the cotton plant, and

The Linaceæ (Flax Family) including the flax, valuable for its fiber as well as the seeds which furnish linseed oil.

The Labiatæ (Mint Family) is quite remarkable for aromatic herbs, and contains basil, balm, sweet marjoram, pennyroyal, lavender, spearmint, peppermint, horehound, hyssop, thyme, summer savory, rosemary, bergamot, cat-mint, motherwort and sage.

Our leading trees belong to about twelve families, including the tulip-tree and magnolias, the basswood, the holly, the maples, the catalpas, the ashes, the elms, the buttonwood, the walnuts, and hickories, the oaks, chestnuts, beech, the birches, and alders, the willows, and poplars, and last, but by no means the least, the cone-bearing trees. The latter contains the cedars, cypresses, sequoias, balsams, firs, spruces, larches, and pines.

Gramineæ (The Grass Family.)—Tufted annuals or perennials, usually herbaceous and evergreen; with fibrous roots, often stoloniferous or with a creeping rhizoma. *Stem* (culm) endogenous, simple or branched, cylindric, rarely compressed, usually hollow, and closed at the joints, sometimes solid, especially when young; the nodes solid, mostly swollen. *Leaves* parallel veined, rarely net veined, narrow, undivided, alternate, rarely two or more at a node, distichous; *petiole* dilated, usually convolute, sheathing the culm, margins free or often united, especially in the lower leaves; *ligule* adnate to the sheath at the base of the blade, scarious, sometimes only a cartilaginous ring or a fringe of hairs. *Inflorescence*, spicate, capitate, racemose or paniculate.

The Spikelets consist of two, three or more, distichous, chaff-like concave scales or bracts (*glumes*), their concave faces towards the axis (*rhachilla*), the 2, or sometimes 1, or rarely 3 or more lower ones, and sometimes 1 or more upper ones empty, the other one or more with one sessile flower in the axil of each.

Floral glume terete or laterally compressed, enclosing a 1-2-sexual flower, and a flat, often 2-nerved scale (*palea*) with inflexed edges. *Perianth* of 2 (rarely 0 or 3 or more) minute scales (*lodicules*), placed opposite the palea. *Stamens* (*andrœcium*) usually 3, sometimes 1, 2, 4, 6 or more, even to 30, one of which alternates with two lodicules, filaments very slender, anthers versatile, 2-celled, linear, pendulous; pollen mostly yellowish-white, sometimes purple or red. In rare cases the stamens are monadelphous, as in *Streptochæta*. *Ovary* simple, free, sessile, sometimes stipitate, 1-ovuled; *styles* 2, rarely 3, free or more or less united; with hairy or feathery stigmas; ovule anatropous.

Fruit (*caryopsis*) erect, free or often adherent to the palea, and sometimes to the floral glume.

Seed usually adnate to the pericarp (free in *Sporobolus*) testa membranous, *endosperm* farinaceous, or somewhat horny. *Embryo* at the inner side of the endosperm at its base; *cotyledon* scutellate; *plumule* well developed; *radicle* thick, obtuse, endorhizal (with a sheath).

The embryo contains one, (often five as in wheat) or more rudimentary roots.

The peculiarities of the styles, stigmas, lodicules, and the caryopsis are of great value for describing grasses, but on account of their small size and the difficulty of always finding grasses in flower, they have not been much employed for that purpose.

It is not botanically correct to call any plants *grasses* unless they belong to this family (*Gramineæ*). There are many widely different plants which in popular language have the name "grass" attached to them, such as knot-grass, rib-grass, cotton-grass, sea-grass, eel-grass, sedge-grass, the clovers, and others, but these do not belong to the family here under consideration.

The plants most likely to be mistaken for grasses are the *Cyperaceæ* (sedges), of which there are two thousand species or more. They are abundant on wet land, and often constitute a large part of what is known as bog or marsh hay. Sedges have three ranked leaves, or leaves spreading in three directions.

The *Gramineæ* (grass family) contains Indian corn, wheat, oats, barley, rye, rice, doura, sorghum, broom-corn, sugar-cane, millet, Hungarian grass, bamboo, Timothy, red top, June grass, fowl meadow grass, blue joint, buffalo grass, orchard grass, meadow fox tail, the fescues, rye-grass, oat-grass, sweet vernal, Bermuda grass, and many more which contribute to the food of domestic animals.

The grass family heads the list of food producing plants, which are the foundation of all agriculture. The cereals, such as sorghum, rice, doura, maize, wheat, rye, oats, barley, furnish a large part of the food of the human race, while the meadow and

9

pasture grasses, together with the cereals, largely feed our domestic animals.

Along the low lands of India and some other tropical countries, many millions of people subsist mainly on rice, while further back on the higher lands, a much larger number feed on several varieties and species of sorghum. Sorghum feeds the most people of any cereal, while wheat outranks them for making the best quality of food.

To assist in comprehending their great importance, let us glance at the figures as they appear in the last census of the United States,—taken in 1890:

Corn, bushels	2,122,000,000
Oats, bushels	809,000,000
Wheat, bushels	468,000,000
Barley, bushels	78,000,000
Rye, bushels	28,000,000
Total, bushels	3,505,000,000

To these figures must be added, as belonging to the grass family:

Hay, tons	67,000,000
Rice, pounds	129,000,000
Broom corn, pounds	39,000,000
Sugar, pounds	301,000,000
Molasses, gallons from (cane and sorghum)	50,000,000

To these figures we might add the immense products of grazing, including meats, horses, cattle, swine, hides, wool. Live stock alone foots up $2,208,767,000. The estimate for hay most likely includes the clovers.

Of the staple crops of the United States, the grass family contains about five-sixths of the total value.

The cereals and the pasture grasses the world over, are of more value to man and his domestic animals than all other plants taken together! Vastly more than half of the value to man of all vegetation belongs to one family, the grasses. "Grass is king. It rules and governs the world. It is the very foundation of all commerce, without it the earth would be a barren waste, and cotton, gold, and commerce all dead."— Solon Robinson.)

Grasses are remarkably and evenly distributed in nearly all parts of the habitable globe, in every soil, in society with others and alone: under the equator or in Greenland, wherever moisture and sufficient heat favors the earth, there grasses are a leading feature of the flora.

In the whole world the family ranks fifth in size: the *Compositæ* ranking first, the *Leguminosæ* second, the *Orchidaceæ* third, the *Rubiaceæ* fourth, the *Gramineæ* fifth. This does not convey an adequate idea of the value or number of grasses in unwooded regions, because the number of individuals of many of them is exceedingly large. Doubtless there are more individual plants in the grass family than are found in all of the others named above.

In the words of Dr. Bessey, of Nebraska:

"When we come to the inquiry as to what proportion of plants of a given area are grasses, we find the number varying very much from those just given. For example, in forest regions the actual number of grass plants is much lower than it is in the same region after the forests have been partially cut off, and if again we compare the latter with the prairies we observe a still greater increase. There are many great tracts in Nebraska, miles and miles in extent, over whose whole surface the grasses constitute fully nine-tenths of the actual vegetation."

"Of the individual plants on the great prairies of the Northwest, 90, yes, 99 per cent are grasses!"—(Scribner).

There are about 300 genera and 3,100 to 3,200 species. They are the most abundant in the temperate zones, where they often clothe large tracts with a growth which is fine, soft, and thick. Here the *Poaceæ* predominate. In the tropics the *Paniceæ* are prominent; many of which are more isolated, growing singly in tufts or small groups. In the temperate zones grasses are slender, and seldom more than a few feet in height; in the tropics they become giant bamboos, 60 or even 170 feet high, and eight or ten inches in diameter, approaching a tree in size.

In the United States, east of the Mississippi, the grasses constitute about one-twelfth of all the species of flowering plants. A State like Michigan, Illinois, or Massachusetts, has been found to contain not far from 130 species of grass. It is not uncommon in the northern States to find 60 species within a distance of a few miles of each other.

Although very few grasses possess brilliant or aromatic flowers, and look so humble and so much alike, there are few plants which are more beautiful in "all stages of growth, whether examined one by one or in masses; with the naked eye or with the microscope."—(American Agriculturist, 97, 1852.)

Who has not admired a gently rolling field as the wind swept over the even tops of thick grain? What view surpasses a field of waving grass, or a closely shaven lawn? Grass is "a thing of beauty and a joy forever." It even beautifies the grave, spreading a green carpet over the remains of friends gone before.

"Here I come creeping, creeping everywhere;
By the rusty roadside,
On the sunny hillside,
Close by the noisy brook,
In every shady nook,
I come creeping, creeping everywhere."—(Sarah Roberts.)

In very early spring, some of the grasses begin to put forth their green leaves, which are soon succeeded by the culms, which carry the flowers. In the northern States *Poa annua* is the first to unfold its spikelets, and spread its flowers, which are soon followed by *Oryzopsis*, *Poa sylvestris*, sweet vernal and meadow foxtail. In central Michigan, these and perhaps a few more, blossom in May, while June is as profuse of grasses as it is of roses. Quite a number flower in July, and some delay till August and September. The annuals are usually late in flowering, though some of the perennials are very late; for instance the Muhlenbergias, Andropogons and *Chrysopogon nutans*.

The Name of a Plant consists of two words, first the *generic* second the *specific*. These must have the Latin form. The generic name is substantive and singular, and very nearly answers to the surname of a person; the specific name is most generally an adjective (rarely a noun), and agrees with the noun in gender and number. The specific name is followed by a name or the abbreviation of the name of the person who applied that name to the plant under consideration.

Latin names are often objected to by persons who cannot see why the common English names will not answer every purpose. But suppose the Germans and the French should say the same thing of the plants they described? German and French names are not so easy for us as those in the Latin form. Those in Latin are often short and easy, and have been adopted as the common names, such as Dahlia, Crocus, Ixia, Orchis, and Iris. There is certainly an objection to using such names as *Kraschenninikovia*, *Andrzeiofskya*, and *Pleuroschismatypus*, names which have actually been thrust upon plants.

In the use of common names, many take the liberty of making their own name, and then the same name is applied to more than one plant.

A certain well known tree in some parts of the United States is called "White-wood," in others, "Tulip-tree," in others. "White Poplar," in others, "Yellow Poplar." *Acer rubrum* is known by several names in various States, as "Red Maple," "Soft Maple," and "Swamp Maple." *Poa pratensis* goes by the name of "Blue grass" in Kentucky; in other places it is called "Kentucky blue grass," "green grass," "green meadow grass," "June grass," "spear grass," and very likely several other names. One species is known as "Timothy," "Herd's grass," "Cat's tail;" another is known in different places as "red-top," "herd's-grass," "Burden's grass," "red bent," "summer dew grass," "small red-top," "fine red-top," "fowl-meadow grass." Other illustrations appear in connection with the consideration of the species most cultivated. It is quite necessary, for accuracy, that only one name be used for each kind of plant.

Collecting and Preserving Grasses.—No person can make satisfactory progress in this study without frequent reference to species with which he has formerly met. For this purpose specimens can easily be pressed, poisoned, dried, labelled, classified, and arranged in order, convenient for future reference. A very simple, and in some respects quite desirable method, is to preserve the grasses in the form of dried bunches. In this way they become brittle, and cannot well be handled without damage, but they preserve their appearance, better than those which are submitted to pressure between folds of paper. Even though these dried bunches may not be kept very long, their use is to be recommended.

If it be within the means of the student, he will find it of great advantage to transplant bunches or grow seed in his garden. In this way he can often see the several kinds and study their peculiarities at every stage of growth.

On account of scattering seeds which produce young plants that cannot be identified till they have become well established, the writer has found it best to separate kindred species by some others which are quite different. It is very natural and seems quite desirable to the systematic worker to want his Poas all in adjoining plats, and his Fescues in other adjoining plats, but if he can succeed in keeping them separate when thus planted, he will do better than the writer has ver been able to do with his twenty or more years of experience. Another word of advice: plant the seeds in rows, never broadcast, as this will much facilitate weeding when the plants are yet small. No agrostologist will be satisfied to study mere dried specimens, as they will not reveal many points to best advantage, but for permanent use at all seasons of the year, the mode adopted in the herbarium will be the best for preserving grasses. To save time in arranging thoughts for the following account in reference to preserving grasses, the writer has taken hints from an article of L. H. Hoysradt in the bulletin of the Torrey Botanical Club for 1878.

Have a tin case made of oval cylindrical shape, 17 inches long, four by six inches wide. It is provided with a light strap to throw over the shoulder, and so attached to the box near the front narrow side so as to have the lid open from the person when hung on the shoulder. The lid opens nearly the whole length of one of the flat sides,—15 by 4½ inches, with ¼ inch lap,—made to fit as tight as possible, and fastens with a simple spring catch.

Procure some thin, unsized paper, without printing on it, in the form of folded sheets, about 11 by 17 inches. A poor quality of printing paper is suitable. The driers are half sheets 12 by 18 inches and are thick and free from sizing. The specimens are left in the thin sheets through all changes of driers, till they are thoroughly dried. Change the driers every 12 hours or oftener at first, and submit them to fire heat or direct sunlight. Press the pile with a weight of fifty to seventy pounds. Several pieces of thin board 12 by 18 inches will be needed.

Always preserve some of the roots, lower leaves and rhizomas, if there be any, as a part of a specimen. Some part of the plant should be in flower. If too long to preserve in its natural form, double the culm in a zigzag manner, so that it shall not be too large. Bite the angles with the teeth to make them submissive. The stem of June grass, or a small thread, or the angle pressed through slits in pieces of paper, are convenient to keep the grass from spreading till it is dried. Short pieces of fine annealed wire are still better.

It is of first importance as grasses are collected and placed in the portfolio or press, to be particular to write on a label the name of each species if known, but by all means, the locality and date of collection.

Instead of changing the driers, plants can be well dried by binding them in a portfolio with sides of wire-netting.

A plant is well dried when it rattles, will not stay bent, or does not feel cool when applied to the cheek.

For safe and long keeping the grasses must be poisoned. Use 95 per cent. alcohol nearly saturated with corrosive sublimate, and apply with a flat brush, and place the specimens again in driers for half a day or more. Common white arsenic in alcohol is just as good and perhaps better.

Procure some white paper 11½ by 16½ inches, which is thick enough to stand on edge without doubling up. On this paper fasten a specimen of one species only, by pasting over it several narrow strips of gummed paper. Many prefer to fasten the plants to the paper with a fine quality of glue. Fasten the specimen directly above the label which is stuck fast to the bottom of the sheet. Specimens of the same species from different localities may be placed on the same sheet, each over its own label. The label contains the generic and specific name, perhaps the common name also, the locality, date of collecting, and name of collector. There are as many styles of labels as there are persons who preserve plants.

The sheets of all the species of the same genus, when not too numerous, or of a section of it, should be placed in a *genus-cover* which consists of a folded sheet of firm manilla paper, 12 by 17 inches. The generic name should appear on the cover at the lower left hand corner after it is folded, with the folded portion to the left.

The sheets are placed on the sides in pigeon holes, arranged in the order as treated or described in our standard books. Begin with the first genus at the upper left hand pigeon hole and go downward to the bottom of the case, then begin at the top of the next column of pigeon holes.

When a sheet of the herbarium is wanted for study, open the door, find the name on the genus-cover and draw out for a few inches the one needed with all above it in the same hole, then remove the genus cover needed, leaving those above in the position described, to show at a glance just where to replace the specimens.

Grasses found in certain Localities.—The following grouping of the grasses is founded on their habits and localities, and will hardly rank as a classification. It is essentially the one proposed in a prize essay by Prof. James Buckman.—(Jour. Royal Agrl. Soc. 1854.)

1. *Jungle or Bush Grasses* are those which generally grow isolated, in bunches or a few plants here and there. The bamboos are examples for the tropics. In the northern countries many of the smaller grasses have a similar disposition of growing in tufts; of this kind we have *Deschampsia* (*Aira*) *cæspitosa*, L. (hair grass), *Avena pratensis* (Narrow-leaved Oat-grass). *Festuca elatior*, L. (Taller Fescue).

A few others, if sown alone and not very thick, not closely fed or rolled, will assume the same habit. Of such we have *Festuca ovina*, L. (Sheep's Fescue), *Festuca duriuscula*, L. (Hard Fescue), *Dactylis glomerata*, L. (Cocksfoot or orchard grass.)

2. *The Aquatic or Water Grasses* are those which elect to grow by the margins of rivers, in brooks or ditches, or around the edges of ponds. With few exceptions they are of little value agriculturally. Their presence is a sure indication of lack of drainage. The following are examples: *Phalaris arundinacea*, L. (Reed Canary-Grass), *Phragmites communis*, Trin. (Reed-Grass), *Glyceria aquatica*, Smith. (Reed Meadow-Grass) *Glyceria fluitans* R. Br. (Floating Meadow-Grass), *Leersia oryzoides*, Swartz. (Rice Cut-Grass), *Zizania aquatica*, L. and *Z. miliacea*, Michx. (Indian Rice).

3. *The Marine or Sea Side Grasses*, which are chiefly found near salt water or the Great Lakes. They are generally very coarse and distasteful to cattle. A great proportion of the plants in these situations, which are cut and cured by the farmers of New England and Long Island, and the Jersey coast, and known as salt marsh hay, are not grasses, but belong to other families of plants, such as the *Juncaceæ* and *Cyperaceæ*. The rhizomes of some are very useful in preventing the water from washing away the soil. Of marine grasses the following are examples: *Spartina juncea*, Willd. (Rush Salt-Grass), *Glyceria maritima*, Wahl. (Goose-grass), *Psamma arenaria*, R. & S. (Sea Mat-weed.)

4. *The Meadow or Pasture Grasses*.—Most of the grasses of much value to agriculture belong to this section. They are the leading grasses of our best meadows, pastures, and the grazing lands of the prairies, and have received the most attention in this work.

Some of these are especially suited to,—

a. Upland pastures, thin soils.

b. Poor, stiff soils, hungry clays.

c. Rich, deep loams.

d. Meadows on the banks of rivers, subject to perennial floods.

e. Irrigated meadows, in which the water can be entirely controlled. Long lists have been given, but there is much risk in prescribing for such a great country as the United States.

5. *The Agrarian Grasses* are more properly those which occur in land under tillage. They are weeds such as *Bromus secalinus*, L. (Chess). *Agropyron repens*, L. (Quack or Couch Grass), *Setaria viridis*, Beauv. *S. glauca*. Beauv. (Fox tails). *Panicum sanguinale*, L. (Finger Grass).

The various Uses of Grasses.—These are,—

1°. For the grain as food, cereals, or drink as whisky, etc.

2°. For pasture.

3°. For hay.

4°. For manufactures, paper, substitute for lumber, sugar, mats, hats, etc.; bamboos for many things.

5°. For fuel; the tops in close ovens, Indian corn in stoves.

5°. For preventing the washing of banks and drifting of sands.

How to begin the Study.—We will suppose the learner has but a very limited knowledge of botany and is possessed of some enthusiasm, a good stock of patience and perseverance. The latter quality is of the greatest importance, and without it, success cannot be attained. A diligent pursuit of the subject is sure to crown the student with success, and this, for several good reasons, is worth all it costs.

This book contains many good illustrations, but a study of pictures alone, with a few superficial glances at plants, will never make a botanist. There may be a hundred names, rather uncommon and technical, to become familiar with, but this should discourage no person of good ability. All of these words are defined in the glossary at the end of the volume.

To have the use of a good, simple miscroscope magnifying ten to twenty diameters is absolutely essential. This should be mounted on a stage or block on which the flowers or other small parts may be laid, while both hands are free for dissecting with needles mounted in handles. Common number five needles, broken in two, and with forceps pushed blunt end first into the pith of a one-year-old stem of European larch, are cheap, nice, and durable. There should be some means of adjusting or varying the focal distance of the microscope. A small, sharp knife, and a pair of fine-pointed forceps will be very useful.

Take in hand a complete specimen of some grass, the name of which is well known. We will suppose it is a sample of Timothy (*Phleum pratense*, L). If dry, the flowers or top can be made soft by soaking in water, if warm all the better. The roots are fibrous; the stalk, *culm*, has solid joints, *nodes*, from each of which starts a leaf. Towards the base of the stalk, the nodes are close together, and one or more may be enlarged or swollen into a simple *tuber* or *corm*, sometimes improperly spoken of as the *bulb*. For some distance above each node, the sheath of the leaf rolls like a scroll around the stem, one edge covering the other closely, but usually not growing fast together. At the upper end of the sheath, the blade of the leaf spreads away from the culm. Just where the blade leaves the stem, at the throat of the sheath, is a delicate ring, fringe, or often a thin, scarious appendage. This is the *ligule*, the form of which is usually constant in all the plants of one species.

The leaves are alternate, one at each node, and two ranked, i. e., there are two rows of leaves alternating with each other on opposite sides of the stem. The leaves are *parallel veined*, and may be stripped or torn lengthwise into narrow pieces. The beginner may consult the chapter which treats of leaves.

The top of the stem bears a cylindrical *spike* of spikelets, some of which on close examination, it will be seen, have very short branches. Select a small portion of the material from the spike and place in a drop of water, while it is seen with the microscope. With one needle hold a portion fast, and with the other or with the forceps separate the parts of the specimen.

Figure 62 illustrates what should be seen. At *2* are the *outer glumes*, looking much alike in shape and size. At the base one appears to be a trifle inside of the other. They are tipped with a *mucronate* awn, and are *ciliate* on the back. If cut in two crosswise, the section of a glume appears in shape much like a broad letter V. At the base of the letter is a rib. Such glumes are *keeled*.

At *3* is the floral glume, covering the *palea*, which is rather smaller. Still within the palea, if the specimen be in flower, may be seen three slender *filaments*, each bearing an *anther* at the extremity. At the center is a small ovary, from the top of which spread two feathery *stigmas*. Turn to the page where *Phleum pratense*, L. is described, if you have not already done so.

Do not hurry, but try hard to understand everything as you proceed, and whenever you come to a word which is not understood, consult the glossary or some other portion of the book. In a note book make a list of all the new technical words, perhaps with their definitions, and frequently study them over like a spelling lesson, till they become familiar. With this thoroughness on the start, you will very soon master the difficulties and progress will be certain and satisfactory. Review often and thus become well acquainted with the first lessons. This review may not be so interesting as advanced lessons, but it is time spent in a very profitable manner.

The beginner is likely to hurry and run over too much ground. He is almost certain not to understand what he looks over. He becomes superficial, and often fancies he is learning a good deal, when in reality he possesses scarcely any definite information of value.

The following are suggested as desirable grasses for the beginner to study: Ray or rye-grass, *Lolium*, quack grass and wheat, barley, rye, wild rye. It makes no difference which is taken up first. The reader will soon see that those last named agree in several respects, and belong to the same tribe, *Hordeæ*. In each case, free use can be made of the excellent illustrations, but the careful examination of each species *must not* be omitted.

Another lot of closely related species are sweet vernal, canary-grass and vanilla grass. One genus contains June grass, wire grass, fowl-meadow grass and a few others which are common. Orchard grass is of fair size and well suited to the beginner.

It is an excellent plan, where possible, to take up in connection with each other, especially in reviews, grasses which are nearly related. Any two such species may very profitably be critically compared.

"There is no way for the student to do but to take the thing described in his hand, and patiently compare it with the definition given, until he distinctly sees the application of every part. He must, therefore, take a cornstalk or some other grass, and study its structure until he has made out every statement in the definitions given."—(Gould).

CHAPTER V.

NATIVE GRAZING LANDS.

Effects of Over-Feeding Dry Districts.—The grazing of sheep and cattle often change the character of vegetation for the worse instead of better.

Every farmer knows the value of sheep to exterminate wild raspberries, blackberries and most other bushes, but many times they also introduce troublesome weeds as well.

Dr. A. Gray, in Am. Jour. Science in 1874, notices a contribution by Dr. Shaw to the Linnean Society, in reference to the ill effect of overstocking the dry grazing districts of Southern

Africa with Merino sheep. Troublesome burrs are introduced, which crowd the grasses besides injuring the wool.

When first introduced, the sheep fed mainly on the grass, which in this dry, hot country, began to fail. There were too many sheep for the moderate supply of grasses. Soon the sheep fed on the brush and scrub, and the ground left to them, and to obnoxious and poisonous herbs. As the vegetation became scarce, bitter and nauseous plants of the neighboring region came in and helped to extirpate the indigenous flora, and render it more and more unfit for sheep. As these were forced to eat diagreeable food, it greatly injured the mutton. What is true of Southern Africa is proving true in many parts of the dry, native pastures of the United States. Numerous herds will soon over feed and "stamp out" the native grasses.

Continuous manuring of any kind, continuous mowing or pasturing,—a continuous treatment of any kind will soon produce a change in the plants.

Dr. Samuel Aughey, in Science, 1883, in speaking of the Nebraska flora, says: "A remarkable peculiarity is its changeable character. This is conspicuous among the grasses. In 1865, much of the uplands of Lancaster county was covered with buffalo-grass. By 1871 nearly all of this species had disappeared, and its place was taken by blue-joints (*Andropogon furcatus*, *A. scoparius*, etc.) interspersed with *Boutelouas*, *Chrysopogon nutans*, *Sporobolus*, etc. In 1878 the blue-joints disappeared, and the *Boutelouas* usurped their place. Similar phenomena were observed in almost every county in the State. During the last two years *Chrysopogon* (*Sorghum*) *nutans* has been gaining in Eastern Nebraska over all others. This tendency to change is common in other States. When old Fort Calhoun, above Omaha, was occupied by the military, twenty-five years ago, Kentucky blue-grass was brought in baled hay to that post from the South. It spontaneously took root and spread in every direction, and now it can be found in prairies thirty miles away. Under favorable conditions the wild, native grasses produce from one to three tons or more of hay per acre."

Professor Shelton, for Central Kansas, says: "Our prairie grasses cannot endure close pasturing or heavy tramping. Notoriously, the most promising wild pastures, after three or four years of even moderately close grazing, become permanently occupied by coarse, rank-smelling, worthless weeds."

"In Nebraska," says Dr. C. E. Bessey, in 1885, "There have been notable migrations of plants within the past twenty or thirty years. The buffalo grasses of various kinds were formerly abundant in the eastern part of the State, now they have retreated a hundred to a hundred and fifty miles, and have been followed up by the blue-stems (*Andropogon* and *Crysopogon*). The blue-stems now grow in great luxuriance all over great parts of the plains of Eastern Nebraska, where twenty years ago the ground was practically bare, being but thinly covered by buffalo grasses. In Dakota it is the same, the tall blue-stems are marching across the plains, and turning what were once but little better than deserts, into grassy prairies."

Native Pastures.—With reference to grazing in Colorado, R. A. Cameron, in the National Live-Stock Journal, 1872, says: "The rainfall is precipitated mainly in the spring as rain, and in the winter entirely as snow. The summer months are dry, with rare rainfalls, and these are short, followed immediately by cloudless skies. The grasses grow rapidly in the spring, but are cut short by the drought, and ripen and dry up in June. It is the absence of moisture in any quantity during the warm weather that not only completely cures the native grasses, but which preserves them unfermented, sweet and nutritious during the summer and winter. They assume a brown color, and give a sombre aspect to the great plains, striking the eye of the farmer from the New England States very unfavorably. But, short and brown as they are, they are no doubt the richest in the world."

Some of the leading grasses which form the native pastures of Texas are: Gama—grass, (*Tripsacum*). *Panicum virgatum*, a kind of Panic grass, Indian grass, *Chrysopogon nutans*, *Andropogon scoparius* and *A. provincialis.* The last two are known as blue-stems, and the latter as broom grass, or broom-sedge. *Tricuspis (Triodia) sesleroides*, fall red-top, is prominent in places. A vast number of smaller species help make up the pastures, but they are less widely diffused or less prominent than those named above.

In the Report of the Department of Agriculture for 1870, J. R. Dodge states: "The relative value of these species as forage grasses differs very widely, a few of them being entirely worthless. The largest number of the species could be dispensed with without manifest disadvantage to the grazing interests of the country. The relative value of the twelve most important species is exhibited in the following table of per centum estimates, one hundred representing the aggregate value of the twelve:

SPECIES.	Missouri River Region.	Rocky Mountain Region.
	Per Cent.	Per Cent.
Andropogon (furcatus) provincialis	40	16
Andropogon scoparius	20	10
Chrysopogon (Sorghum) nutans	20	12
Sporobolus heterolepis	12	1
Buchloë dactyloides	5	5
Bouteloua oligostachya	0	10
Spartina cynosuroides	2	2
Festuca ovina	0	20
Festuca macrostachya	0	5
Bromus Kalmii	0	8
Poa serotina	0	8
Stipa viridula	0	5

These estimates can only be approximate for that time. The first three are quite tall, and make the main bulk of hay in the wild regions referred to.

I have taken the following from General Alvord's Bulletin, as quoted in the Agricultural Grasses of the United States, by Dr. G. Vasey:

"In the arid Rocky Mountain plateaus, the grasses, as they stand on the soil, are cured in the sun during the summer, the action of heat retaining and concentrating in the stalks the sugar, gluten and other constituents of which they are composed. It is so cold and so dry in those elevations that there are neither heat nor moisture to rot them. And the snows are so fine (save in some exceptional seasons) in that cold atmosphere, that they are so blown by the winds into drifts, that four-fifths of the soil is never covered by them.

"The difficulties in lower altitudes than those I have described, have been, that after a warm spell and a thaw, the snow freezes to a crust and the grass is matted down by the ice, and kept from the stock.

"In Texas the grazing grounds are mostly at so low a level above the sea that the grasses rot in winter. Hence, in the latter part of winter, the animals there are often poor. The region higher than 3,000 feet above the sea, fit for winter grazing, includes nearly all up to the timber line, of Montana, Idaho, Wyoming, Utah, Nevada, Colorado and New Mexico, and five-sixths of Arizona, one-half of Dakota, one-third of Nebraska, one-fifth of Kansas, one-fourth of Texas, and one-sixth each of California, Oregon and Washington Territory. This embraces about one-fourth of the area of the whole United States."

The Native Grasses of the Pacific Slope.—The following are free extracts from the notes furnished by C. G. Pringle:

One going into the Southwest from New England, where all

deforested areas are closely sodded with perennial grasses, is struck with the insignificance of permanent grasses there and the almost entire absence of sod.

To speak of Arizona and Southern California: In the bottom of the valleys and along the line of the water-courses, though water may not flow over the surface except during the period of summer or winter rains, and in soil more or less impregnated with alkali, the traveler occasionally meets with natural meadows.

Distichlis maritima, with its creeping roots forming a close network in the soil, and *Sporobolus Wrightii*, growing in great clumps, chiefly form these meadows. The former has wiry stems, and its foliage is tough, but animals accustomed to subduing spring opuntias and thorny shrubs thrive on it. The latter is a rigid, coarse grass, its culms often four to five feet high and as thick as a goose quill. When its stems are but recently grown animals browse away their upper portion, and cull out somewhat from amongst the bristling stumps of the stems of former years, standing dense and stiff, some two feet in height, the long radical leaves of the plant. To arrive after nightfall and a long forced drive to reach grass and water upon such a meadow, and to be compelled to picket our horses on such pasturage, closely gnawed away by the herds of ranches far and near, seems hard, but from May till August the valleys and plains afford nothing better.

Sporobolus cryptandrus var. *strictus* has much the habit and value of *S. Wrightii*. *Sporobolus asperifolius* occupies patches of wet soil with a fine herbage, and its abundant and leafy sterile culms yield forage more easily appreciated by animals. *Panicum obtusum* growing in low lands, particularly in the partial shade of shrubs, contributes a trifle of forage by its long, wiry, but leafy creeping stems.

In low lands scattered tufts of *Andropogon saccharoides* and *Trichloris fasciculata* contribute a better food to animals, as acceptable, probably, as any afforded by the perennial grasses. *Panicum lucophæum* and *Andropogon contortus*, in their scattered tufts on the mesas and foot hills, are of similar value.

Hilaria rigida on sandy plains has hard stems and tough leaves, but animals are forced to consume it. *Panicum fuscum*, *P. capillare* var., and *P. colonum* are rather weeds of tilled fields, and as forage plants probably equal *Panicum Crus-galli*, *P. sanguinale*, *Setaria glauca* and *S. viridis*. With them may be classed *Helopus punctatus*, *Eragrostis Purshii* var. *diffusa*, *Chloris alba*, *Leptochloa mucronata*, as they are tender and eaten with avidity.

Agrostis verticillata, on the margins of water courses, is a tender and nutritious morsel; so also *Eatonia obtusata*, less abundant in Arizona, *Agrostis exarata* by brooks, and *Phalaris intermedia*, more widely scattered along streams and in wet, cultivated soil.

To cattle straying over miles of arid wastes, nibbling at the leaves of thorny trees and shrubs, or pulling here and there a bitter weed, such grasses as *Setaria caudata*, *Tricuspis pulchella* and *mutica*, *Muhlenbergia debilis*, and even *Aristida Americana* and *A. Humboldtiana*, and *Bouteloua aristidoides* and other species, all scattered in thin tufts over hill and mesa, furnish dainty bits seized upon with avidity. When the summer rains fall abundantly these species renew their growth, or spring up from seed, and grow rapidly, so as to cover the soil with a pretty close growth of herbage, which furnishes an abundant pasturage to fatten herds during the autumn months. Only a small part of this is consumed while green; but drying up in the droughts of October and November, and being little weather beaten in that dry climate, it serves to sustain the herds through the winter and early spring months. The more densely covered areas are sometimes mown for hay.

Cotten makes its growth entirely, as far as I have observed, during the summer rains, and this and the two species of *Pappophorum* may be classed in point of economic value with the species of *Aristida* and *Bouteloua*, though apparently less common than these.

Hilaria cenchroides, a perennial, not rare on hills, grows freely, fruits during the dry months, from April to July, and contributes a little to save stock from starvation. So likewise does *Muhlenbergia*, both wiry but nutritious grasses. Under the summer rains they grow more luxuriantly; and the latter growing in bushy clumps, retains in its wiry stems much nutriment, so that it supplies the more common sort of hay in the towns and at the stop stations, being pulled by the Mexicans or Indians, and brought in on the backs of donkeys or on carts, even as late as May, when it is gray with age.

Poa annua var. *stricta* and *Festuca microstachys* furnish few tender bits of food to cattle following up the mountain streams in spring.

Beside streams of mountain canyons, *Imperata Brasiliensis* var. at any season furnishes tall, leafy clumps, to be eaten down eagerly by the animals fortunate enough to attain to them. On the higher slopes of the mountains, particularly in those turned from the direct rays of the sun, and under the partial shade of pines and oaks, I found in May, *Atropsis* (*Glyceria*) *Californica* and *Muhlenbergia virescens* growing in clumps, standing so close together as to remind one of a northern meadow. The former furnishes the tenderest and sweetest of pasturage, and the latter is a soft and leafy grass. These two species largely compose the "deer parks" of those mountains, but unfortunately for our horses, while we were camping on the mountains they began at such an altitude (6000 feet), that we could seldom get our horses up high enough to take the benefit of them.

In Arizona the coarse grass of the valleys was called by a Spanish name, which sounded as I used to hear it pronounced by miners and Mexicans "Saccatone," though I suppose it began with a "Z." The name was applied to *Sporobolus Wrightii* and similar species. This and one other are all the names in use among the Mexicans to distinguish the shorter, softer grasses of the mesas.

Beyond the cereals, notably barley most extensively sown for hay, the agricultural grasses are scarcely employed in California agriculture.

Where permanent pasturage and hay is wanted, and where it is possible to secure and maintain this by irrigation, Alfalfa (*Medicago sativa*) is employed almost exclusively. I saw but a very few fields of Timothy, and those were confined to the higher valleys that could be irrigated by mountain springs. In winter and early spring the hills and plains are green with a species of Crane's-bill (*Erodium*), called by the Spaniards "Alfilerilla." Formerly *Avena fatua* covered the hills and valleys of California, but it has been reduced in extent by sheep.

The native grasses contribute but an insignificant portion toward the maintenance of the flocks and herds of California. On the open ranges, cattle scour large areas, browsing upon every green thing that is not too repellant or too repulsive.

Aira (*Deschampsia*) *danthonioides* offers, on damp mesas, etc., patches of fine, soft herbage, which is eaten with avidity by animals.

Deyeuxia Aleutica, growing in dense tufts on the northern coast and adjacent hills, is often sufficiently abundant to be of importance in pasturage; although it is a coarse grass, cattle eat it readily.

Deyeuxia Bolanderi, sparingly scattered through damp forests, with *Hierochloa macrophylla* and *Phalaris amethystina* (these observed at Mendocino) are most tender and palatable, but are of very slight amount.

Deyeuxia rubescens, a hard grass, grows in small, scattered patches, or thin tufts, on pine barren plains, where there is nothing else to feed deer.

Elymus condensatus grows in thin clumps, or small, scattered patches; its leaves, though tough and hard, are stripped off by hungry animals.

Oryzopsis cuspidata, *Sporobolus airoides* and *Stipa speciosa* are tufted grasses, scattered over the Mojave desert, and furnish an occasional bite, palatable, though tough to chew, to antelopes, and to strolling cattle and sheep.

Glyceria pauciflora, found on ruins of mountain tarns, is a tender and sweet grass to deer or stock coming to drink.

Melica imperfecta and *Stipa setigera*, tufted species frequenting mesas and hills, are tender and nutritious in April and May.

Throughout the mountains where cattle cannot be herded so successfully, sheep are everywhere led by their herders, swarming like vermin, and creeping up to the very pinnacles of rock or to the snow line, nibbling or tramping in the dust all vegetation. No grass at ever so great an altitude, but must contribute its mite towards the sustenance of these flocks.

Thus *Stipa stricta*, *Sporobolus depauperatus*, *S. gracillimus*, *Agrostis varians*, *Trisetum canescens*, *Melica stricta*, *Poa tenuifolia* and *P. Pringlei*, on bare mountain tops and around mountain springs and rills, must all yield a dainty mouthful to the miserable dust begrimed sheep, compelled in their ascent to live on the foliage of shrubs and on brittle herbs.

Deyeuxia equiralcis, a tender and sweet grass, grows on the verge of mountain brooks.

Agricultural Grasses of Montana.—The following notes are from a paper read at the fifth meeting of the Society for the Promotion of Agricultural Science, by F. Lamson Scribner:

"Although located so far north, and at no point less than three thousand feet above the level of the sea, horses and cattle thrive upon the 'ranges' throughout the year without care or shelter. In the valleys the standing grass cures, with all the nutritive properties held within the tissues, affording excellent hay for winter grazing.

The region abounds in a great variety of species, the whole number discovered being one hundred and twelve. Some are rare; many have little value, while one or two can only be treated as troublesome weeds.

Broom-sedge, Broom-grass, or Beard-grass (*Andropogon scoparius*), is widely dispersed from Maine to Texas, and west to the Rocky Mountains. It grows in dry, thin, or sandy soil, and thus serves a good purpose in furnishing fair forage where little else will grow. In some parts of the Missouri river and Rocky Mountain regions this grass is very abundant, and is highly prized, both for hay and for grazing. In the East it is looked upon as comparatively worthless.

Reed Canary-grass (*Phalaris arundinacea*) grows naturally in Montana in wet places, along streams, etc., and adds a little to the grazing.

Mountain Timothy (*Alopecurus pratensis, var. alpestris*). This grass is quite common at elevations of from five to seven thousand feet above the sea, growing in rich soil along mountain streams, and frequenting the so-called 'mountain meadows.' In the large, open park, a few miles west of Neihardt, there are many acres covered with this grass, and when I passed through the place, August 14th, it was being harvested for hay. It yields a large bulk of fine, long, bright-colored hay, and is highly valued. It has tall, slender, leafy culms, three feet high, with an oblong head, similar to that of Timothy, whence its local name, but the heads are shorter, thicker, and conspicuously hairy. For the more elevated meadows of the Rocky Mountain region and for northern latitudes, there is no grass which so highly commends itself as this, both for hay and for summer grazing. It is closely allied to the European Meadow Foxtail.

Feather Grass (*Stipa*). Several species of this genus are common to the region, the most prevalent being *Stipa comata* and *Stipa viridula*. They are often found together, and are usually associated with *Poa tenuifolia* and *Kœleria cristata*. The first named (*S. comata*) is the least valuable, but the more hardy of the two, growing on bench lands in soil too gravelly and thin for even *Poa tenuifolia*. It has very narrow and involute radical leaves, a few-flowered panicle, and smooth, twisted and more or less curled, hair-like beards, or awns, five inches long. Both this and the *Stipa viridula* are sometimes called wild-oat grass. The latter is by far the most valuable of the Stipas. *Stipa spartea*. Porcupine Grass, occurs, but happily in no great abundance.

Bunch Grass (*Oryzopsis cuspidata*) is very abundant on the sandy bench lands along the Missouri and other rivers. It thrives in soil too dry and sandy for the growth of other valuable species, and is much esteemed for grazing.

Alpine, or Native Timothy (*Phleum alpinum*).—This species, which closely resembles our cultivated Timothy, is common in the mountain districts, growing near streams, at elevations of from 6,000 to 8,000 feet. In the mountains back of Fort Logan, I saw this grass associated with *Phleum pratense*, and it was the more luxuriant plant of the two—not so tall, perhaps, but growing to the height of two feet, with stouter and more leafy culms. The common Timothy (*Phleum pratense*) has been introduced, and succeeds well when irrigated. But there are a number of native grasses which would yield equally fine and more abundant crops with less care.

Drop-seed Grass (*Sporobolus*).—There are several species of this genus more or less common, but none of them sufficiently abundant or valuable to have received local names.

Agrostis grandis is a species of bent-grass, common along the rich, moist banks of streams in the mountain districts. This is certainly a valuable grass to introduce into cultivation.

Reed Bent Grass (*Deyeuxia*).—There are quite a number of species of this genus native to the country, all possessing some value for forage, being readily eaten by stock. Grasses that grow naturally on these dry bench lands without irrigation, and hold the ground in spite of excessive grazing, deserve special attention, for these are the species which will best meet the requirements of the farmer when it becomes necessary for him to cultivate grasses on these same lands.

Buffalo Grass, Mesquite (*Bouteloua oligostachya*).—The true Buffalo Grass (*Buchloë*) was not seen, but this Bouteloua, which the ranchmen of Montana recognize under that name, is a no less valuable species for grazing. It frequents the bench lands at elevations of from 3,000 to 4,500 feet, and not uncommonly covers wide areas. Its strong, perennial roots and fine early leaves make a dense turf that yields a large amount of forage, and no other species seems better to withstand the tramping of stock than this.

June Grass (*Kœleria cristata*).—This is one of the most common grasses of the bench lands, disputing possession of the soil with *Poa tenuifolia,* with which it is almost always associated. On the dry benches it is seldom over a foot high, but on irrigated grounds it grows to the height of two feet or more, and makes excellent hay. June-grass is the only local name I heard applied to this species. [It may be needless to say here that this is not the grass called June grass in the east.]

Bunch-grass, Meadow-grass, Spear-grass, etc., (*Poa*).—There are a large number of Poas found throughout the northern portion of our country, and one and all are excellent pasture grasses. Wherever grasses grow at all, from the sea-shore to the mountain-top, from the arctic zone to the antarctic, this genus has its representatives. In Montana, *Poa nemoralis* ascends to the altitude of 9,000 feet. At this elevation it is dwarfed in habit, but lower down the mountain's side it soon becomes taller, and makes a valuable forage plant. Kentucky Blue-grass (*P. pratensis*) is truly indigenous, and grows quite abundantly along the streams and rivers. *Poa tenuifolia* may well be regarded as *the* grass of the country. No species withstands the long summer drought so well, and it constitutes the chief forage upon the dry bench lands. It has several local names, such as Bunch-grass, Red-top, Red-topped Buffalo-grass, etc. In the drier soils the culms are low, less than a foot, and slender, usually of a reddish color, and the foliage is reduced to the short and dense radical tuft; but the plant responds readily to richer soils and better situations, and when growing along streams or on irrigated land it makes a luxuriant growth of foliage and attains the height of two or three feet. As fine a field of natural grasses as I saw in the Territory, or, in fact, as I have ever seen anywhere, included *Poa tenuifolia, Kœleria cristata, Stipa viridula, Stipa comata,* as the leading species, the Poa being the most abundant. In this field the Stipas were unusually fine, overtopping the other grasses.

Manna-grass (*Glyceria*)—Three species are common; Reed Meadow-grass (*G. aquatica*), a well-known grass in the eastern and middle States, grows in similar situations here—wet grounds and along the borders of streams—attaining the height of from three to five feet. *Glyceria nervata* is still more abundant.

Great Bunch-grass, Buffalo Bunch-grass, (*Festuca scabrella*).—This is one of the characteristic grasses of the country. On the mountain slopes and foot-hills, at elevations of over 6,000 feet, it is the prevailing species, constituting one of the most valuable forage grasses of the winter ranges. It often covers many thousand acres of the mountain parks, and during August it is cut in large quantities for hay; it makes excellent feed, both for horses and cattle, but is rather too hard and coarse for sheep.

Sheep's Fescue (*Festuca ovina*).—The name of Bunch-grass is applied also to this species, which, in point of altitude, occupies a belt just below that held by the Great Bunch-grass. Several varieties are recognized, and all afford excellent grazing for all kinds of stock.

There are several species of *Bromus*, one of which is much like Schrader's grass: in general, however, these brome-grasses are little esteemed.

Blue-joint, or Blue-stem (*Agropyron glaucum*, var.)—There is no grass in Montana that the settlers more highly value for hay than this Blue-joint or Blue-stem, so named because of the decided bluish tint of its leaves and stalks. In appearance it resembles our common witch or couch-grass (*Agropyrum repens*), and has by some been regarded as a variety or form of that species. Like the couch-grass, this has creeping roots, making it equally objectionable in cultivated lands. It grows naturally on the dry bench lands and river bottoms, and although the yield per acre is not large, the quality of the hay is judged unsurpassed by any other species.

Fox-tail Grass, Squirrel-tail Grass (*Hordeum jubatum*).—This is a common grass in the low countries, especially where the soil is generally moist. It is considered a great nuisance, for when associated with other grasses it entirely destroys their value for hay. The long and sharp-pointed beards or awns stick fast in the nose and mouth of horses, often penetrating the flesh, and cases are reported where they have caused the death of the animals.

Lyme-grass, Wild-rye, (*Elymus condensatus*), is a large, native grass sometimes cut for hay, but is not very valuable, holding a rank much like the eastern species.

At Jefferson City, June 28, altitude about 5,000 feet, one would rarely see in any part of the country a finer looking or better sodded field of grass than was observed at this place. The strong and luxuriant growth of the grasses, *all native species*, gave sufficient proof of the resources of the Territory in this direction. I venture to say she will not find beyond her borders more valuable species either for hay or for pasturage.

Poa tenuifolia, Kœleria cristata, Stipa viridula, and *Poa pratensis* (three feet high) were the prevailing species; then came *Stipa comata, Agropyrum glaucum, A. divergens, Elymus condensatus, Poa Nevadensis, Agrostis scabra,* and *Hordeum jubatum*; along the streams or growing in the water were *Glyceria aquatica, G. nervata, Beckmannia erucœformis, Alopecurus aristulatus,* and *Catabrosa aquatica.*

Some species extend over many degrees north and south, others range within narrow limits. Some of wide range have their limits of greatest abundance confined to a few degrees. So it is in the matter of elevation above the sea. Some species range from sea level to nearly the line of perpetual snow, others are found only at certain elevations, extending but a little above or below a given altitude, while others again may have a considerabe range, but it is only within narrow limits that they are able to conquer in the struggle for life and gain almost complete possession of the soil.

With a little experience one could tell in Montana with a considerable degree of certainty the altitude of his position by the prevailing grasses about him. *Bouteloua oligostachya* and *Oryzopsis cuspidata* were never abundant above 4,000 feet. *Agropyrum glaucum* ranged a little higher, while *Poa tenuifolia, Kœleria cristata,* and *Stipa viridula* prevailed up to about 5,000 feet. *Agropyrum divergens* became the leading species between 5,000 and 5,500, when *Festuca ovina* took the field and usually held its own up to 6,000 feet, when it in turn gave way to *Festuca scabrella,* which has its line of greatest vigor between 6,000 and 7,000 feet.

Native Grasses of the Great Basin.—For Arizona and New Mexico in this basin, Dr. J. T. Rothrock ventures the assertion that for want of water there will always be much waste land so far as raising crops is concerned. For want of water, neither of these territories have reached anything like the real possibilities of the soil and climate.

Sereno Watson, a very careful observer who has spent much time in the Basin, makes a long report in the United States Geological Exploration of the Fortieth Parallel. He observes that the climate is characterized by a very dry atmosphere, small amount of rain and snow, by a cold winter and a correspondingly hot summer. No portion of this whole district is destitute of some vegetation, even in the driest seasons, excepting only the alkali plats. The vegetation possesses a monotonous sameness of aspect, and is characterized mainly by the absence of trees, by the want of a grassy greensward, the wide distribution of a few low shrubs, and by the universally prevalent gray or dull olive color of the herbage.

The turfing "buffalo" or "grama" grasses, which make the plains east of the Rocky Mountains a vast pasture for the bison, deer, and antelope, are here unknown. The grass grows in sparsely scattered tufts, dying away with the early summer heat. The two or three species that mat into a sward are confined to alkaline meadows and are nearly worthless for pasturage.

Native Grasses of Northern Mexico.—During the summer of 1885, C. G. Pringle collected and studied the flora of this country, mainly in the Mexican State, Chihuahua. By request he has furnished full notes, from which the following are taken:

With respect to the cultivated species, I think I shall surprise you by declaring that though I botanized carefully in the irrigated and tilled valleys as well as on the plains, and on the hills and mountains of every geological formation in that State, and from the beginning to the end of the season, I saw not a single plant of the exotic grasses commonly cultivated in the United States; not a stalk of *Phleum pratense* (Timothy), nor *Poa pratensis* (June grass), nor *Agrostis vulgaris* (red-top), nor any other whose tender and nutritious herbage so largely maintains the flocks and herds of the American farmer.

The only attention which I have seen the Mexican ranchero bestow upon grass is to inclose, rarely, a limited area of valley sod, formed of hard and tough species like *Sporobolus Wrightii*, *Distichlis maritima* and *Panicum obtusum*, and use the field to restrain a few saddle horses and work oxen. He provides scarcely any store of fodder for his animals, so when the growth of vegetation is arrested by the frosts of winter, they must bite shorter the half dead but still nutritious herbage, and must range widely to do this, and when the growths of the spring months, always feeble, have been entirely checked by the withering droughts which reach their worst in June, they must, if they can, maintain life by browsing shrubs, cactuses, etc.

To supply the wants of the animals kept in the cities gives employment during winter to many of the poorer class, who hawk about the streets, in ox-carts and on the backs of donkeys, bundles of dead grass gathered on far away hillsides or plains. By the beginning of March the neighboring rancheros are selling green wheat and barley in the same way, and they plant maize from early till late to succeed these. Great stacks, freshly cut, may be seen walking into town early in the morning with donkey's legs, scarcely more than the feet visible—a droll sight.

The exotic grasses which accompany cereals as weeds of tillage seem to be very few in northern Mexico. Of the 108 species on my list, I count only three such: *Panicum sanguinale*, L., *P. Crus-galli*, L., *Phalaris canariensis*, L.

Nearly all the grasses range northward from Chihuahua to a greater or less extent into the United States. All my species of *Aristida* and *Stipa*, and some species of *Muhlenbergia*, are as yet undetermined.

Paspalum Hallii, V. & S., is confined to moist situations, as the vicinity of streams and the banks of irrigating ditches. Its herbage is tender, its growth strong, and it might be cultivated to advantage in fields capable of profuse irrigation.

Panicum reticulatum, Torr., is a soft and tender annual, growing in low, scattered tufts on rich plains, and contributes not a little to the sustenance of the herds which range over them.

Panicum cæspitosum, Swartz. On rich, moist soil this forms a low, dense mat of tender and leafy herbage, relished by animals. Although only an annual, it might well be employed in irrigated fields for grazing.

Eriochloa polystachya, H. B. K., like *Panicum sanguinale*, L., is a weed in cultivated fields, and often yields large crops of a quality which would be considered good in the southwest.

Hilaria cenchroides, H. B. K. Here this is a plant of much importance to the stock grower. It forms a close perennial sod in patches of greater or less extent on the plains and mesas. As its culms are few and small and its leaves short, its yield is light, but it is a pasture grass of good character and quality.

Hilaria mutica, Benth. called in Arizona "Black Gramma," is considered one of the most valuable grasses in that region. It grows in dense perennial clumps about a foot broad, and these growing close together, to the exclusion of other species, occupy areas of considerable extent, usually in depressions of plains or mesas, sometimes even on hillsides. Except during the rainy season, about August, the plants show few living leaves, but at all times of the year the numerous branching stems contain nutriment. The clumps are detached from the soil by a blow with a mattock directed at their base, and this gives rise to the saying that hay is cut in Arizona with a hoe. As the dead leaves and their sheaths adhere for a long time on the slow growing perennial branches, a patch of this grass presents a dark grey appearance, which gives it the name of Black Gramma. Its stems are very hard, so that I was at first surprised that animals could eat it at all. My horses soon got tired of it, preferring softer grasses.

Heteropogon contortus, R. & S. This is probably the most abundant grass of dry hills of igneous rock thinly covered with soil. It grows in tall, narrow clumps, and is a perennial with numerous leafy branches. Stock show a preference for other grasses if such are to be found. The hay on sale in Chihuahua last spring seemed to be mainly composed of this plant, usually dead when gathered, and blanched by winter weather. During autumn I found its seeds a nuisance. Their long bearded and twisted awns sent the slender and rough seeds into my clothing, and often into my flesh. Sheep, goats, and even donkeys must find these seeds a terrible annoyance.

Andropogon hirtiflorus, Kth., is a fine, soft, leafy plant, growing in dense clumps, but apparently confined to hedges, etc.

Andropogon saccharoides, Sw., grows in clumps three or four feet tall. In valleys, and the moister depressions of the plains, this is a grass of some importance.

Some ten species or more of *Aristida* are mostly bunch grasses of hills and mountains, of average frequency in such situations, and of full average quality, contributing largely toward the upland pastures.

There are two new species of *Stipa* also, tall and tender plants, excellent for forage but not abundant.

There are many species of *Muhlenbergia* of more or less value. *M. Texana*, Thurber, is the mesquite grass of our southwest, and one of the most valuable species of those regions, common over mesas and hills. It is such a favorite with animals that it is exterminated except when growing under the protection of thorny shrubs, usually mesquite bushes (species of *Prosopis* and *Acacia*). Its leaves are short and scanty, but its branching, perennial, wiry stems are nutritious, and at all seasons furnish forage which is greatly relished by all kinds of stock. In Arizona the Indians bring it during winter and spring long distances into the towns to sell, the men tying the bundles behind and beside them on their ponies, and the women carry them on their backs or heads, trundling painfully behind the ponies. How many times I have contended with the horrid mesquite bushes to gather an armful of this grass to carry joyfully to my hungry and jaded horses.

Muhlenbergia gracilis, Trin., thrives on cool, grassy summits of mountains, perhaps the most important element in the pasturage of such ranges. It grows in small clumps about two feet high, and is rather wiry and tough.

Sporobolus Wrightii, Munro, grows along water courses, forming great clumps, nearly contiguous, four to six feet high. These are browsed down by stock within a foot or two of the ground. The culms are stout and stiff and the leaves long and conspicuous. They appear to be acceptable to animals. It is to this species notably that the Mexicans apply the name *Zacaton* or *Zacate grass*, meaning great grass. The same name is given to other species.

Bouteloua hirsuta is a common grass on rocky, dry soil of the hills and plains, with rather wiry culms. The quality is equal to the most of the species enumerated, and furnishes an important proportion of the forage of the region.

Bouteloua oligostachya, Torr., var. *pallida*, Scrib., is the most abundant species of the plains, especially abundant, and forms a close sod in the less arid portions. In amount of yield and in quality it is surpassed by no common grass of the plains, and is the one native species adapted to permanent mowing. I believe it would bear the effects of close grazing in enclosed areas.

Bouteloua Havardi, Vasey, I find to be the most valuable pasture grass of the hills and mesas around the city of Chihuahua. It is perennial and forms a sod more or less interrupted. It is tender and nutritious, and is kept closely cropped during most of the year.

Six or eight other species and varieties of *Bouteloua* furnish more or less pasture.

Diplachne dubia, Benth. sends up here and there, over hills from perennial shoots, a few late, succulent stems, especially relished by animals. It seems probable that this grass would succeed under good cultivation without irrigation, and, if so, no species native to Mexico would be likely to yield crops of greater amount or of superior quality.

Diplachne imbricata, Scrib., is similar to the last in habit and quality, and would probably succeed as well, but only on rich soil with copious irrigation.

Arundo donax, L., grows on the banks of streams, and is stripped of its broad leaves by cattle, which crowd upon the tall canes, straddling them to bring the leaves within reach.

Eragrostis erosa, Scrib., is a tall, soft, leafy bunch grass of the mountains, than which none can be more acceptable to stock. *Eragrostis lugens*, Nees, is a closely related species, of similar habit and quality.

In these notes I have said little about the possibility of the species mentioned for cultivation, because it does not to me seem possible that Mexican agriculture can in this generation, or in several generations, attain to the cultivation of grasses. I cannot say in what way their methods are in advance of those of the ancient Egyptians and Syrians; certainly one is astonished to find numerous parallels between their customs and practices and those of the ancients.

In regard to other forage plants which interest stock growers, I have seen a little lucerne or alfalfa grown there, but only a little. The place it occupies in the American southwest is there filled by barley, wheat, and corn.

The clovers, native or introduced, are almost entirely wanting in the southwest. Of course there are a large number of plants of many natural orders which help to sustain animal life, and I can attest from observation that there are few plants so repulsive to taste or so spicy that they are not occasionally appropriated by animals, according to the extremity of their hunger. I used to think that nothing but starvation could induce cattle to nibble at horrid opuntias, as I have seen them doing during drought.

The mesquite tree, *Prosopis juliflora*, of the southwest is worthy of especial mention. It is a godsend to those regions. Its abundant and nutritious pods, resembling those of our field bean, begin to fall in August before the grass has made much advance under the midsummer rains, and afford much relief to the half famished stock. I have gathered sacks full of them for my horses as I have journeyed through those arid districts.

Seeds are Distributed in a great variety of ways—through the agency of wind, water, snow, animals, including man himself, who purposely or unintentionally accomplishes more than any of the other agents.

The small size of the seeds of most grasses is a great help in their distribution. Many of them remain attached to the glumes and palea, or even to some of the branches, and others are provided with beards, hooks, or awns, all of which make it easy for them to be carried about by the wind, water, snow, or animals.

"It would seem that nature has appointed every animal as a special disseminator of the plants which furnish it with food. We have seen the activity of the rodent in scattering the fruits of the Nuciferæ, and of birds in sowing broadcast the seeds of fruit-bearing plants, and the ruminants seem no less active in performing a similar work for their favorite grasses. The great efficiency of animals as disseminators of seeds appears more marked when we regard them in masses. The herds of reindeer and elk on the plains of northern Europe and Asia, the bison on the prairies of North America, or the herds of naturalized horses and cattle on South American pampas, migrating from place to place in immense masses, cannot fail to sow as they pass along a host of seeds which adhere to their coats, or which they have swallowed and drop uninjured upon the soil."—(Prof. A. N. Prentiss, in Prize Essay.)

A few examples may be given to illustrate the distribution of grass seeds. The panicles of *Panicum capillare*, when ripe, easily separate from the culm and are freely tossed about and carried by the wind, scattering seeds as they go for long distances, even leaping over fences and bushes.

When snow arrives its surface becomes slightly packed, and seeds, with their chaff or branches still left on the dead culms, are occasionally torn away and drifted for long distances before the wind.

Small seeds in the mud adhere to the feet of many kinds of animals, and are thence transported from one place to another.

The elongating and spreading root-stalks of some grasses and clovers enable them to spread and occupy different ground or more ground.

The chapter on the geographical distribution of grasses will be given in the second volume.

CHAPTER VI.

GRASSES FOR CULTIVATION.

PHLEUM, L. TIMOTHY.

Spikelets in spike-like panicles, 1-flowered, rachilla very short and jointed above the empty glumes, extending beyond the floret, rarely bearing a rudimentary flower. Flower perfect. The empty glumes persistent, nearly equal, membranous, much compressed laterally, keeled, awned, or mucronate. Floral glume much shorter, broader, hyaline, truncate or toothed, 3-5-nerved. Palea narrow, hyaline. Lodicules 2, hyaline, toothed on the outer margins. Stamens 3. Styles distinct, long, slender, hairy. Caryopsis ovoid, enclosed in the floral glume, and palea, free.

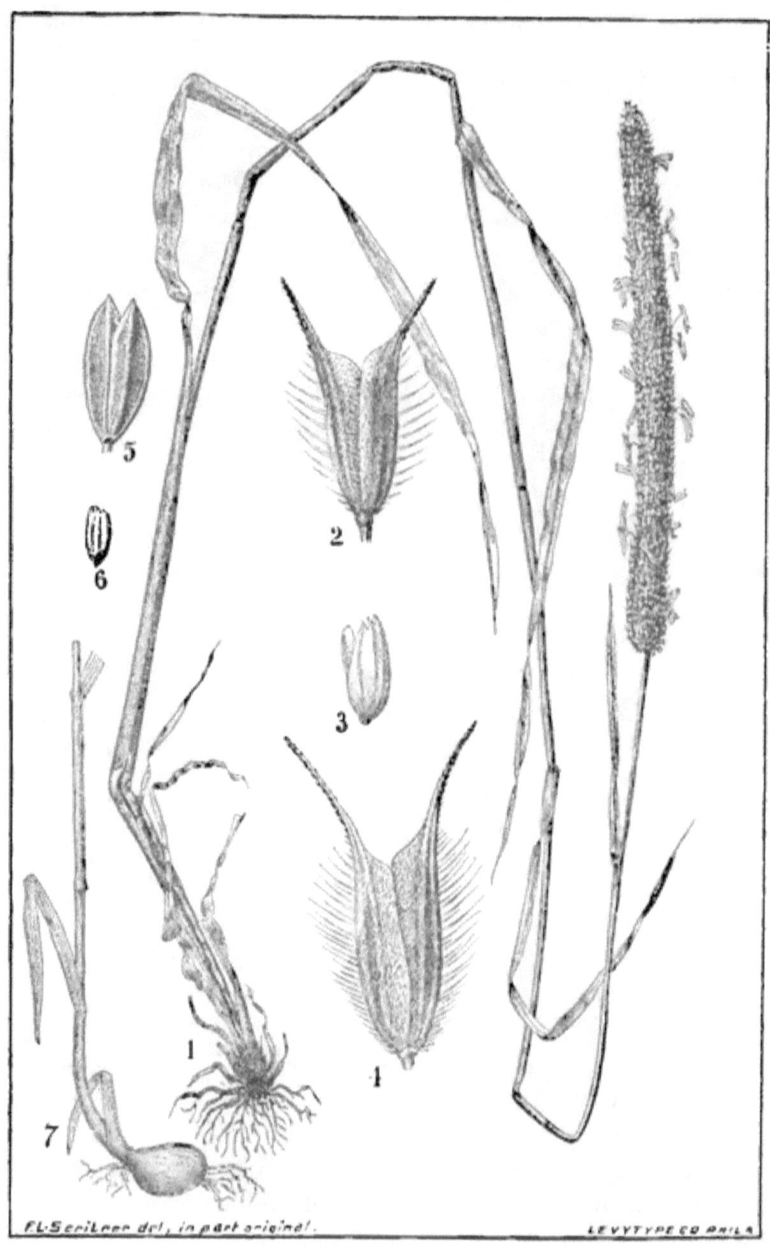

Fig. 62

Erect annuals or perennials with flat leaves. Ten species in N. and S. temperate and arctic regions.

P. pratense, L., Timothy, Herd's-Grass, Meadow Cat's Tail.—Panicle cylindrical. Empty glumes truncate with a scarious tip, and a hispid keel.

This is the best known, most extensively sown, and one of the most profitable grasses of any in the United States. In Pennsylvania and some other States, *Agrostis alba* is called "herd's grass," while at the north this is known as "red top." There are several other grasses called "cat's tail" in different portions of our country. The first common name comes from Timothy Hanson of Maryland, who introduced the grass from England about 1720. The next name comes from a man by the name of Herd, who found it growing in New Hampshire and began its cultivation.

In 1760 or '61, Peter Wynch took seeds of it from Virginia to England. It is a native of Europe, and very likely also indigenous to some portions of the United States. It is widely distributed in north Africa, western Asia, and other portions of the world. Timothy is a perennial not likely to be mistaken for any other grass, and in fact this is about the only one that is generally known by all farmers.

The leaves are short and flat, and on good soil the stem is from two to four or more feet high, each bearing one stiff, erect, rough spike as long and as thick as a lead pencil. The plant is rather coarse to the touch and sight.

Having a large bulk of stems, with few leaves, the hay wastes but little in transportation. The grass stands up well, the hay is easily cured, heavy for its bulk, presents a good appearance and suffers less than many other grasses when allowed to go to seed before cutting.

Fig. 62.—*1*. Plant of *Phleum pratense*, L.; *2*, spikelet; *3*, floral glume and palea; *4*, spikelet of *P. alpinum*; *5*, spikelet of *P. arenarium*; *6*, floral glume; *7*, base of culm of *P. pratense*, showing one enlarged solid internode, tuber, or corm, improperly called a "bulb."—(Trinius and Scribner.)

Everyone in town and country knows the grass as soon as he sees it and can distinguish it from all others, hence a leading reason why it is raised, fed, and sold. Consumers buy Timothy and fear to buy anything else, even though it were better, because they do not know what it is. They will buy even if it is dead ripe.

The same remark applies to a well known and popular grass, perennial rye grass, generally raised in England. After a long time if a grass or fruit becomes well known, and it has good qualities if not the best, people buy it because they know what they are getting.

In this country Timothy is often sown alone, at the rate of about eleven pounds to the acre. The sowing usually occurs in autumn with wheat or rye, or in the spring with oats or barley. It is often sown as the only forage crop on moist land or on strong, clay loam, but on lighter land it is usual to sow on some red clover also. If quite sandy, clover without any true grass is generally sown. Timothy is two to four weeks behind red clover in coming into flower ready for the mower. Among its other good qualities, Timothy seeds very freely, yielding 6 to 10 or more bushels of cleaned seed to the acre; and this is easily saved and threshed with a flail or a machine, can be easily cleaned and separated from seeds of weeds, and can be put onto the market in abundance and sold cheaply. It only takes from one to two pecks to sow an acre, and this costs but little.

While Timothy has many good qualities to recommend it, it has many marked defects. When sown with clover, it makes but a small growth and must be cut young, if the clover is secured in good season. It starts very slowly in spring, is a long time in coming into flower, and after cutting the second growth is slow, feeble, and of little consequence, seldom large enough to cut a second time or to afford much pasture. If cut early the tuber at the base of the stalk does not become sufficiently matured to keep the plant alive and healthy. If cut close, the tuber is cut off, and the plants suffer and become feeble, and perhaps perish.

It is hardly suited for pasture at any time, unless it is kept quite large. Horses, sheep, and especially hogs, must not be allowed to eat it close to the ground. In England it stands pasturing in spring without much injury to the hay crop. Besides these objections, Timothy is likely to be short lived; the tubers are easily trodden out by cattle, killed by drought or frost, or eaten by mice or gophers. It sometimes rusts badly. It is not hard to kill when cultivating for another crop; it starts quite readily from the seed, and is well suited for one good crop of hay in a season, but is not well adapted for pasture. It is not as well liked in Kansas and vicinity and in the south as it is at the northeast.

Timothy is one of the five grasses in the list recommended by Mr. De Laune for permanent pasture and meadow in England.

For Kansas, hear what Professor Shelton says:

"Of this favorite eastern sort, we shall say but little, believing that over a large portion of the State it is of little value. We have obtained good yields upon the college farm, and have seen good crops of Timothy grown west of this point. Still it suffers much from drought, and from the attacks of chinch-bugs, and it rarely savives the ravages of the grasshoppers."

For Nebraska, read from a lecture by Dr. Bessey:

"My inquiries were very generally answered, and in a most satisfactory manner. They all indicate that throughout the greater portion of the eastern half or two-thirds of the State, Timothy is an exceedingly valuable grass for farm use. It is invariably doing well, and in many instances producing crops of hay far beyond the most sanguine expectations of those who sowed it. It is of course not to be expected that it will succeed as we pass far into the northwestern portion of the State."

From Howard's Manual we learn that: "At the south it does not thrive on upland."

Major H. E. Alvord, of New York, in *Rural New Yorker*, reports as follows:

"Timothy is not a favorite of mine. Its hold upon the land is too slight, and, as a rule, it falls off in yield too fast after the first crop. My preference is to treat it like a grain crop—sow alone on well prepared land, in August, a half bushel of seed to the acre. After cutting the first crop turn over at once, manure and re-seed; or cut once, top-dress well, cut a rowen crop, then one crop the second year, plow and re-seed. I know of no suitable mixture for Timothy, if for hay, and do not consider it as desirable as a large part of any mixture for pasture."

Waldo F. Brown, of Ohio, writes in these words:

"I think that land seeded to Timothy and with three or four pounds of Mammoth Clover seed sown to the acre, will produce one-half more hay than Timothy alone, and the clover cures beautifully with the Timothy.

"In sowing Timothy for hay, I use a bushel of seed to three acres, and think the quality of the hay much better than when sown thin. There are many farmers of my acquaintance who sow a bushel to ten acres, and then allow it to stand till dead ripe before cutting."

With reference to saving the seed of Timothy, the following was written for *The Prairie Farmer* by Hon. Samuel Dysart:

"It is very difficult to fix any specified time for harvesting this crop, because a change in the weather may make a great difference in the ripeness in a single day. When the amount to be harvested is not large, a better yield of seed will be had by letting it stand until all the heads are ripe, and a few of the early ones shelled off. But in doing this there is much risk. A windy day may thresh half the crop. A shower of rain, followed by a warm sun, will change the color of a field in a few hours. Of late years I have harvested from 75 to 100 acres of this crop annually. I make it a rule to start the harvester into it when the early heads begin to shell at the tips. The straw of most of it is then quite green, and if carefully put up, makes fair feed for stock after threshing. If cut before fully ripe, much care must be taken in shocking, or there will be a great loss of seed in threshing, for this reason: When Timothy is ripe, the cell which holds the seed opens. If cut too green and the bundles are left exposed to the sun, the straw dries like hay, these small cells do not open, and no machine can knock the seed out of them. If cut before fully ripe this difficulty may be largely overcome by putting in round shocks as soon as cut, packing the bundles close together to exclude the air. In this condition the natural process of ripening will go on; but if set up two and two, as many set the bundles, it will dry and stop at the same stage as when cut. A good crop of Timothy should give eight bushels to the acre. I have had more, and also less. As a farm crop there is more uncertainty in saving it than others grains. It must stand in the shock at least two weeks to be dry enough for threshing. During that time, if heavy rains and high winds occur, there will be considerable waste in the shock. The less the bundles are handled after drying, the less waste. Hence I thresh it directly from the shock. All separators are now made with sieves for cleaning this crop."

DACTYLIS, L.

Spikelets several-flowered, laterally compressed, nearly sessile, crowded in dense one-sided fascicles, at the end of the branches, forming a one-sided panicle. Flowers all perfect, or the uppermost one staminate. Empty glumes unequal, membranous, keeled, the upper one larger, 3-nerved. The floral glume larger than the empty glumes, cartilaginous, keeled, 5-nerved; awn short, scabrid. Palea 2-fid, nerves ciliate. Lodicules 2.

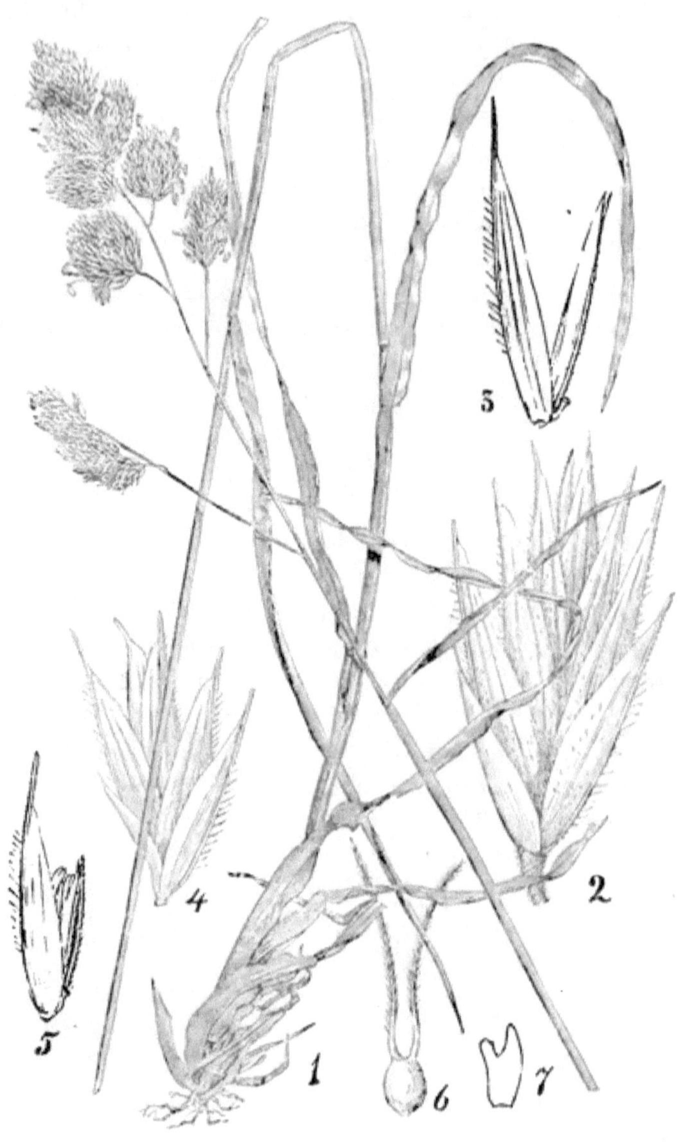

FIG. 63.

acutely toothed. Stamens 3. Styles distinct, stigmas feathery Caryopsis compressed, loosely inclosed in the floral glume, and palea free. A perennial grass with broad leaves. One species, found in cold and temperate regions of Europe, Asia, and Africa.

D. glomerata, L. Orchard Grass, Cock's-Foot.—Leaves long, keeled, conduplicate when dry, culms stout, rough, 2–5 feet. Ligule long, panicle 2–6 inches, often tinged with violet spikelets 3–5-fld., ⅓in. long.

For the past twenty years or more the writer has been accumulating notes and making observations and experiments in reference to our most noted grasses, and concerning none of the true grasses has there been more said or written or more inquiries made than about the one above named. Like every question capable of dispute, this one has two sides, and shrewd men of the same neighborhood often differ very much in their estimation of orchard grass.

The grass is perennial, lasting for many years, two to three, or even five feet or more in height, rather large, coarse, rough, of a light green color, and grows in dense tufts unless crowded by thick seeding. The lower leaves are sometimes two feet or more in length. The clustered spikelets make dense masses on the small spreading panicle; the flowers appear with those of early red clover.

It is a native of Europe, and is also now found in North Africa, India, and North America, and perhaps in other countries. Although it came to this country from Europe, it did not attract much attention in England until sent back there in 1764 from Virginia.

So far as quality is concerned, if cut in season or pastured when young it stands well the test of cattle and the chemist.

FIG. 63. *Dactylis glomerata*, L. Orchard grass; *1*, entire plant; *2* and *4*, spikelets; *3* and *5*, florets; *6*, young pistil; *7*, a lodicule.—(Spikelets by Scribner.)

It is very nutritious, the seeds start quickly and make a vigorous growth, and if the grass is not a very valuable one, it is certainly not for the lack of good testimonials.

The stems are not very abundant when compared with the leaves, hence the plant is more suitable for pasture than for meadow.

James Hunter, of England, considers: "For permanent pasture for alternate husbandry, or for hay, there is no more valuable grass, and its liberal use for all these purposes is strongly recommended."

Mr. De Laune estimates it as "By far the most valuable of all grasses because it grows in all soils; it produces the greatest amount of keep; it is the must nutritious grass, and seems to grow faster and stronger in extremes of weather, either wet or dry, than any other grass." This is one of the five which he recommends for permanent grass lands.

According to Sir J. B. Lawes, "It is very abundant and productive on good soils and is much improved by cultivation. It is really prominent only with a liberal supply of ammonia, associated with a correspondingly liberal supply of mineral constituents. It is a formidable opponent to other grasses, where it has once got possession."

The following from Alexander Hyde of Massachusetts, is excellent and to the purpose. "We have found it one of the most luxuriant and nutritious, both for grazing and for hay. It never says die. It is the first to furnish a bite for the cattle in spring, is little affected by the droughts of July and August, and continues growing until the severe cold of November locks up the sources of nourishment. When cut or grazed it starts up with the vigor of the fabled hydra. We advise no man to sow it on his lawn, for it would need cutting every day before breakfast. If cut while in blossom, both cattle and horses are exceedingly fond of the hay, and do well on it. If left to stand till the seeds are matured, it becomes more tough and woody than even Timothy, and cattle will need to have their teeth sharpened to eat it in this stage of its growth.

"Orchard grass loves a deep, rich, moist soil, and in such a soil no other grass yields such an abundant harvest. Why it is so much neglected among us we cannot divine, unless it is the fashion of sowing Timothy and clover, and fashion is as much a tyrant among farmers as among the ladies, though showing his power in a different mode."

A. W. Cheever, a most successful farmer and editor of Massachusetts writes. "I have now cultivated this grass some ten or twelve years, and feel that I can speak of it understandingly. It is a grass that must be understood to be appreciated. Grown on poor, dry land, by a poor, lazy farmer, who is always behind hand with his work, it will not give satisfaction; but on rich, moist land, capable of cutting two or three crops in a season, sown thickly with a mixture of clover and June grass, or other kinds ripening at or about the same time, and under the management of a wide-awake farmer, I can confidently pronounce it the most valuable grass known in this country at the present time. It may be cut two or three times a year, producing large crops of the very best of fodder, just as long as the fertility of the land can be maintained by top dressing. It is the earliest grass in the spring and the latest in fall."

E. H. Libby, in 1883, wrote me that "A little while ago the *New England Homestead* contained numerous letters from farmers week after week, speaking in the highest terms of this grass."

T. D. Curtis, of New York, says, Orchard grass is a most excellent hay plant, but it requires a rich soil. A well sodded pasture of this grass is a thing to admire as well as for use and profit.

Wm. Crozier, of New York, speaks as follows:

"Heretofore the base grass in all the Northern States has been

Timothy: but experiments that have been carried on for a period of twenty years have led me to believe that orchard grass is much better fitted to be the leading kind in mixtures, whether for pasture or for hay, or used alone or otherwise; and I place it far in advance not only of Timothy, but of any other grass we have thus far in cultivation. It is very early. The advantage of this earliness is not only that it gives three weeks longer for the aftermath to grow, but another reason, far more important is, that at this date the white ox-eye daisy (*Chrysanthemum leucanthemum,*) and other troublesome weeds are not yet in a condition to seed, so that should any of them happen to be in the fields, they are destroyed by being cut before they have ripened their seeds."

The following is by Prof. I. P. Roberts, of New York:

"Orchard grass is hardy with us, and gives an abundant yield of good hay, if cut early and carefully cured. Where we have used it as the principal grass in pastures, it becomes patchy; that is, some portions of the field the cattle will eat close, while other portions, where the grass gets a little start, will go to seed, after which all growth ceases till the next season. I have frequently mowed the pastures as the grass was heading out; sometimes the cut grass was left on the field, sometimes cured for hay. It grows in hummocks to such an extent that evaporation from the soil in dry weather goes on so rapidly that the other grasses perish for want of moisture, and then, too, orchard grass is always 'dry' and takes the lion's share of the moisture. Except for timber lots, and for mixing with a variety of grasses for permanent pastures, its value is not great with us."

A writer in the Connecticut Report of the Board of Agriculture for 1868, remarks: "Orchard grass does well on dry land, giving a large yield of coarse, black looking hay, very sweet and palatable to cattle, but it must be cut early, suffering more from standing too long than any other grass with which I am familiar.

On moist places it runs to tufts. No grass does better in the shade than this, and none gives so quick a second growth, or so strong aftermath. With me it ripens precisely with red clover, and I always sow them together. Clover and orchard grass I sow together in the spring, using 12 pounds of clover and two bushels of orchard grass per acre."

L. F. Allen, of Buffalo, New York, approves of a favorable article in the *New York Tribune*, saying: "I have had it in continuous mowing and pasturage for upwards of forty years without disturbing it. As a market hay, I admit that Timothy is more salable, because town's people do not know the value of the orchard grass, which is just as good for any kind of animal."

In the *Country Gentleman* for 1883, the same man of wide experience and observation, remarks: "Why it is that farmers are so dull in the use of orchard grass, passes my comprehension, when, on a single trial of its virtues, mixed with red clover when sown, equal in proportion for a hay crop, it is better for any class of stock than Timothy."

The late Hon. George Geddes, of New York, said: "It is a very valuable grass, but unless thickly sown it is inclined to grow in tussocks or bunches."

T. A. Cole, Madison Co., New York, in *New York Tribune*, remarks:

"After twenty years of experience, I have settled down upon orchard grass as possessing greater merits than any other, for both pasture and meadow, for fattening animals or for dairy stock. When cut for hay, just before its bloom, and cured with as little sun as possible, it will make more milk than any other variety known to me; if left to ripen it is worthless. When sown thick enough it does not grow in tussocks and will crowd out white daisies, and in five or six years I have seen it crowd out quack-grass. Hundreds of farmers in this region are raising it, and in every instance consider it superior as a forage plant."

The following is from the pen of Major H. E. Alvord, of New York, and was written for the *Rural New Yorker:*

"Orchard grass is a variety which has no superior for pasture or hay, and it matures so early that the crop may be easily got out of the way before Timothy or red top is fit to cut. But orchard grass must have a good strong soil, and can be made most profitable by keeping land thus seeded in sod for a series of years. If cut twice a year or three times, as is often possible, it must be liberally top-dressed. With the land previously in good condition and a well prepared seed-bed, orchard grass is very satisfactory, grown by itself. For this purpose, I would sow it as soon as the land can be put in order in the spring, or in the latter part of August, using at least two bushels of seed to the acre, put on with the greatest care, as it is a difficult material to handle. In 1884, orchard grass was in bloom in May, at Houghton Farm, and good hay was made the first week in June. The period of cutting as to maturity of plant, should be regulated according to the use to be made of the hay. It can be cut so as to make hay as fine as any rowen or coarser than any heavy Timothy. If a mixture is desired for hay, tall meadow oat-grass and clover are the best for maturing with the orchard grass. Although orchard grass is hardy, furnishes the first green bite in the spring, and the last in the fall, and usually provides good protection with its own aftermath, it will winter kill where not well covered with snow, if the land is moist. It prefers a location rather high and dry, naturally or artificially well drained."

Prof. J. R. Page, of Virginia, says: "It does well and yields one and a half to two tons per acre."

W. F. Tallant, of the same State, in the *Country Gentleman* remarks: "It will grow more in one week after cutting than blue grass will in a month. It makes a larger aftermath, and makes it quicker than any other grass I know of. It is ready to cut before harvest and after planting. Timothy is too near wheat harvest, so that it is often left until that is over, when it is entirely too ripe. I have tried it on rich land and poor land with good results."

Orchard grass is much raised in Kentucky, where it has been grown since 1817.

Richard Waters, of Oldham county, in *The Tribune*, says: "Orchard grass grows best in good, strong loam, reasonably dry, not on sandy land, nor in wet land. It will graze more stock to the acre, and can be grazed ten days earlier in the spring than any other grass. It makes good winter pasture, and during one recent winter I kept 800 ewes on this grass all winter without any other feed."

On the same subject, we learn from Dr. J. B. Killebrew, of Tennessee: "It likes a soil moderately dry, porous, fertile, and inclined to be sandy. It withstands hot, dry weather better than any other valuable grass."

A prominent writer in *The Rural World*, of Missouri, states:

"When suitably located and properly grown, it is one of the best of our cultivated grasses, but when not so located and grown, it is of indifferent value. Sow two bushels of seed to the acre, if sown alone."

Prof. D. L. Phares, of Mississippi, prefers to sow this grass in the spring without a grain crop, and on well prepared land. It thrives well without renewal on the same ground for thirty or forty years, and is easily exterminated when the land is desired for other crops. The growth in clumps may be obviated by thick seeding.

"Altogether and from every standpoint, I am compelled to say still, as I did many years ago, that I prefer orchard to any other grass. I could fill volumes with testimonials more strongly expressed than my own in favor of this grass over all others.

"It produces seed freely, and they germinate with certainty, a bushel weighing twelve to fifteen pounds."

In *Howard's Manual of Grasses*, we read: "This valuable grass ranks next in importance to the tall meadow-oat-grass for hay and winter pasture. The second growth after cutting should be reserved for winter grazing. Where hay is an object, meadow-oat and orchard grass should be sowed with red clover and white, as each of the four blossoms at the same time and is simultaneously ready for the scythe. The cultivation of these two grasses at the South cannot be too strongly recommended on soils adapted to them."

Prof. S. A. Knapp, of Iowa, after looking the ground all over, concludes that orchard grass is valuable for early and summer use, but not superior for late fall pasture upon the open prairie.

Prof. G. E. Morrow, of Illinois, in *Rural New Yorker*, says:

"Both for pasture and for hay, I think we have underrated the value of orchard grass, if sown thickly and not allowed to become harsh and woody by standing too long."

Those living on the dry prairies will be interested in the following from Professor Shelton, of Kansas:

"Two years ago, in giving our experience with this grass, we stated that it had proved to be 'one of the very best and safest of all the pasture grasses that we have tried.' It has proved with us but an indifferent hay plant, yielding moderately upon ordinary soils; and the hay, when well secured, is not relished by our stock. In our experience the hay is hardly equal to that cut from the prairie. Our experience is totally against this grass as a hay plant; but, in grazing, its valuable qualities soon become apparent to the farmer. We feel confident that it will yield fully twice the feed that can be obtained from the same area of blue grass or Timothy, and in nutritive qualities is certainly greatly superior to blue grass. Orchard grass is one of the earliest grasses to start in the spring, and the last to succumb to the frost in the fall. By giving it a good start in the fall, it will furnish good pasture far into the winter. It is consumed with great relish by stock of all kinds, especially if the grass is cropped short. It seems to do equally well upon heavy clay and sandy soil; and any rich and well drained soil seems suited to it. It germinates about as easily as oats; and, with good seed, no difficulty is experienced in getting a 'stand' that will endure moderate cropping the first fall after seeding. As might be inferred from its common name, it does better when moderately shaded, and is admirably suited to orchard culture; yet there are few grasses that will so well endure the prolonged sunshine of our dry seasons. For these reasons we feel safe in recommending this grass to the farmers of central Kansas for the purposes of pasture.

"Clover has always thrived with orchard grass, besides furnishing to animals that variety of food so agreeable to the taste. We have found that orchard grass is relished even by swine, and therefore it makes excellent 'hog pastures.' In our experience, too, no amount of tramping or close grazing at any season has been able to injure a well-rooted sod.

"Orchard grass will endure late seeding better, perhaps, than any other sort: but this operation ought not to be delayed much beyond the middle of April."

Still later on he concludes as follows:

"Of all the large number of grasses that have been tested at the College Farm during the past twelve years, this has proved the most generally useful, because: 1, a 'stand' is easily and quickly obtained; 2, it yields wonderfully of pasturage and hay if the land is good—indeed orchard grass is such a gross feeder that it is not worth a trial upon very poor land; 3, it does not winter-kill, does not 'tread out,' is not injured by too close cropping, and will survive an uninterrupted four months' drought. It winter-killed badly in 1885-6."

Hear a few good words from Dr. Bessey.

"The nutritive value of orchard grass, as determined by chemical analysis, shows it to rank well up toward the high value of blue grass. It is much more nutritious than Timothy, and very nearly as valuable as red-top. It is shown by trial to grow well in many parts of Nebraska, and is considered by many to be one of our best grasses."

Hon. L. N. Bonham, of Ohio, among other things says: "If the land is not too strong, orchard grass is an improvement sown with clover intended for hay. In strong, black land, however, I have found the culm incased by the several folds of the leaves, so thick and sappy that it does not cure before the leaves are so dry as to crumble into powder. On poorer land I prefer orchard grass with clover. Where pasture is desired, orchard grass adds greatly to the value of the clover field. It furnishes a variety, recovers quickly after mowing or being eaten down, and comes early. It is not appreciated, and is neglected by farmers because the seed is more difficult to sow and is more expensive than clover or Timothy per acre. Its chief value is for pasture."

As we might expect where a grass has been so long in cultivation, it varies much in vigor and size. In England some attention has been given to selecting vigorous varieties. Like Indian corn, it is well to select seeds from large, thrifty, well grown plants.

B. A. R., of Bowling Green, Kentucky, thus describes the mode of saving seeds of orchard grass:

"About the time the seeds are ripe, and before they commence to shatter, take a reaper and set the sickle about one foot above the ground, so as to be above the leaves or blades, and cut, bind and shock as wheat, only make the bundles and shocks smaller. Leave the shocks uncapped for three or four weeks, exposed to the action of the sun and rain. This is necessary to make it thresh clean from the head. At the end of three or four weeks, as above stated, place a canvass in the bottom of the bed or frame in which it is to be hauled (to avoid waste, as it shatters very badly at this time), and haul to the place of threshing. If not ready to thresh right away, you must cover it with something, stack it, or put in barn, as too much dampness will prove injurious to the seed at this time. Remember to handle over a canvass as much as possible whenever you move it, for otherwise the loss will be considerable. As for the yield, that is very variable—all the way from five to fifteen bushels per acre, according to the age of the meadow and fertility of the soil. Orchard grass increases its yield every year from the second to the sixth or seventh after sowing. But even at this seemingly small yield it is very profitable, as the labor is not very great and there is an abundance of good hay left to be mowed after the seed are saved."

Of the producers or of country merchants of Kentucky and Tennessee, the seed may be obtained for much less than is usually paid to the seed dealers of the northern States. It is usually put up in eight-bushel sacks, 14 lbs. being allowed to the bushel.

These long quotations have been selected from wide awake, observing men living in remote parts of our great country. I have neglected to quote much from those who speak against it, believing that they do not understand the grass and consequently make mistakes in its management.

As a rule it blossoms but once a year, and then about a month ahead of Timothy and red top.

It is often mentioned as very suitable for growing in the shade, but June grass does as well, comparatively. It will not spread and make a fine, handsome turf.

Many cut it too late, when the hay will be of poor quality.

A farmer should not have too much of it for meadow, because it all comes on at once, and then it should be cut; the weather often controls the time of cutting. If rainy when the grass is in flower we must wait often a week or more. In such cases most grasses endure the delay better than orchard grass.

Some object to its use as the seed costs too much, from one to two dollars a bushel, making the seed for an acre cost from two to five dollars. Others sow on stiff, poor soil, where it makes a feeble growth.

Fig. 64.

ARRHENATHERUM, BEAUV.

Spikelets subterete, 2-flowered, panicled; rachilla jointed above the empty glumes, extending above the upper flower; the lower flower staminate, the upper perfect or pistillate. Empty glumes persistent, membranous, unequal, mucronate; the floral glumes firm, 5-7-nerved, the lower one bearing a long, bent awn below the middle, the upper one bristle-pointed near the tip or awnless, or rarely bearing a stout, bent awn. Palea narrow, hyaline, 2-nerved. Lodicules 2-fid. Stamens 3. Styles short, distinct, stigmas feathery. Caryopsis ovoid, free. Perennials, leaves flat or convolute when dry. Three species, found in Europe, northern Africa, and western Asia.

A. avenaceum, Beauv. Tall Oat-Grass, False Oat-Grass, French Rye-Grass, Evergreen Grass [at the south].—(*Avena elatior*, L.) Panicle narrow, long, nodding. Spikelets ⅓-in., floral glume, with bristly hairs at the base, palea shining; introduced.

Within the past few years this grass has become somewhat prominent, and has won many notes of praise from the farmers, especially from those living in the south and west. It has long been grown in some portions of Virginia.

Tall oat-grass is a hardy perennial, growing from three to six feet high and bearing a loose panicle somewhat resembling one of the common oats, only more slender in every way. It is common in Europe and western Asia, and has some peculiarities which ought to be well understood by those attempting to grow it for meadow or pasture.

In place of much experience by our best farmers, the writer will quote the somewhat conflicting views of several eminent authorities.

Dr. Lindley, of England, said: "It is bitter and ungrateful

Fig. 61.—*Arrhenatherum avenaceum*, Beauv. (Tall Oat-Grass): *a*, plant; *c*, spikelet nearly closed; *b*, rather more enlarged and the outer glumes removed.—(Details by Scribner.)

to animals, and there is no reason why this grass should be regarded as fit for cultivation. The variety *bulbosum* is apt to become a troublesome weed, difficult to extirpate."

William Gorrie, of England, says: "It is most extensively cultivated on the continent; speedily attains to maturity from seed, yields continuously from early spring till winter frosts a large bulk of produce, yet it contains a small proportion of nutriment and possesses a very disagreeable, bitter taste, which causes it to be avoided by horses, cattle, and sheep. It is very subject to rust and black smut. It abounds chiefly on light, dry, arable soils. Its cultivation under any circumstances would not fail to create suspicions of lunacy against the grower. Its extirpation alone demands attention."

Sir J. B. Lawes says: "The endowments favorable to this grass are its hardiness, its comparative indifference to the character of the soil, its particularly ample root growth, both deep and superficial, its strong, tufted habit, and its early flowering tendency. It yields a considerable quantity of foliage on the culms, which affords a good deal of leafy feed in the spring. It produces rapidly after cutting; its taste is bitter, but it is not disliked by cattle. It does not grow abundantly except upon poor soils, and is, upon the whole, of somewhat questionable value. It is much grown in France."

The late Professor James Buckman, also of England, a good botanist who had given much study to the grasses, said: "This is exceedingly bitter, uniformly refused by cows and sheep unless starved to it by want of something better. We think it would be better to discourage its growth. We have two forms, one of which is the variety *bulbosum*, growing in sandy lands. In this the bulbs become enlarged and look like a string of onions on a small scale, which gives it the name of 'onion couch.' The only way to get rid of it is to hand pick it after repeated plowing and harrowing."

Even the English seedsmen, who recommend the use of many sorts, the value of which is questioned by farmers, do not include tall oat grass in the list of valuable grasses.

But the reader doubtless cares less about what the English think of tall oat grass, than he does about what some of the best American farmers think of it.

Judge Jesse Buell, of Connecticut, in 1823, quotes the opinion of Dr. Muhlenburg and Mr. Taylor, of Virginia, who place this at the head of good grasses. It possesses the advantages of early, late, and quick growth, for which the orchard grass is esteemed, and is well calculated for a pasture grass. Dickenson, quoted by Buell, says: "It makes good hay, but is most beneficial when retained in a close state of feeding."

Prof. D. L. Phares, of Mississippi, says: "It has a wonderful capacity of withstanding the severest heats and droughts of summer and colds of winter. It admits of being cut twice a year, yielding twice as much hay as Timothy, and is probably the best winter grass that can be obtained. To make good hay it must be cut the instant it blooms. For green soiling it may be cut four or five times, with favorable seasons. Along the more southerly belt it may be sown in November and onward till the middle of December. It is one of the most certain grasses to have a good catch."

The late Mr. C. W. Howard thought, this grass deserved to be placed at the head of winter grasses for the South. It does not answer well on moist land. Seed sown in the spring will produce seed in the fall.

Prof. E. M. Shelton, of Kansas, says: "This grass has within a few years been extensively sold in the West under the name of 'evergreen grass.' We have tried it for a number of years upon a considerable scale. No grass that we have yet tried has, during its first season, made such a vigorous growth as meadow oat-grass did last year. In this respect it has greatly surpassed our old favorite, orchard grass. It made a much better stand than did orchard grass growing beside it, and endured the severe and protracted drought of the latter part of the season better, retaining its intense green throughout. This grass, although sown late in April, gave a heavy cutting of hay in July, a feat that we have never before accomplished with any other sort. It makes excellent pasturage early in the spring and late in the fall, but as a hay plant, and for general pasturage, it is greatly inferior in Kansas to orchard grass."

The Students' Farm Journal, of Iowa Agricultual College, sums up its merits as follows:

"It vegetates earlier in the spring than any grass we have ever seen, producing pasture for cattle by April 15. It stood five feet ten inches May 1, started April 5. This is a great item to the farmer, for hay and corn are worth something in April and so is the time required in their feeding. It grows strong and even throughout the entire year and very late in the fall. It is best for pasture but makes coarse hay, but of fine flavor if cut early. It will blossom twice in one season if cut early. Its flavor and smell are good. By chemical analysis this grass contains some more flesh or muscle forming material than Timothy. More fiber and less fat. But chemical analysis is not the most important element used in judging of a plant's value. It is better than Timothy in not being so hard on the soil, and produces nearly twice as much hay. No grass in the college experimental grass garden is more promising than this. It ripens earlier than Timothy and is therefore better mixing with clover."

Lieutenant Governor Sessions, of Ionia, Michigan, has given this grass a good trial and reports: "In a very dry season the newly seeded clover and Timothy disappeared, but the oat-grass sown with it grew well. It more than holds its own with clover and Timothy. It is rank and early and will seed twice each season. It makes good pasture and good hay, and is very pro-litle. I want a permanent grass, so I have not tried to destroy it."

The writer has raised this grass on rather light, sandy soil at Lansing, Michigan, for twenty years or more, has seen it in some other localities in the State, and thinks he can tell why there are such conflicting opinions in relation to its value. In England the climate is moist, and the finer succulent grasses thrive well, while tall oat-grass does better in a hotter, dryer climate. He has had occasion to kill several plats and has had no more trouble with it than in killing so much Timothy. There are some bulbs on the sort raised in Michigan, but they are not hard to kill. Like orchard grass, it ripens very quickly after blooming, and to make good hay there must be no delay in cutting. As it blossoms rather early, many let it go too long before cutting, when the stems become woody and of poor quality. Again, bad weather often interferes with the cutting just at the right time, and poor hay is the result. A man doesn't want a very large quantity of this grass to mow, unless he is prepared to cut it all in a day or two. It makes a fine growth the first season after sowing, and if sown alone will cut a good crop of hay.

I find that stock eat the grass well, though most likely they would prefer to have some grass not so bitter for a part of the time. The seed is rather light, weighing fourteen pounds to the bushel in the chaff. About two bushels to the acre are usually sown. Only half of the flowers set seed, as every other one is staminate. The seed is rather large, starts early, and soon makes a vigorous plant. This fits it for alternate husbandry and for dry countries.

In saving the seed, care needs to be used to cut the grass just as soon as the top of the panicle is ready. Not a half day should go by or seed will be lost. It is cut high, bound in small bundles, shocked till well cured, when it is drawn to the threshing floor on a wagon supplied with a canvas to save the shelling seeds. It yields from ten to twenty bushels of seed to the acre.

FESTUCA, L. FESCUE.

Spikelets 3 or more flowered, subterete, in a compact or slightly spreading panicle, rachilla jointed above the empty glumes and between the flowers; flowers perfect or rarely staminate, empty glumes, persistent, unequal, shorter than the lowest floret, acute, keeled, the outer 1-nerved, the inner larger, usually 3-nerved. Floral glumes narrow with 3-5 obscure nerves, acute, mucronate or awned at or near the tips. Palea shorter, 2-nerved. Lodicules 2, notched. Stamens 1-3. Styles short, terminal, distinct; stigmas feathery. Caryopsis oblong or linear, more or less adherent to the floral glume and palea. Many are tufted perennials; leaves flat, covolute when dry, or narrow and permanently conduplicate. The glumes are longer and more pointed than in *Poa*, otherwise the two genera blend together.

About 80 distinct species, many of which are quite variable. Found in arctic, cold, and temperate regions.

F. elatior, L. Tall Meadow Fescue, Randall Grass, Evergreen Grass. A perennial, 2-4 ft. hi., usually tufted. Leaves broad, flat, panicle narrow, erect or nodding, 5-9 in. Spikelets 3-7-flowered, about ½ in. Floral glume pointed, 5-ribbed. When the panicle is much branched the glumes are narrower and more pointed, and the ribs less distinct.

This is a very variable perennial, two to four or even five feet in height, generally growing in tufts or bunches, which from year to year creep slowly upward, as the new growth springs from the side of the old culms, a little above that of the previous year.

The roots are stout and woody, with a slight tendency to sucker. The leaves are rather firm, flat, varying much in length and breadth, but often one to two feet long.

The nodding, spreading panicle somewhat resembles that of a

slender top of chess, with which every farmer is familiar. It flowers about a week or ten days before Timothy.

This valuable grass, with several kindred species and varieties, is found throughout Europe, western Asia, and has been introduced into North America.

Tall fescue has long been in high favor with the best farmers of Great Britain, as it is well liked by all domestic herbivorous animals.

Mr. Gorrie, a competent British authority, speaks of it as the most important species of the fescues, highly valuable for permanent grass lands, both for spring and autumn, but not the best suited for alternate husbandry, as it does not attain to full productive powers till the third year from sowing. It is very nutritious, making excellent hay as well as pasture.

This grass is seldom sown in a pure state, but is frequently met with, in the northern States especially, where the soil is heavy and inclined to be moist. It seeds freely, and the seeds germinate quickly and make strong young plants. If used alone, sow two bushels to the acre.

For the South, Dr. Phares considers it one of the best winter grasses, and says it is much prized as far north as Virginia, where it furnishes cattle good grazing in mid-winter. To the writer this seems to be one of the most promising grasses for the dry prairie lands of the west.

Festuca pratensis, Huds. Meadow Fescue, Randall Grass.— By many botanists this is considered a mere variety of the former species, or the former grass a mere variety of this one; some choosing one name, some the other. The one now under consideration much resembles the tall fescue. It is a little earlier, considerable smaller, with shorter, thinner leaves, less inclined to grow in tufts, top narrower and simpler. The reader should consult the remarks on *Festuca elatior*, which mostly apply to this one also. Most of the seeds sold for tall meadow fescue are

Fig. 65.

those of meadow fescue or perennial rye-grass, and most of those sold for meadow fescue are all seeds of rye-grass, or they are very extensively adulterated with those of rye-grass, to the extent of fifty to ninety per cent.

In the words of James Hunter, a seedsman of England: "So closely to the naked eye do the seeds of meadow fescue resemble the seeds of perennial rye-grass, that abundant opportunities for adulteration are afforded and are certainly not neglected. The fact that the average price of perennial rye-grass is only about one-fourth or one-fifth that of meadow fescue, sufficiently explains the motives of those who mix these seeds."

The writer knows well that the frequent adulteration of the seeds of meadow fescue is one of the chief causes why so little is raised in this country. A farmer not knowing either grass, orders seed of this one and gets seeds of rye-grass, which produce plants not satisfactory to his needs. This is one of the five grasses recommended by Mr. De Laune, of England, for permanent pastures and meadows, the others being tall fescue, orchard grass, Timothy, and meadow foxtail. His valuable experience is noticed under the head of "testing seeds," and "what to sow."

The writer at present would advise no one to buy seeds of meadow or tall fescue unless he is a good botanist or employs a good botanist to examine the seeds for its identity.

For the South, Prof. Phares thinks "Randall, evergreen grass, or meadow fescue is a magnificent winter grass; in fact it may be grazed from September till June. Or taking stock off in April, it will make a large crop of seed, and a heavy crop of hay, as the seeds mature while the stems and leaves are still green. This may be made to yield two to four tons of hay per acre and of high nutritive rank. It grows well on wet or dry bottoms, hillsides and tops, gravelly and loamy

---

FIG. 65.—*Festuca elatior*, L. (Taller Fescue; Part of plant, *a*, spikelet enlarged; *b*, floret enlarged. (Scribner.)

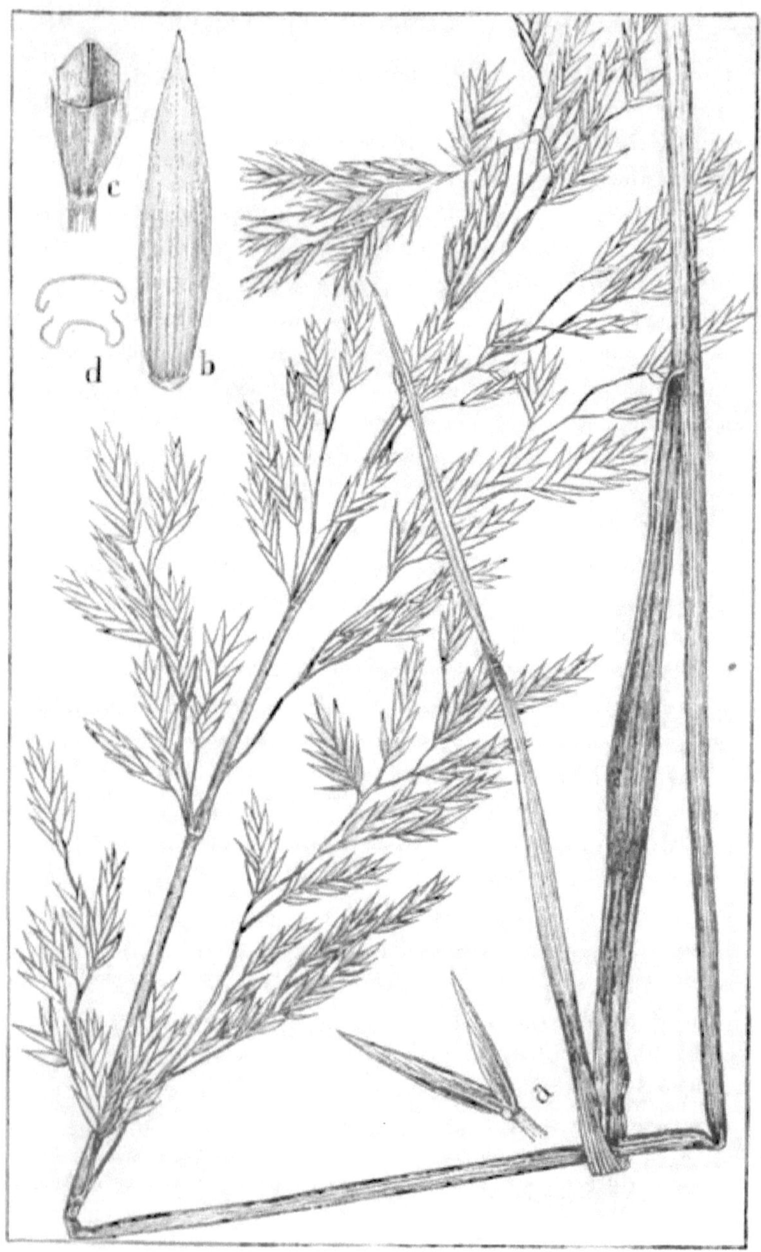

Fig. 66.

lands and clays, and having many fibrous roots running down eight to fifteen inches, resists the droughts."

For Kansas, read what Professor Shelton writes: "After experimenting for twelve years, I have often wondered that the cultivation of this grass has not been more widely extended. It gives a good amount of early and late feed of good quality, and yields heavily, of good hay. It endures dry weather, in strong lands, without injury. People east and west can afford to give this Festuca a trial. Sow two to two and a half bushels of seed to the acre."

Professors Latta and Troop, of Indiana, say that "Meadow Fescue and Taller Fescue do remarkably well at Lafayette, and we look upon them as the coming grasses of this section. The first seems to give better satisfaction as hay, while the second furnishes more pasture after cutting. Its leaves are too rough and harsh for hay."

As before said the various samples of this grass already vary much in size and vigor, and this shows what might be done with a little time and care in selecting certain types and in raising each by itself. Like Indian corn, they seem ready to break up into permanent varieties. Prof. James Buckman, of England, tried, side by side, the two fescues above named, and another called *Festuca loliacea*, and found all intermediate stages passing from one into either of the others, but under certain circumstances each maintained its distinct characters.

Festuca elatior var. arundinacea, Tall Meadow Fescue.—Leaves longer, broader, firmer, culm stouter and taller, panicle more erect, roots larger and stouter than those of *F. elatior*.

For many years the writer has had three separate forms or *races* of the larger fescues, each of which came from seeds of distinct selected plants. The mixed seed at first was received

FIG. 66.—*Festuca elatior* var. *arundinacea*; part of plant; *a* outer glumes; *b*, floral glume; *c*, section of floral glume and palea; *d*, a cross-section of same.—(Sudworth).

from the Kew Gardens. Of all the races this one seems the best adapted to the dry prairie regions of the Central United States.

Festuca ovina, L., Sheep's Fescue.—A small perennial, densely tufted, leaves chiefly radical, very narrow, conduplicate, appearing cylindrical, the upper more or less flattened. Panicle one-sided, short; spikelets, 4-10-fld. Glumes faintly nerved. Dry, hilly pastures, very variable. Of little value, but here mentioned because it is so common and likely to be found.

Festuca duriuscula, L., Hard Fescue.—Compared with the preceding, less densely tufted, taller, larger, sheaths downy. Panicle more open, varying much in color. All intermediate forms can be selec'ed from this to the preceding, of which many consider it a mere variety.

For dry pastures this seems to be worthy of some attention.

POA, L.

Spikelets, 2-6-flowered, compressed, in loose or close panicles; bunches, 2-nate or in ½-whorls. Rachilla jointed between the flowers which are perfect, rarely imperfect. Empty glumes unequal, shorter than the lowest floral one, keeled, acute or obtuse; the lower 1-nerved, the upper larger, 3-nerved. Floral glume often webbed below, keeled, acute or obtuse, 5, rarely 7-nerved, tips hyaline. Palea 2-nerved, ciliate. Lodicules tumid below. Stamens 3. Styles 2, short, terminal, distinct, stigmas feathery. Caryopsis avoid, oblong, grooved, free. Annuals or perennials, low or tall, leaves flat or conduplicate. About 80 species, chiefly in cold or temperate regions, nearly related to the fescues, having shorter and more compressed glumes, without awns.

P. pratensis, L. June Grass, Spear Grass, Green Grass, Smooth-stalked Meadow Grass, Blue Grass, Kentucky Blue Grass.—A perennial, 1-2 ft. hi., with creeping root stocks. Culm, smooth, terete. Leaves narrow, keeled, tip closed, ligule short, obtuse. Panicle pyramidal, 2-3 inches long, with slender, spreading branches, 3-5-nate. Spikelets ovate or oblong, 3-5-flowered. Floral glume silky-hairy on the keel, 5-nerved. See Fig. 51.

This is one of the most common and most useful grasses in the Northern temperate zone; especially valuable in North America for lawns and permanent pastures.

It is not so highly esteemed in Great Britain as in this country, as there it is objected to on account of excluding other grasses which are considered more valuable in that climate. It is found also in Asia and Australia, varying considerably in size and appearance.

June grass varies in height, from a few inches to a foot, and in rich ground, where the stems have not yet become crowded, samples may be found which exceed four feet. It is noted for root stocks which spread rapidly and fill the ground near the surface with a close mat of turf, much like quack grass. This makes the grass very tenacious and hard to kill, especially in moist land or in wet seasons when the land is used for a hoed crop. The crowding of these root stocks weakens the stems above ground and soon a large amount of vegetable matter accumulates near the surface.

It flowers about the same time as the earliest red clover and orchard grass, and nearly all comes on at once. The seed soon matures, and, unlike *Poa compressa* and *Poa serotina*, the culms soon turn yellow and die, and the leaves become feeble or perish.

It flowers but once a year. The leaves are slender and when dry fold up like the two halves of a book when closed.

As stock feed off the leaves, many of them, and some new ones, continue to elongate by growth at the base of the blade near the apex of the sheath. The apex of the blade is the oldest portion; the base the youngest.

In a wet season, in a hedge, the writer found some leaves still green and thrifty where they were almost (5½ ft.) five and a half feet long.

Although this grass is so very common, yet frequent inquiries are made in reference to its value. Are June grass and Kentucky blue grass, or blue grass of Kentucky, the same?

Frequent experiments and careful study by the botanists prove that they are without question identical—one and the same.

We have a rather thin, short, late grass, with short leaves, a small top, and a flattened stem. This one noted in the last sentence is very rich, of a dark bluish-green color, and is often called "blue grass," a name which it richly deserves. It is *Poa compressa*, wire grass or flat stemed poa, an account of which should be read in this connection.

June grass starts quickly in spring, after mowing or feeding, unless the weather be quite dry. It is very rarely injured by the cold, and very hard to kill by dry weather, hot sun, the tramping of hoofs, or close mowing. It is a perennial, living on and on almost indefinitely.

In most soils the stalks are too short for a large yield of hay, but if cut early, in flower or a little before, and well cured, the hay is very rich, and will go a great way, considering its bulk.

It is too frequently condemned for its single crop of short stems and leaves.

It does not get a great name on account of its value for meadow, but on account of its endurance and great worth for permanent pasture and lawn.

The leaves keep growing and make much feed, if the soil and season be not unfavorable.

Like all other grasses for feeding in cold weather or in a very dry spell, it should be allowed to get a good start before this trying time arrives.

But few sections of country are suited to a permanent and fine growth of June grass. Such land is always deep, rich, and valuable for many other crops. The forest trees in such sections of the United States are usually large, tall, thick, abounding in sugar maple, black walnut, hickory, white, black, and blue ashes, red elm, black cherry, and burr oak.

A Kentucky farmer says: "Whoever has lime-stone land has blue grass; whoever has blue grass has the basis of all agricultural prosperity, and that man, if he has not the finest horses, cattle, and sheep has no one to blame but himself."

Besides some portions of Kentucky, there are also a few counties or parts of counties in Ohio, Michigan, and Indiana.

It requires three years or more to become well established, and on this account should not be sown for one or two crops of grass or hay.

Among the numerous plats of grasses, clovers, and other experimental plants of the Michigan Agricultural College, not one is so persistently omnipresent as June grass. The seeds push up and make young plants at all growing seasons of the year; these cannot always be certainly detected until their tops appear. In Michigan it is certainly a good fighter. The spreading so rapidly by root stocks, and its tenacity of life, account for the fact that it soon appears in pastures or old meadows when the other grasses die out. June grass is not very often sown for pasture or meadow, yet it abounds in most of our pastures, especially if they have not been plowed for some time.

Read Professor Phares as to its success in the South: "Kentucky blue grass grows as well in most parts of the Gulf States as in Kentucky or any other State. In these States this grass is perennial and excellent for hay and grazing through a large part of the year."

In the famous experiments on meadow grasses by J. B. Lawes, in England, *Poa pratensis* on the permanently unmatured land made a very poor fight, amounting to one-quarter of one per cent or even less of the herbage, while on plats receiving a large quantity of potash, phosphate of lime, and salts of ammonia, in fact everything that is necessary to grow luxuriant grass, it managed to fight its way onward, so that in ten years it was credited with twenty-two per cent of the whole herbage.

With a still more generous diet, it had to give way to cock's foot [orchard grass], which in turn gave way to meadow foxtail.

Concerning this grass for Kansas, Professor Shelton writes: "What we said five years ago in writing of this grass has been fully borne out by recent experience. It can be grown almost anywhere in the now settled portions of the State. We have never failed to secure a good stand, and ultimately a good sod,—even during such very dry seasons as 1875,—when good seed was sown upon well prepared land, and at the proper season, which is early in the spring. However, our experience with the grass,—a very extended one by the way,—has convinced us that, for all useful purposes except lawns, in central and western Kansas, this is one of the most worthless of the tame grasses. It starts early in the season, and for a short time yields a small amount of quite inferior feed; but in May it ripens its seed, the grass becomes brown, dry, and fibrous, and in this dormant condition it remains until fall, and often until the following spring. We have invariably found, too, that, in a field containing other sorts, cattle will not touch blue grass until all these others are consumed. Moreover, dry weather will almost certainly injure blue grass sod seriously, when no damage is sustained by orchard grass and clover growing in the same field. On the other hand, in the eastern portions of the State, particularly in the counties bordering the Missouri river, we know from personal observation that blue-grass thrives abundantly, and is very profitable grass.

"We can easily see that this grass possesses great value for a region like Illinois and Kentucky, where winter rains abound, enabling it to make a slow and continuous growth; but the

Kansas winter is generally our dryest season, and for this reason we doubt much if this old favorite sort has any place in our agriculture. To obtain a good stand quickly, blue grass seed should be sown in the early spring; and in amount not less than three bushels per acre of ground."

The following in reference to the quality of this grass, is from the pen of of Dr. Bessey, of Nebraska:

"Passing now to the nutritiousness of blue grass, we find that it stands at the head of the list of cultivated grasses, as shown by repeated chemical analysis. It is very nearly twice as nutritious, weight for weight, as Timothy. As compared with red top, the latter has about five-sixths the value of that of blue grass. Orchard grass likewise, has about five-sixths the nutritious value of blue grass. From what has been said, it is clear that the high rank held by blue grass for pasturage is well merited."

June grass, in a cool climate, is one of the best of all our grasses for a lawn, and when sown on rich land and cut often, it makes a a soft, thick green turf, which is a delight to the eye and to the feet which tread upon it. In many places nothing else need be sown. For a good lawn sow four bushels to the acre.

Much of the seed is saved in Kentucky, where it is tied in bundles and set up in cocks till the tops decay sufficiently to break up easily. It needs care to prevent this seed from heating and injury when piled up in the chaff.

In some cases the tops, when ripe, are taken off with a stripper with a box behind it, the whole kept on wheels.

Poa compressa, L. Wire Grass, Blue Grass, Flat-stemmed Poa, Flat-staked Meadow Grass.—A perennial, about 1 ft. hi., with creeping rootstocks, and smooth compressed culms. Leaves short, with flattened sheaths, and a short, obtuse ligule. Panicle oblong, 2-3 in. long, slightly spreading branches, 2-3-nate. Spikelets ovate-oblong, 4, 6, or even 9-flowered. Floral glumes

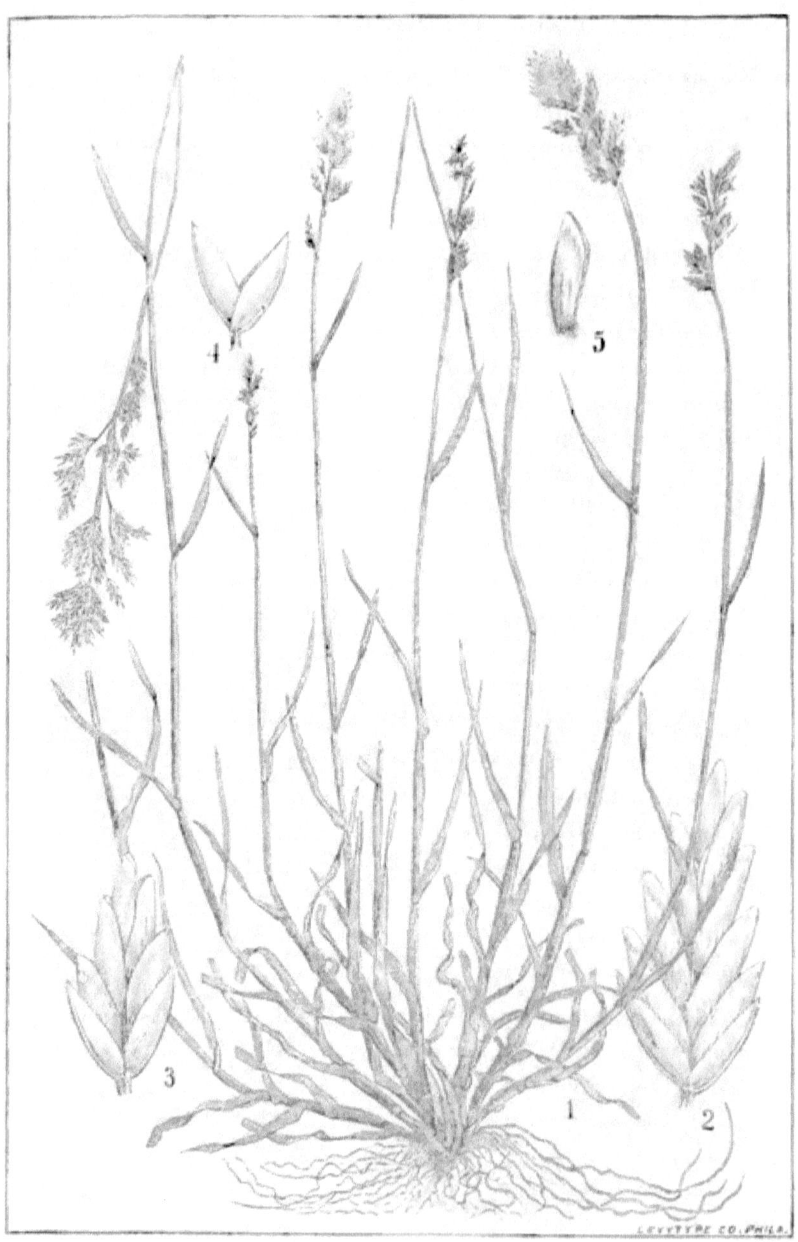

FIG. 67.

with minute silky hairs along the keel, margins hyaline; nerves obscure.

Although not purposely sown anywhere, so far as the writer can learn, it deserves notice because so often found in rather dry, thin pastures on sand, gravel, or clay soil, in company with June grass, which it somewhat resembles.

The grass is a perennial, a foot or more high, with a stem nearly solid, hard to cut, soon gumming the knives of the mower. When compared with June grass, it flowers several weeks later, the panicle is shorter, narrower, more compact; the leaves shorter, the stem much flattened, and the whole plant of a much darker color. In this country it does not spread rapidly by root-stalks, as is the case with June grass and quack grass, but in England it does spread rapidly.

It well deserves the name "blue grass," by which it is often known, as the whole plant has a dark, bluish, glaucous-green color. It is to be regretted that the name "blue grass" was ever applied to *Poa pratensis*, as is commonly the case in Kentucky and vicinity.

Prof. D. L. Phares, in his manual of grasses for the Southern States, says: "*Poa compressa* is blue, the 'true blue' grass, from which the genus received its trivial name. It has priority of claim to the name *blue* grass, and justly too, as the leaves have a deep bluish tint."

Like *Poa serotina*, fowl meadow grass, it may be allowed to get ripe before cutting, as its stalks remain green and nutritious. No grass makes richer pasture or richer hay

Gould says, "It never forms a close turf, and is rarely found intermixed with other grasses. It never yields a great bulk of hay, but this bulk weighs very heavily, frequently a ton or a ton and a half to the acre, where one would not expect to get half a ton.

FIG. 67.—*1*, Plant of *Poa compressa*, *2*, *3*, spikelets; *4*, empty glumes; *5*, floral glume. (1 from U. S. Agricultural Report. 2-5, F. L. Scribner.)

"It is certain that cows that feed upon it, both in pasture and in hay, give more milk and keep in better condition than when fed on any other grass. Horses fed on this hay will do as well as when fed on Timothy hay and oats combined. Sheep fatten astonishingly when fed upon it.

"The crops are remarkably even; it rarely suffers from excessive wetness or dryness. By manuring, we have increased the size of the culms, and caused them to grow two feet high. It is one of the hardiest grasses known. It is perhaps, rather better suited to moist, gravelly clays. It keeps green even until the heavy frosts of winter. It loses less weight in drying than any other species. Although this grass is spoken of by most writers on the subject in terms of contempt, we must differ very decidedly from them, and adhere to the opinion which we have formed after much observation and experience, that it is one of the most valuable and nutritious of them all."

Poa serotina, Ehrh. Fowl Meadow Grass, False Red Top.—Culms rather weak, 2-3 ft. hi. Leaves narrow, smooth, ligules $\frac{3}{16}$ in. Panicle, 6-14 in., slender, open, branches mostly 5-nate. Spikelets numerous, acute, short, pedicelled, often purplish. Floral glumes obscurely nerved, webbed at base.

The name "Fowl" meadow is said to have been applied to this grass because ducks and other wild water birds were supposed to have introduced the grass into a poor low meadow in Dedham, Massachusetts.

This is a native grass, found on bottom lands in the eastern half of the Northern States. It flowers about the same time as Timothy. It makes a soft, pliable hay of excellent quality. The stems in damp weather branch at the lower joints, and thus the grass inclines to spread. On account of the large top, and the slender stem, this grass when sown alone is rather inclined to

FIG. 68. *1*, Plant of *Poa serotina*; *2*, *3*, spikelets; *4*, floral glume. (The first from U. S. Agricultural Report, 2-4, F. L. Scribner.)

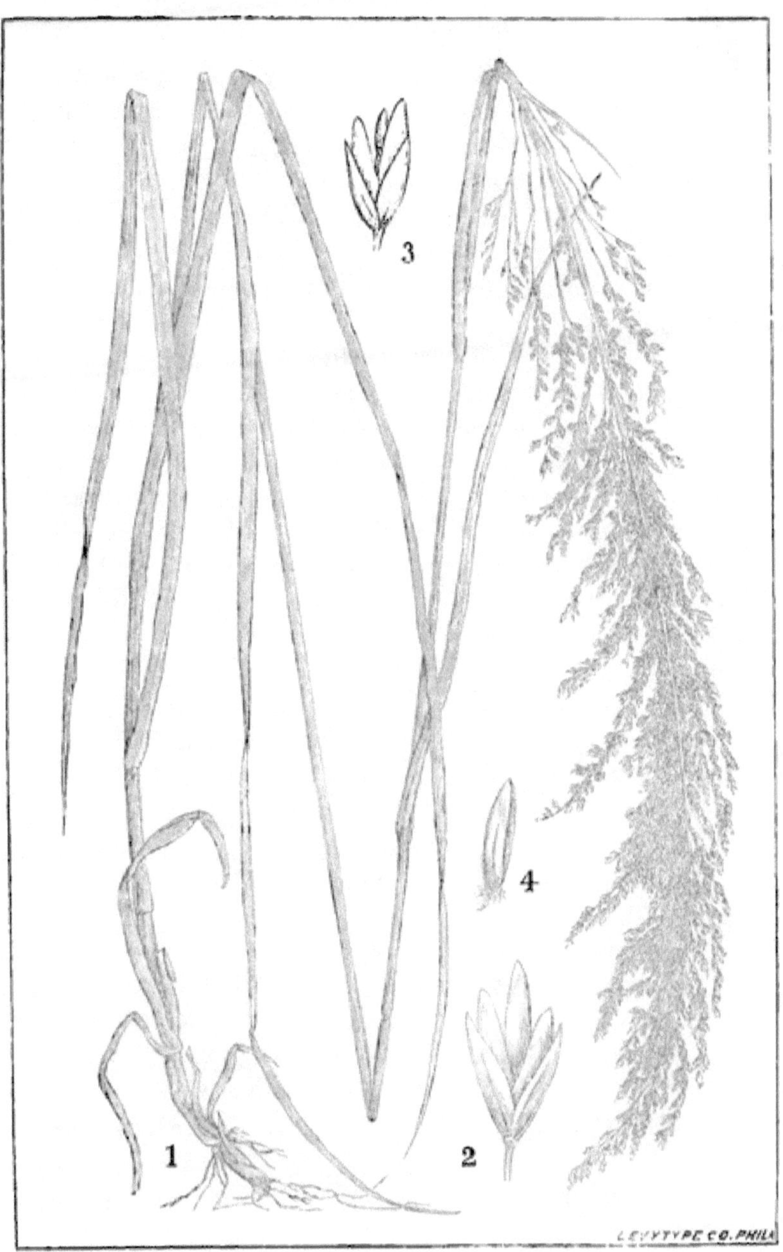

Fig. 68.

fall over or lodge. This is one reason for growing it with other grasses, like red top, which has larger and stiffer stems. Like *Poa compressa* or wire grass, it flowers rather late, has a dark green stem, which remains green and nutritions for a long time after the plant has gone to seed. It does not spread by root-stocks, like June grass. Owing to the fact that the stems remain green and succulent after flowering, there is not so much need of cutting this grass when in flower as there is of cutting most other grasses at that time. It may be allowed to go to seed before cutting, then threshed, and the straw fed out. In this way the hay is not so good, but answers very well, makes a profitable crop, as we get both an abundance of good seeds and forage. The second growth, after feeding or mowing, starts quite slowly, and like Timothy, it is not well adapted for pasture. The grass will grow on almost any rich, arable land, making a fair crop, but it likes moist land. The seeds are small and require more than one year to make strong plants, hence it is not suitable for alternate husbandry.

Although grown in the Eastern States for 150 years, pure seeds are not often found in market. They are difficult to identify by seedsmen and farmers, and both are liable to be misled. This is true of many other grasses, and constitutes one of the many "practical" reasons why farmers stick to a few well known sorts.

Although tried in Europe, its culture has not met with much favor there.

Poa trivialis, L. Rough-stalked Meadow Grass.—Culms decumbent at base, without rootstocks, taller and more slender than *Poa pratensis*. Culms and sheaths usually rough, ligule oblong, acute. Panicle 4-6 in., slender, spreading, 5-nate. Spikelets mostly 3-fld. Floral glumes acuminate, nerves distinct. Found in Europe, N. Africa, Siberia, and introduced into America.

This perennial is employed in Great Britain for meadow and

pasture, and is there usually much preferred to June grass, which it much resembles. It is to that country what June grass is to the eastern part of the United States. The grass is suited to deep, moist loam, to sow with red top and fowl meadow grass, but poorly suited to dry soils.

The late Prof. James Buckman, of England, said: "*Poa trivialis*, is a month later than June grass and inferior to it." In Europe the seeds of June grass are often sold for those of *Poa trivialis*.

In my plats of grasses in several places, this has always proved a slow grower, and has soon been crowded out by June grass.

P. arachnifera, Torr. Texas Blue Grass.—This plant is well supplied with creeping rootstocks, and is taller than *P. pratensis*. The leaves are long and slender; ligule short and obtuse. Panicle 4-6 in. by ¾ in., light colored. Floral glumes prominently ciliate on the keel below the middle; at the base usually are very long, webby hairs. Palea ciliate on the nerves, slightly adherent to the caryopsis, which is twice as long as that of *P. pratensis*.

For the South, as a pasture grass, this one seems very promising. It is hardy and a more rapid grower than Kentucky blue grass.

As far north as Kansas, Professor Shelton says, it endures the winters and resists drought perfectly, making *three or four times* as much pasture or hay as does its near relative, Kentucky blue grass. He is very hopeful of this grass.

AGROSTIS, L.

Spikelets small, 1-fld., panicled, flower perfect, empty glumes persistent below the joint, keeled, acute; floral glume shorter, broad, hyaline, frequently supplied with a slender awn below the middle. Palea very slender, hyaline, short or none. Stamens mostly 3. Styles distinct, very short, stigmas feathery. Caryopsis included in the floral glume, free. Annuals or perennials,

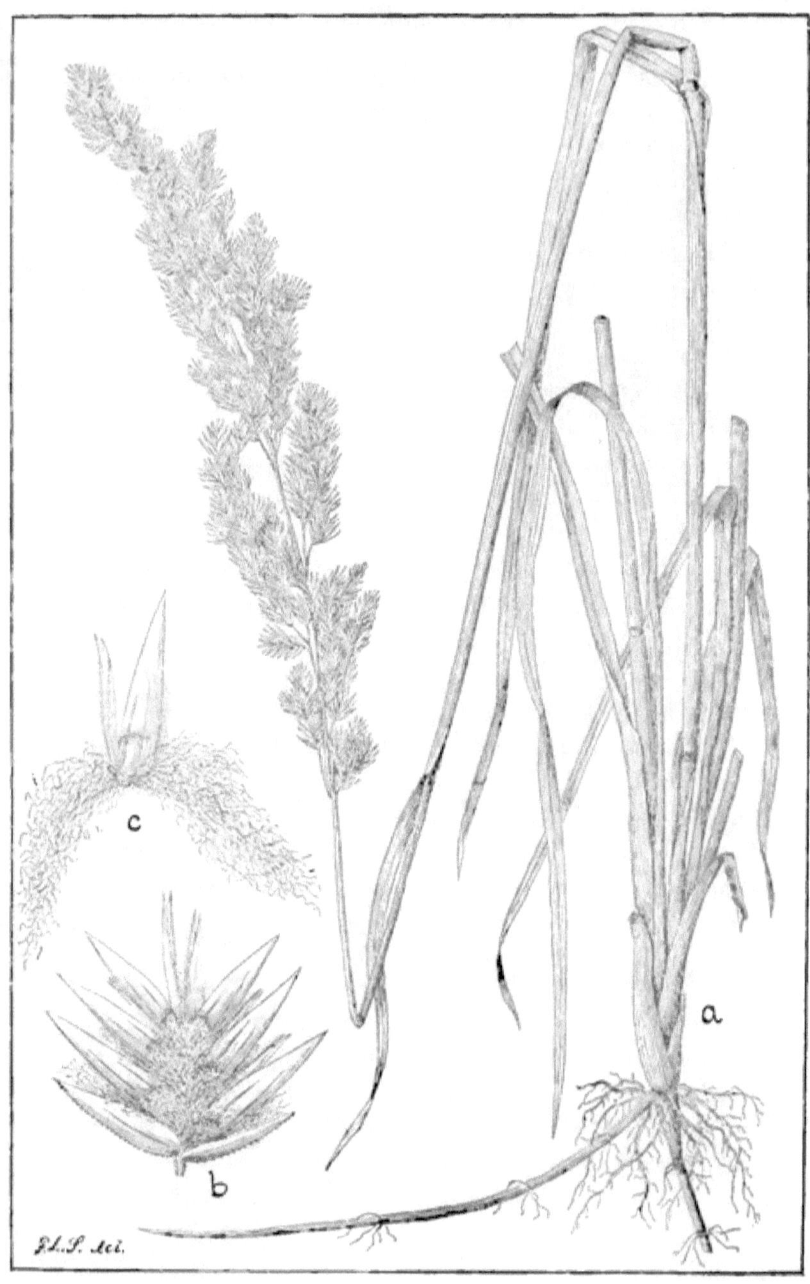

Fig. 69.

tufted, leaves flat or bristly. Panicle terminal, usually slender, much branched; branches slender, spikelets numerous.

About 100 species, mostly found in temperate regions.

A. alba, With, Red Top, Herd's Grass (of the South), **Burden's Grass, Summer Dew Grass.** Culms 1-2 ft. hi., ascending, smooth, from creeping rootstocks. Leaves short, flat, ligule oblong. Panicle oblong, 3x8 in, branches spreading. Empty glumes subequal, or the lower longer, ovate or lanceolate, acute, often purple. Floral glume shorter, truncate, 3-nerved; awn short or none. Palea about one-half as long as the floral glume.

This is erroneously sometimes called "fowl meadow grass." In England it is also called "red bent, "purple bent." "creeping rooted bent," "black twich"

It is a well known, common, native, perennial grass, found on moist bottom lands, where it flowers with Timothy or later. The spreading panicle varies considerably in appearance, but is usually tinged with purple.

It starts rather late in spring or after cutting, affording very good pasture, remaining green for a great part of the year. It yields from one to two tons of hay to the acre, is of good quality and rather light for its bulk. Chemical analysis shows it to rank next to June grass, very high in nutritive qualities.

Red top in this country is often sown on marshes, too wet for some of the better grasses. It is not well adapted to alternate husbandry, as it takes several years to become well established.

Gould says: "Its interlacing thick roots consolidate the sward, making a firm matting, which prevents the feet of cattle from poaching. It is generally considered a valuable grass in this country, though by no means the best one. Cattle eat hay made from it with a relish, and as a pasture grass it is much valued by

FIG. 69. *a*, Plant of *Poa arachnifera*, Torr. (Texas Blue Grass); *b*, spikelet, enlarged and spread out; *c*, floret.—(Scribner).

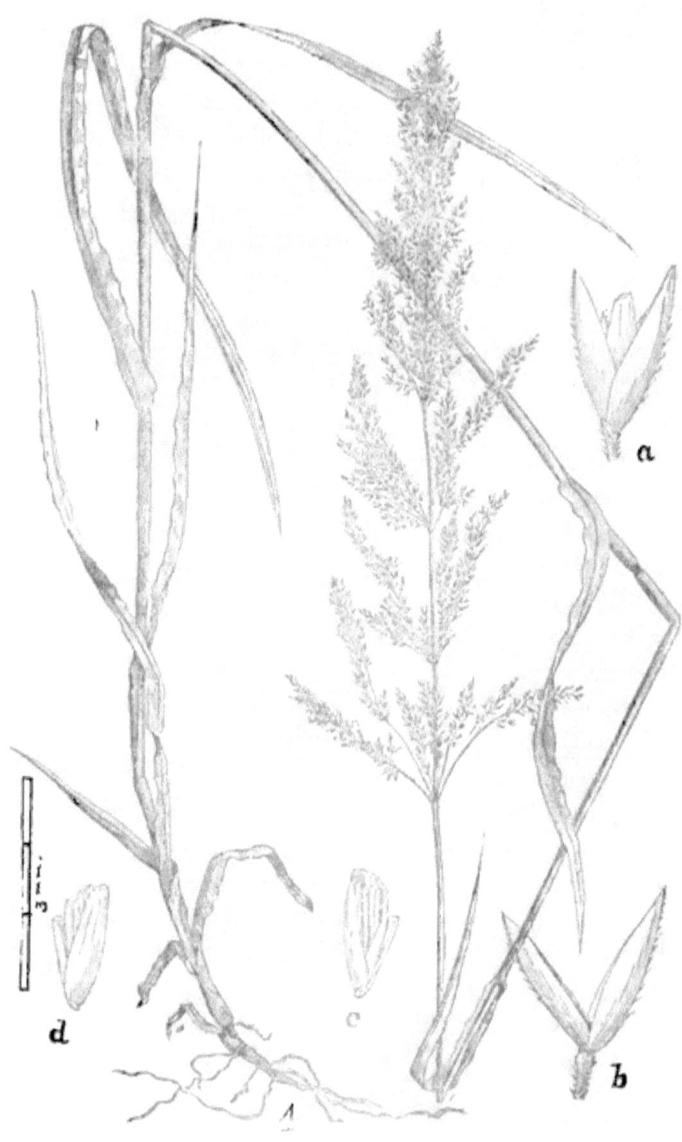

Fig. 70.

dairymen, and in their opinion the butter would suffer much by its removal "

Professor Phares says "It furnishes considerable grazing during warm 'spells' in winter, and in spring and summer an abundant supply of nutrition. It will continue indefinitely, though easily subdued by the plow. It seems to grow taller in the southern States than it does farther north, and it makes more and better hay and grazing. It does well with Timothy, but will finally root out the latter Sow about two bushels (24 lbs.) per acre if alone.

"Red top may be pastured here through most of the year, furnishing considerable grazing even along through winters, growing on almost all soils if not kept too long submerged in water It is very hardy, and in mixed pastures exterminates, after a few years, most other grasses."

Killebrew, of Tennessee, says. "Red top is next in importance to Timothy as a meadow grass. Grazing is necessary to its preservation, as, if allowed to go to seed a few years, it dies out.

It is the most permanent grass we have, and by means of its long, creeping roots will, even if sown too thin, quickly take possession of the ground. On uplands it is not a good producer. It stands the effects of drought much better than Timothy. For stopping gullies in old fields it is superior to blue grass. The seed is usually sold in the chaff It is probably better adapted to all the soils of the State [Tennessee] than any other grass."

Howard of Georgia, says. "It will grow almost in running water. It yields a valuable return on thinner land than, perhaps, any other of the cultivated grasses. Timothy and red top should be sown together, as they are ready for the scythe at the same time. This mixture is better than either grass singly."

In England *Agrostis vulgaris* differs somewhat from the same

---

FIG 70.—*Agrostis vulgaris* — *alba*. (Red Top); number *1*, a plant: *a*, spikelet: *b*, empty glumes: *c*, *d*, florets.—(Scribner.)

grass in this country. In that country it is not given in the lists of grasses recommended for cultivation.

Dr. Lindley says: "They are little better than weeds, except in soils where better grasses cannot be obtained. It grows in *dry*, gravelly, sandy places, and is a troublesome weed."

Mr. Gorrie, of England, says: "Remarkably variable in habit and appearance, too common and disliked by cattle. It starts late in spring."

Sir J. B. Lawes, says: "It flourishes most on dry soils, and is a troublesome weed on arable land, disliked by cattle and sheep. It is reported as useless, and should be discouraged as much as possible. In manuring the land, the proportion of this grass was very much reduced in every instance, a result certainly not to be regretted."

Agrostis vulgaris. Creeping or Marsh Bent, Fiorin, White Bent, White Top, Bonnet Grass.—A perennial, 6-24 in. hi., often prostrate below. Leaves flat, sheaths smooth, ligule long, acute. Panicle contracted, narrow, 3 in., many small branches in a whorl. Palea with two tufts of hairs at the base. Very variable.

By some this is equivalent to *Agrostis stolonifera*, by others it is thought to be a mere variety of red top, or red top a mere variety of this grass. Although not considered very valuable, yet it is often recommended in Great Britain in mixtures for permanent pastures. It starts early and holds out very late in autumn. A creeping habit makes it much like June grass, difficult to kill out on wet land. It is not suitable for dry land, but for wet, bottom lands or for permanent irrigated meadows, where it often produces large crops.

Along the Connecticut river, the straws are cut for braiding to make bonnets. In this country, so far as the writer has seen,

---

FIG. 71.—*Agrostis canina* L. (Brown Bent. Rhode Island Bent.) *a*, plant; *b*, spikelet *c*, back of floral glume with awn. (Reichenbach.)

Fig. 71.

FIG. 72.

florin does not seem to be as large, vigorous, productive, or as valuable as our own native red top.

Agrostis canina, L. Brown Bent, Rhode Island Bent, Fine Top, Furze Top, Burden's Grass.—A very variable perennial, much like small plants of *A vulgaris*. Culms 6-18 in. hi., stoloniferous. Ligule oblong. Panicle 2-4 in., contracting in fruit, usually purple. Floral glumes shorter than the empty, 5-nerved, awned on the back, near the middle or below. Palea minute or none.

Widely distributed in cool regions.

J. B. Alcott says: "There is as much difference between this and red top as there is between the Tom Thumb pea and the marrowfat. It will make beautiful, close, fine sod upon quite sterile soils. This, red top will not do. It is especially satisfactory for lawns, which in strong soils is apt to overgrow."

It makes very good pasture, though it is too small and grows too closely to afford much of a bite. For fifteen years the writer has watched it in Michigan, on thin soils and on rich soils, on moist land and on dry, sandy land, and he unhesitatingly recommends it as one of the very best grasses to mix with June grass for producing a fine lawn. If sown alone, four bushels of seed in the chaff is none too much.

This grass, with considerable variation, is often found on mountains in Europe, Asia, Australia, and North America.

A small *Agrostis*, probably *A. vulgaris*, of Europe, has been much used for lawns, and by some it has passed for *A. canina*.

ALOPECURUS, L. FOX TAIL.

Spikelets 1-flowered, flat, crowded into a head or cylindrical spike-like panicle, jointed at the apex of the enlarged pedicel, flowers perfect. Glumes 3 or 4, the 2 outer empty, acute, awnless or short awned, often connate below, flat-keeled, the keel ciliate

Fig. 72.—*1*, Plant of *Alopecurus pratensis*, a little reduced; *2*, spikelet; *3*, floral glume.—(Trinius and Scribner.)

and sometimes winged; the floral glume obtuse, hyaline, 3-5 nerved, with a short awn on the back, or mucronate, the margins joined at the base into a tube inclosing the flower, the palea sometimes present, narrow, hyaline, keeled, acute, partly included by the floral glume. Stamens 3. Styles distinct or rarely joined at the base or the middle, stigmas short, hairy. Annual or perennial grasses, erect or decumbent at the base; leaves either flat or convolute, upper sheaths often inflated. Spikelets or panicles spike-like, terminal.

About 20 species in temperate and cold countries.

A. pratensis, L. Meadow Foxtail.—A soft erect perennial, 1-3 ft. hi. Leaves flat, upper sheath inflated, longer than its blade; ligule oblong truncate Spikelets 3-8 in., $\frac{1}{4}$ in. or more in diameter, dense, obtuse, soft, pale green. Spikelets $\frac{1}{8}$-$\frac{1}{4}$ in long Empty glumes, membranous ciliate on the keel only, ovate lanceolate, acute, connate at the base. Floral glumes ciliate, as long as the empty glumes, awn near the base and projecting half its length.

Found in Europe, North Africa, Western Asia, introduced into America.

In Great Britain and other parts of Europe with a like climate, this is one of the best known and highly esteemed grasses which is cultivated for permanent meadow and pasture.

It bears considerable resemblance to Timothy, though the culm and leaves are shorter, the spikes shorter, broader, and softer, the whole plant less firm and rough, and it starts much earlier in spring, flowering three or four weeks before that well known grass.

Meadow foxtail is not well adapted for alternate husbandry, as it requires three or four years to become well established, but on deep, rich, moist, or irrigated soils, in a cool climate not subject to droughts or very hot weather, it is a fine grass and peculiarly well adapted for permanent pasture. It makes a quick growth in spring or after feeding or mowing. It is fine, nutritious, and palatable for all kinds of stock.

Like Timothy, it has no tendency to spread, as is the case with June grass, quack grass, and white clover. Mr. Lawes' experiments show that it thrives best with high manuring, supplying much nitrogen. In this respect it comes into competition with orchard grass.

There is considerable difficulty in procuring good seed, which is very light, and sold in the chaff, only weighing five pounds to the bushel. If sowed alone, three bushels to the acre is none too much. The seeds are often adulterated with those of *Holcus lanatus, Alopecurus agrostis* and rye-grass. The first two grasses are of very poor quality, the latter is much cheaper and costs not over one-eighth part as much per pound. Other seeds are also often found with those of meadow foxtail.

The seeds ripen unevenly, some beginning to fail while much is yet immature or even in flower.

Many glumes are empty: insects, blight, or something else causing the failures. The small seed produces a small, feeble plant, which requires a favorable chance for a long time before it becomes well established.

It will not likely ever be popular over a very large portion of the United States. It is well adapted to parts of New England, New York, Canada, and mountain districts farther west and south. It is a native of Europe, and one of the five grasses recommended for permanent grass lands by Mr. De Laune, of England.

ANTHOXANTHUM, L. SWEET VERNAL-GRASS.

Spikelets 1-flowered, narrow, slightly compressed, crowded into a cylindrical spike-like panicle, rachilla jointed above the lower glumes, often hairy. Glumes 6, the 2 lower persistent below the joint, acute, mucronate, or very short awned, the second longer

Fig. 73.

than the others, the two intermediate glumes much shorter, empty, narrow, keeled, with an awn on the back or very near the base, the two upper glumes much shorter, hyaline, obtuse, awnless, of which the fifth is very broad, including the narrow 1-nerved sixth (or palea ?) and the flower; other palea 0. Stamens, 2. Styles, distinct, with long, feathery stigmas. Caryopsis oblong, included by the inner glumes, free.

Aromatic or sweet scented annuals or perennials, with flat leaves. The terminal panicle spike-like, pedunculate, dense or rather loose, with many very short dense branches.

Species, four or five, European.

A. odoratum, L. Sweet Scented Vernal Grass.—A rather slender, erect perennial, 1-2½ ft. high. Leaves slightly hairy, ligule oblong, obtuse, blade of upper leaf about 1 in. long, sheath slightly inflated. Spike-like panicle, 1½-3 in. long Spikelets ¼-⅓ in. long, linear, oblong. First glume ovate, acute, 1-nerved, half as long as the second which is 3-nerved, elliptical when spread out. The third and fourth empty glumes emarginate, obscurely 5-nerved below the apex; the straight awn of the former above the middle, projecting ½ its length The twisted awn of the latter below the middle, projecting twice the length of the glume.

Native of Europe, widely dispersed in temperate Asia, North America, Australia, often sown for pastures and lawns

Sweet scented vernal grass is a pretty name, and suggestive of something agreeable, and is one of a very small number of grasses which possess this peculiar odor. The grass is perennial, with a culm one to two feet high. It starts very early in spring and soon flowers. It has often been recommended for lawns and pastures, but for the latter purpose some of the best farmers of

Fig. 73.—*Anthoxanthum odoratum*, L. (Sweet Vernal Grass); *1*, plant; *a*, spikelet; *b*, the same with the outer glumes removed; *c*, the same as *b* with the empty glumes removed; *d*, the stamens and pistil; *e*, the pistil with one style removed; *f*, one of the hairs from the stigma.—(*e*, *f*, from Kunth, the others by Scribner.)

England now omit sowing this grass. Its fragrance when wilted, bruised or dried, is its chief recommendation, and about the only one for its use on the lawn. It is too apt to kill or be crowded out, a little coarse, bunchy, and uneven for a velvet and elastic turf. It likes rich, moist soil and cool summers. T. M., in Treasury of Botany, says: "The fragrant resinous principle which occurs in this grass, and is called coumarin, is a widely diffused natural perfume, being found in the Tonka Bean, the Faham tea-plant, the sweet-wood-ruff, melilotus, and the blue or Swiss melilot."

"In Northern Michigan, and probably in other places, the Indians raise this grass and use the stems for mats, card baskets, and other small articles. It is sometimes used for bonnets, hats, and bouquets. The culms and flowers possess the strongest perfume, which remains for a long time after drying."

"The Italians," says Dr. Lindley, "Are said to employ the distilled water as a perfume."

The writer and one of his students made some experiments in feeding sweet vernal when in flower. A young horse kept on dry feed, preferred fresh June grass to sweet vernal, but ate all of both, small bunches.

Some meadow foxtail and sweet vernal had been cut in flower, and dried for some days. One short-horn cow rather preferred the foxtail, but ate both readily. Another ate both alike, another ate the foxtail and refused the sweet vernal An Ayrshire cow ate both greedily without preference Two of the farm horses ate both alike, while a third preferred the foxtail, but ate both readily. The cows had been in pasture some weeks, the horses had been kept on dry feed.

Gould says: "It is nowhere considered a very valuable species for hay, as the culms are wide apart, very thin, and bear but few leaves; hence it gives a light crop of hay, at best not over three-quarters of a ton of hay from an acre. The seeds are not very abundant, nor easy of acquisition. Sheep are not fond of it. It was once thought to give a sweet flavor to butter and to mutton; but these notions are now exploded. It may be beneficial medicinally, as cattle will eat a little of it mixed with other grasses, but when in any considerable bulk they always refuse it."

Sir J. B. Lawes says: "Upon the whole this grass takes rank somewhat low in the scale of the better grasses for permanent purposes. The growth is much discouraged by highly nitrogenous and farm-yard manures, such as greatly increased the amount and proportion of the graminaceous hay plants, as a whole. It only becomes prominent under conditions which do not induce special luxuriance in its competitors, and it seems to be more injured by association with more luxuriant grasses than by the direct action of manures."

The seeds are sometimes adulterated with those of *Anthoxanthum Puelii*, a much smaller and insignificant plant, which is an annual. The grass is a native of Europe, and is extensively naturalized in North America.

LOLIUM, L.

Spikelets many flowered, sessile, distichous, compressed in a simple spike, placed with one edge to the rachis. Rachilla jointed between the flowers; flowers perfect or rarely imperfect. Empty glumes, firm, 5-7-nerved, convex on the back, obtuse, acute, or awned; the empty glume next to the rachis wanting, except in the terminal spikelet. Palea shorter than the floral glume, narrow, 2-keeled. Lodicules ovate, ciliate. Stamens 3. Styles distinct, very short, stigmas feathery Caryopsis oblong, smooth, adherent to the palea. Annuals or perennials, with flat leaves. Spike terminal, elongated, spikelets placed on alternate sides of a jointed rachis.

Species about 20, found in north temperate regions.

Fig. 74.

L. perenne, L. Perennial Rye or Ray Grass, Darnel. An erect or slightly decumbent perennial, 1-3 ft. hi., smooth, culm slightly compressed. Leaves flat, shining; ligules short. Spike 4-10 in., slender. Spikelets 8-16, obtuse or pointed, rarely awned, ⅓-½ in. long. Empty glumes strongly ribbed, linear-lanceolate, floral glume linear-oblong. In the terminal spikelet the second glume is usually empty.

In many portions of moist, temperate Europe, this grass has for over 200 years taken the rank among the farmers that Timothy has in the United States. Rye grass is termed a perennial, though it can hardly be relied on to last for more than two to five years, and especially after one seeding, the plants mostly soon perish. Self-sown seed, ripening on the ground, help supply a continuous crop of plants.

It seeds bountifully, frequently producing forty bushels or more to the acre; these are of a large size and make strong plants on a great variety of soils soon after sowing. The stems are one to two or even three feet high, including the straight spike at the top, giving it somewhat the aspect of quack grass. The leaves are abundant, dark green, flat, glossy, succulent, and the whole plant is nutritious.

This grass, with its simple spike, is easily recognized, and people can soon become acquainted with its peculiarities. Its short life and vigorous habit make it a grass especially suited to alternate husbandry, and not for permanent pasture or meadow, although, through ignorance and long precedent, it has very generally been recommended for the latter purpose.

The plants start early, flower early, and repeatedly during the growing season. Seeds are cheap, easily obtained, always in market, and well advertised by the dealers. Above we see stated

Fig. 71.—*Lolium perenne* (Perennial Rye Grass). *a*, Empty glume pulled back from the rest of the spikelet; *b*, floral glume cut above the middle; *c*, *d*, cross sections of floral glume and palea; *e*, spikelet with a portion of rachis of *L. perenne* var. *Italicum*; *f*, ovary and styles; *h*, cross section of an anther with a few grains of pollen.—(Sudworth.)

some of the main reasons why rye-grass has been so popular.

Rye-grass was one of the first if not the first grass ever cultivated in Great Britain, as long ago as 1677, nearly 100 years before the cultivation of Timothy or orchard grass.

Owing to this long cultivation, under varying conditions of soil and climate, as we should expect, the grass has broken up into many varieties, which are more or less permanent and well marked.

As early as 1823, one experimenter pointed out sixty varieties. A few of these became permanent enough to reproduce themselves quite true from seed. Soon after this, a few of these races of rye-grass became quite noted under various names, such as "Pacey's," "Russell's," "Whitworth's," "Dixon's," and now several English seedsmen claim extra varieties and name them after some member of their firm.

Mr. Lawes writes. "It stands at the head of the list as to the quantity in culm. It is obviously a plant of relatively weak habit. It did not flourish where ammonia-salts were used, but where nitrogen was supplied as nitrate of soda, it was much more able to maintain some, though still a rather low position in the struggle."

Only 25 to 30 pounds of seed are required to sow an acre. This grass is often recommended for a lawn, but owing to its short life in most places, it is by no means adapted for that purpose.

Rye-grass has been quite extensively tested in various portions of the older States. It is emphatically a lover of rich land and a moist climate, without very great heat. In many portions of the interior of our country, subject to great extremes, this grass has not proved of much value. It is not well suited to the Southern States, especially the dryest portions. Further remarks concerning this grass may be found on a later page of this volume.

Lolium perenne, var. Italicum. Italian Rye Grass.—This is one of the most distinct, well known, and valuable of the permanent varieties of rye grass. It differs from the species in having short awns to the spikelets. It is larger and more vigorous, makes a quicker growth, but is only an annual, and of course cannot be relied on for more than one season. Where the ground is favorable, and especially if irrigated, it produces immense crops of valuable feed for live stock, yielding as high as seven and a half tons of dry hay per acre. For this crop it is cut four or five times. The seeds are sown in autumn or in early spring at the same rate as that of rye grass.

As was said on a former page of this work, annual grasses, like rye, and rye grass, may often be induced to live for two years or more if kept cut short and not allowed to seed.

For most parts of the Northern States, however, rye grass perishes with the cold winters, and except in some of the cooler and more moist portions of our country, has proved of little value. It will make little growth on dry ground.

CYNODON, PERS.

Spikelets small, 1-flowered, sessile on one side of a flattish rachis, alternately 2-ranked, rachilla jointed above the empty glumes, extended into a small stipe beyond the flower, flower perfect. The empty glumes persistent or deciduous, slender, keeled, acute, or obtuse. The floral glume broader than the empty glumes, membranous, ciliate, keeled, transversely pilose near the apex, awnless. Palea scarcely shorter than the floral glume, hyaline, 2-nerved, ciliate. Stamens 3. Styles distinct, clothed with short hairs. Caryopsis oblong, smooth, included, free.

Perennials, creeping or stoloniferous, with short, narrow, flat leaves. Spikes slender, 3-6, digitate at the apex of the culm, straight, erect, or spreading.

Four species in warm regions.

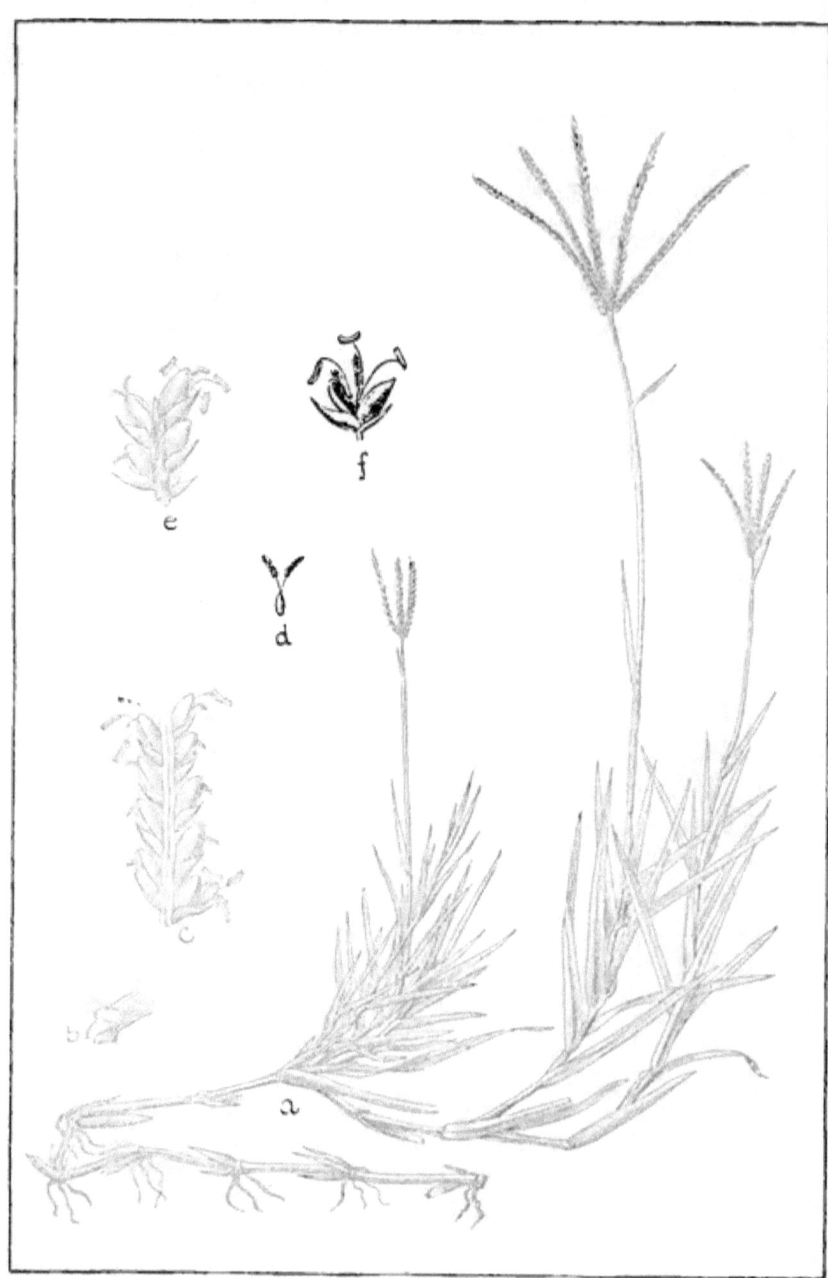

Fig. 75.

C. Dactylon, Pers. Bermuda, Wire or Scutch Grass. Spikes 3-5, seldom seeding, but spreading rapidly by stout rootstocks. This grass belongs to southern Europe and to many other warm climates, and is a perennial, thriving from Michigan southward. The stems are low and come from extensively creeping rootstocks, which also penetrate the ground to the depth of three to six inches. The top spreads into several branches, somewhat resembling crab grass, an annual too common in neglected gardens. The leaves are short.

This grass is said to be celebrated in the sacred Vedas as the shield of India, and preserver of nations, as without it the cattle would perish.

For the Northern States it is of no value, starting very late in spring, with the leaves barely an inch high when meadow foxtail is in flower, but for permanent pasture in warm countries it is highly prized, standing heat and dry weather remarkably well. It rarely ripens seed in the United States, but may be propagated by washing the rootstocks, running them through a cutting machine and then sowing broad-cast.

Like quack grass, it is a terrible pest in field crops, where its deep, stout rootstocks make it hard to kill. Thorough cultivation will kill, if not pastured. June grass, cow peas, or other rapid growing plants will shade and choke it out.

The following is from Killebrew: "In the South it has been the chief reliance for pasture for a long time. It revels on sandy soils, and is used extensively on the southern rivers to hold the levees and the embankments of the roads. It forms a sward so tough it is almost impossible for a plow to pass through it. It will run down the sides of the deepest gully and stop its washing. It has the capacity to withstand any amount of heat and droughts, and droughts that are so dry as to check the growth of

FIG. 75. *Cynodon Dactylon* (Bermuda Grass); *a*, Plant with rootstock; *c*, dorsal view of spike; *e*, front view; *f*, spikelet; *d*, pistil and lodicules; *b*, ligule. (Rieschenbach.)

blue grass will only make the Bermuda greener and more thrifty."

Mr. Elliott, quoted by Professor Phares, says: "The cultivation of this grass on the poor and extensive sand hills of our middle country would probably convert them into sheep walks of great value."

Here, Professor Phares remarks: "As a permanent pasture grass, I know no other that I consider so valuable as this, after having transplanted it from near the mouth of Red River to my present residence thirty-five years ago, and having started it on hundreds of other farms, commons, and levees for a longer period. As hay this grass has been cured and held in high esteem by many farmers in Mississippi for more than forty years.

"It does not bear dense shade, but grows best where most exposed to the intense heat of the sun. To make good pasture it must be kept well trodden and grazed to keep it tender, and to suppress other objectionable grasses and weeds. To make good hay and the largest yield, this grass must be mowed from three to five times every summer. Thus briars, broom grass, and other weeds are also repressed and prevented from seeding, multiplying, and ruining the meadow. Properly managed this grass grows from ten to fifteen inches high."

The following is from Howard's Manual. "Upon our ordinary upland I have found no difficulty in destroying it, by close cultivation in cotton for two years. Work the land in the dry, hot months of summer. When not pastured, broom grass or briars soon destroy it. I think it very doubtful whether there is an acre of land in the South thoroughly set with Bermuda grass, that is not worth more than any other crop that can be grown on it. The Bermuda and crab grass are at home in the South. They not only live, but live in spite of neglect, and when petted and encouraged, they make such grateful returns as astonish the benefactor. While grazed, neither Lespedeza, broom sedge, blue grass, or any other growth will oust it."

Some accounts are given of very large crops of excellent hay made from this grass. Although short, it is thick, fine, and heavy.

The following is from Professor Shelton, of Kansas:

"Bermuda grass has been quite extensively introduced into Kansas from southern Missouri and Arkansas during the last two or three years; and the most extravagant claims have been made for it by interested parties. After five years' experience with this grass upon a considerable scale at the College farm, we have no hesitation in saying that for this section, and we are confident generally throughout the State, these claims are totally unfounded. Bermuda grass has shown itself with us to be quite worthless either for hay or pasture. Of all the tame grasses it is the latest to appear in the spring, and the lightest frosts cut it off level with the ground. Until the hot weather of June had set in, our Bermuda grass showed scarcely a sign of life and growth. But even then the amount of feed which it furnishes is quite insignificant. Moreover, our stock of all kinds showed no great fondness for it, leaving it always for orchard grass growing near by. We are confident that our farmers will do well to keep their farms clear of a grass which, like the Bermuda grass, has some of the worst qualities of the most pernicious weeds."

The latest is from Prof. F. A. Gulley, of Mississippi: "Blue grass, white clover, and orchard grass do well in certain places, but our best grass on most soils is Bermuda. In future, on the College farm, I shall plant Bermuda to the exclusion of other grasses, except on rich soil in small pasture lots near the barn. We plant it like corn and potatoes in strips across a field, where it soon spreads and will remain for all time to come, worse even than quack grass for persistence. On good land it will cut two to four tons of nice hay per acre which is easily cured. It fur-

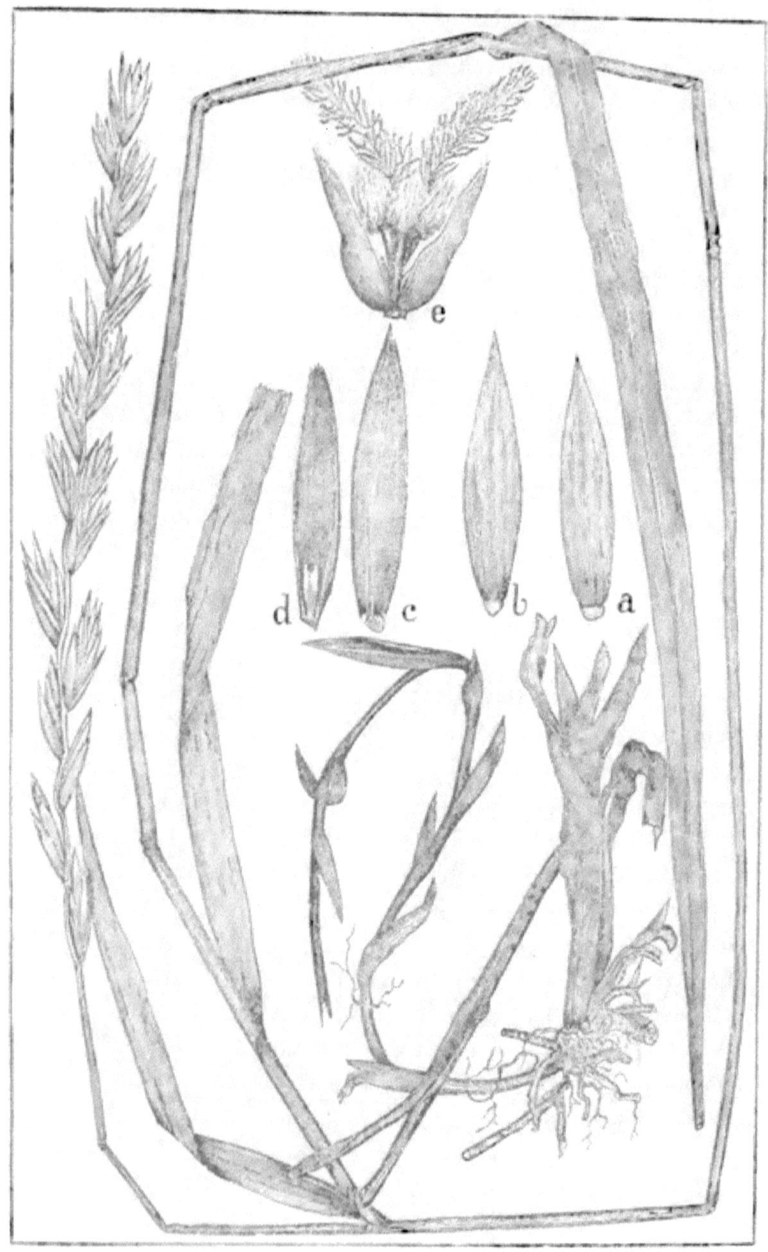

FIG. 76.

nishes a good deal of grazing on rather poor land, and grows right along through summer, when blue grass will dry out entirely. It is improved by breaking up every three or four years, and planting a crop

"I am beginning to believe that in this and Johnson grass, we have for this latitude, for hay and pasture, two plants that are not excelled by anything that grows in the north. They do not fraternize with cotton, so planters are very much afraid of them."

AGROPYRON, J. GAERTN.

Spikelets many flowered, compressed, sessile, one at each joint of the zigzag rachis, distichous, placed with one side to the rachis, flowers perfect, or the upper ones imperfect. Empty glumes narrower than the floral glume, few nerved; floral glumes firm, convex on the back, 5-7-nerved, obtuse, acuminate, or awned, the upper one often empty or enclosing an imperfect flower. Palea shorter than the floral glume, 2-keeled. Lodicules, ovate, entire, ciliate. Stamens 3. Styles very short, distinct, stigmas feathery, subsessile. Caryopsis oblong, grooved, more or less hairy at the apex, adherent to the palea or free. Perennials or annuals, with leaves flat or convolute. Spike terminal.

About twenty species found in temperate climates.

A. repens, Beauv., (Triticum repens, L.) Quack, Quitch, Quick, Twitch, Couch, Dog, Scutch, Rye, Durfee, Chandler, Witch, Quake, Squitch, or Fin's Grass or Creeping Wheat. —A very variable perennial with long, creeping rootstocks. Culms 1-4 feet, smooth, glabrous. Leaves flat, sheaths terete, ligule short, spike 2-10 in., straight or curved, spikelets 4-8 fld., ¾-1 in long. Empty glumes 5-7-nerved, rigid, cuspidate, acute or awned, floral glumes much the same, with an awn nearly its length, or sometimes awnless.

This grass is well known to most of the older portions of our

Fig. 76. *Agropyron repens* (Quack grass); part of a plant; *a*, lower empty glume; *b*, upper empty glume; *c*, floral glume; *d*, palea; *e*, lodicules, including the base of stamens, an abortive ovary and the feathery stigmas.—(Sudworth.)

country. It comes from Europe, though something very much like it is common on the great western plains, where it affords an excellent pasture

The rootstocks fill the soil, and much resemble those of June grass, only they are larger. The leaves near the ground much resemble those of Timothy; the stems are one to three feet high, and each is terminated by a slender spike from two inches to a foot in length. It seldom produces seed till the plants become somewhat dwarfed or stunted by crowding or exhaustion of the soil.

Gould tells the truth when he says, "The farmers of the United States unite in one continuous howl of execration against this grass." They generally dread its presence, which most of them are ignorant of till it has become well established, often in many places on their farms. It is a clean, sweet grass, and affords much good pasture. In good soil, if not too old and crowded, it will cut a fine crop of hay of excellent quality, not surpassed in value by that of Timothy.

The editor of the Rural New Yorker says, 'It will endure the severest droughts of the North, it will thrive in sandy or clayey soils; it is early to appear in the spring; it is the first to carpet a field with green after it has been mown or closely cropped; it makes a compact sod for the door-yard or lawn, and will become as 'velvety' under the frequent use of the lawn-mower as the bent grasses, red top or poas. Its merits are many. We do not know of any true grass about which more may be said in its praise. The great fault with quack is that it seems to be too much of a good thing. A field recently plowed for corn next spring, which had been in grass eight years or more, was nearly all quack—Timothy having disappeared entirely, and the rest forming a small percentage of blue grass and red top. The cultivation which we shall give the corn will practically subdue the quack unless the next summer should prove unusually wet, so that when Timothy and clover are again sown upon the wheat at the end of the rotation there will be scarcely any to dispute their possession. But quack would reäppear in several years and if the land, as in the above case, were retained in grass for six years or more, the quack would again largely predominate. As our lands are sandy with gravelly sub-soil, they need frequent rains, so that a season rarely passes without a drought of lesser or greater severity prevailing. It is then that quack is easily destroyed. The shallow corn cultivator, always here used, exposes the quack roots to the parching air and sun and destroys them. Upon this farm quack is a blessing, though perhaps a troublesome one. We do not believe there is another grass which, when plowed under, will furnish a greater amount of suitable food for Indian corn, while the cultivation given for suppressing its summer growth is no more than that which a full corn crop needs.

"Hence it is that any disturbance of the roots during wet weather, or when the ground is at all moist, serves as much to spread the plant as to suppress it. These rootstocks grow rapidly and persistently, preferring to grow through any permeable obstacle rather than turn aside."

On making the best of quack grass, the *Country Gentleman* says: "When hoed crops are not too prominent or common, quack is not so bad. It is neither killed by drought, hard freezing, nor close feeding. When cut early it makes the best of hay. Where it has a foothold, docks, thistles, whiteweed, and other weeds are unable to put in an appearance. Land intended for permanent fields must be broken often, as the roots form such a close sod it soon binds out. When this is the case, plow and harrow well every third or fifth year after cutting."

In the same paper above named, Henry Ives says: "There are three ways to manage quack. One is for the timid man who thinks he cannot subdue it, and who works accordingly. He gives it just about tillage enough to renovate and keep it thrifty. Another way is to cultivate enough to get a very good crop of something else; a third way is to kill it entirely. To do this, many summer fallow by thorough cultivation all summer, others plow late in the fall and next spring put in a crop. The cheapest way to clear land from quack, is to plow in the fall, then harrow in the spring, cultivate or gang-plow until rather a late planting time for corn, then plant, when the corn will come up quick, cultivate early and often. It cannot be killed by any process of raking and picking it off the ground."

As to the mode of killing, the writer has often tried, with excellent success, the plans named by the last writer. Plow late in the fall, and go on to the ground as soon as possible after thawing out—not waiting for the soil to settle. Cultivate well every three days till no traces are seen, which will usually leave time for a late crop of potatoes, corn, or rutabagas in the same season. It must not be allowed a breathing spell, as it then recuperates rapidly. Do not wait for a leaf to show itself. Give it no peace.

FIG. 77.—Rootstock of quack grass which has grown through a potato. Reduced one-third. (Sudworth).

It thrives in the South as well as at the North.

The apex of a rootstock is quite sharp and stout, and not unfrequently grows through tubers of potato.

SORGHUM, PERS.

Spikelets in threes, panicled, the central one hermaphrodite, sessile, 1-fld.; the lateral ones pedicellate, male or sterile, with sometimes 1-3 pairs of spikelets at the nodes below. Glumes of the sessile spikelet 4, the lower larger than the others, empty, lanceolate or ovate, hard and shining, obscurely nerved; the second empty, narrower, keeled, firm, acute or awned; the third much smaller, hyaline, empty; the fourth or floral glume very slender, hyaline, 2-lobed, awned. Palea minute or none. Stamens 3. Styles distinct, stigmas feathery. Caryopsis included, free. Annuals or perennials, often tall with broad, flat leaves, panicle terminal, large. Species now reduced to two, (*S. halapense and S. vulgare*). Extensively cultivated in warm and temperate climates.

S. halapense, L. Johnson Grass, Means' Grass, Cuba Grass, Egyptian Grass, Green Valley Grass, Arabian Millett, Egyptian Millett, Syrian Grass, Saint Mary's Grass.—From several sources I learn that in 1835 Gov. Means of South Carolina obtained the seed from Turkey. A few years later William Johnson of Alabama obtained the seed of the Governor, and was quite active in advertising its good qualities, hence the popular name of "Johnson grass."

FIG. 78. *a*, Portion of panicle of *Sorghum halapense*, L.; *b*, two spikelets, the other having been removed; *c*, lower spikelet with fertile flower; *d*, one upper spikelet with staminate flower. 1×6. (Sudworth.)

It has sometimes been called *Guinea grass*, though this name has more generally been applied to another, *Panicum jumentorum*.

Johnson grass is a coarse perennial, with large, stout rootstocks often half an inch in diameter. These penetrate the ground in every direction, and each joint may send up a stem after the manner of June or quack grass, only on a much larger scale.

The stems are three to six or more feet in height, and are amply supplied with long, broad leaves. The branching panicle somewhat resembles that of barn-yard grass. For the Southern States this grass has been highly praised by some and tried cautiously by others. It bears great heat and severe drought, and may be cut once a month during the growing season. It affords fine pasture, if any coarse grass can be said to furnish such a pasture, and the rootstocks furnish food for swine nearly equal to that of artichokes.

As might be expected, it is next to impossible to turn up these rootstocks with a plow; hence it is difficult to eradicate, though if no tops are allowed to grow, the parts beneath the ground will soon become exhausted and perish.

Those who have tried it say, that if cut in blossom, or earlier, the hay is most excellent, and on good land the yield is enormous.

Dr. Phares says, "During the recent long drought in north-east Mississippi, on one farm at least, this grass was mowed three times; and on the first of October, when from eight to twelve inches high, the cattle were turned in it and there remained feeding and fattening on its abundant, rich, rapidly-growing foliage to the last of December."

Prof. F. A. Gulley says: "Johnson grass stands first in quantity and quality for permanent meadow, especially on rich, well-drained, heavy land. This and Bermuda for the South are equal to anything at the North. It is improved by breaking up once in a while."

Mr. Montgomery, of the same state, has no hesitancy in saying that it will produce more nutritious hay per acre on rich land than any meadow grass we can grow. To insure a fine quality of hay it should be mowed when the first seed stems appear. Overflows and standing water are death to it. A good plan to propagate this grass is to drop roots between the hills of corn and cultivate with the corn crop.

Here follows the statement of Professor Shelton, of Kansas: "We have had Johnson grass in cultivation upon the college farm for four years, and every year's experience with it makes its total worthlessness the more conspicuous. It never makes its appearance with us much before the first of June, and the first frost in the fall cuts it even with the ground. During the summer's heat it makes a coarse, scattering growth of herbage which our cattle persist in disliking. I notice that our patch slowly increases in size from scattering seeds and rootstocks. You can safely advise your readers in Kansas and the southwest, to keep entirely clear of Johnson grass."

Dr. Vasey says: "It has been tried in Kansas with very promising results. Probably no grass gives better promise for the dry arid lands of the West."

It may be propagated by pieces of rootstocks or by seeds. The writer has tested it on a small scale in Central Michigan, but many of the rootstocks are killed by winter while a few usually remain. It has produced some seed even in the coolest summers. The seeds start slowly, and no sprouts from any source appear above ground till the weather becomes warm and settled.

SETARIA, BEAUV.

Spikelets ovate, jointed with the persistent pedicel, which bears one to many bristles, collected into a cylindrical spike-like

Fig. 79.

or narrow panicle. Glumes 4, the three outer membranous, the lower very small, the second shorter than the third, both empty, the third usually longer, empty or rarely inclosing a palea or male flower, or sterile; the terminal inclosing the perfect flower, shorter, obtuse, indurated as well as the inclosed palea, shining or transversely wrinkled, or simply dotted in lines. Stamens 3. Styles distinct from the base, elongated, stigmas feathery. Caryopsis included in the hard floral glume and palea, free. Annual grasses, often tall with flat leaves. Panicle terminal. Species about ten, found in tropical and temperate climates.

S. Italica Kunth, Hungarian or Bengal Grass, German, Italian, Mammoth, Golden or Cat-tail Millet.—A stout, quick-growing grass, 2-3½ ft. hi., with numerous broad, flat leaves and a nodding panicle 4-9 in. long by ¾-1¼ in. in diameter. Bristles two or three in a cluster.

The term "Millet" is also applied to various other species of plants, and is about as indefinite as the name "blue joint" or "bunch grass" or "pig weed."

The variety of millet which is principally grown as a hay crop in America was distributed through the United States Patent Office in 1854 under the name of *Panicum Germanicum*. There are many races, which, like those of Indian corn, are mixed up in hopeless confusion. It is much cultivated in the West and Southwest.

The millets are among the most ancient of cultivated grains, as is evinced by the variability in the species as well as by ancient mention, and their wide distribution. It is said that a third part of the inhabitants of the globe feed upon the different millets, especially in Africa, Turkey, Persia, India, and Japan. It is mentioned by Pliny as one of the cereals of his time. *Setaria Italica* has an Asiatic origin and a high antiquity, as is evinced

FIG. 79.—*Setaria Italica*. (Hungarian Grass): *a*, portion of plant; *b*, spikelet with the pedicel of a second; *c*, another view; *e*, fertile floret showing palea; *d*, dorsal view of same.—*a* Redrawn from Trinius. *b*, *c*, and *d*, Scribner.

by its Sanscrit name "kangu" and "priyangu." In the old world one variety is grown on watered land, another in palm gardens, and another in dry fields.

The seeds of this or another species are even now sold in London shops as a substitute for rice in making puddings. It requires a dry, light, warm land or medium soil for its best production, and has a remarkable power of resisting drought. It will not grow till the weather becomes settled and warm. It is sensitive to cold and is a shallow feeder, and will bear crowding without injury. Its seeds will germinate under conditions of considerable dryness.

When cut it parts with its moisture very slowly, and cures into hay with difficulty.

When forced to grow fine through crowding, and grown on rich and suitable land, this plant makes from three to four or even five tons of fine-appearing fodder, sweet-smelling if cut early and properly cured, and is relished by stock. If cut early it is certainly quite equal to ordinary hay. If grown thinly the forage is coarse, and is not so well relished by animals. If not cut early its value is greatly impaired. After the seed is ripe it is said to be unhealthy for horses. It is ready for hay when the spikes begin to appear generally over the field. One bushel of seed is sown to the acre, broadcast, or less when sown in drills. Sow only on rich land.

Its rapidity of growth in six or seven weeks after sowing, shows its availability as a catch crop in case there is a failure of the hay crop. As it is a shallow feeder it is well adapted for surface manuring.

The previous account is selected and adapted from an article by Dr. E. Lewis Sturtevant in *The National Live Stock Journal*, p. 522, 1881.

Dr. Armsby says: "The chemist gives it about the composition of fair meadow hay. It is deficient in protein and rich in non-nitrogenous nutrients, and should be supplemented with oil cake."

Major H. E. Alvord, of Mass., in the *Rural New Yorker*, speaks as follows: "Hungarian grass is a valuable auxiliary. Where a piece of grass or grain, which looks well in the autumn or even in early spring, shows in May that it will not produce a profitable crop, its fragments may be depended upon to do most good as green manure. Then plow late in May, turning well, harrow two or three times at intervals, sow Hungarian grass the latter part of June, cut it in August and re-seed the land. Hungarian, according to age at harvesting, may be adapted to any class of stock. It makes quite a draft on the land, and, either when it is sown or with the following crop, a dressing of cheap fertilizer is no more than fair, like agricultural salt, kainit, or the raw ground Carolina phosphate. Knowledge of the facts in every case must determine what can be most economically used."

Waldo F. Brown, of Ohio, in the same paper, writes: "In a season when wheat and clover have been generally killed over a large area of country, many farmers are asking what can we substitute for hay? We have two good substitutes—millet and corn fodder. Either may be put in, in this latitude, as late as June 10, with a good prospect of a crop. Millet will yield largely on good land, but the land should be finely pulverized. It is best to sow as soon after a rain as the land can be worked, as if sown just before a rain there is more danger of weeds coming up with it. The seed should be covered lightly, and I prefer a plank drag for the purpose, as it presses the earth to the seed, and retains the moisture till it sprouts. When sown for hay, from three pecks to a bushel of seed per acre should be used."

DEYEUXIA, CLARION.

Spikelets 1-fld. in a close or open panicle, rachilla jointed above the lower glumes, often extending beyond the floret into a

Fig. 80.

bristle-like or smooth rudiment of a flower: flower perfect. The empty glumes persistent below the joint, slightly unequal, awnless, keeled, membranous; the floral glume often with a ring of hairs at the base, 5-nerved, entire or 2-4-toothed, bearing a short awn on the back. Palea slender, 2-nerved, thin. Stamens 3. Styles distinct, short, stigmas feathery. Caryopsis obovoid or oblong, often oblique, included by the slender floral glume and the palea, free, or slightly adherent.

Grasses with various habits. Panicle terminal. Nearly related to *Agrostis*.

About 120 species in temperate and cold regions.

D. (Calamagrostis,) Canadensis, Beauv. Blue Joint.—A perennial with creeping rootstocks, found in low grounds, 3-6-ft. high. Leaves flat, glaucous. Panicle open, 2-6 in. Spikelets purplish with the rachilla continued behind the palea as a short, hairy pedicel. Empty glumes, ovate-lanceolate, acute, the upper with an obscure nerve each side the middle one. Hairs numerous, as long as the floral glume, which bears a very slender, straight awn near the middle. Palea hyaline, two-thirds as long as its glume.

This native perennial grass is widely distributed in the marshes of the Northern States clear across the continent, where it attains a height of four to six feet or more. The narrow panicle somewhat resembles that of red top, only it is more slender.

Unfortunately, the common name is a very indefinite one, as many other and widely different grasses in various parts of our country have been called "blue joint." It is not much cultivated, but is quite common, and if cut rather early, while in flower, or sooner, it affords a very large yield of good hay. Blue joint will grow on land rather too wet for red top, and for such places, if they cannot be drained, we know of no grass more suitable for cultivation.

FIG. 80. *Deyeuxia Canadensis* (Blue Joint): *a*, upper part of a plant; *b*, empty glumes; *c*, *d*, back of same; *e*, floral glume, palea to the left, and at base a rudiment of a floret; *f*, ovary and styles.—(Scribner).

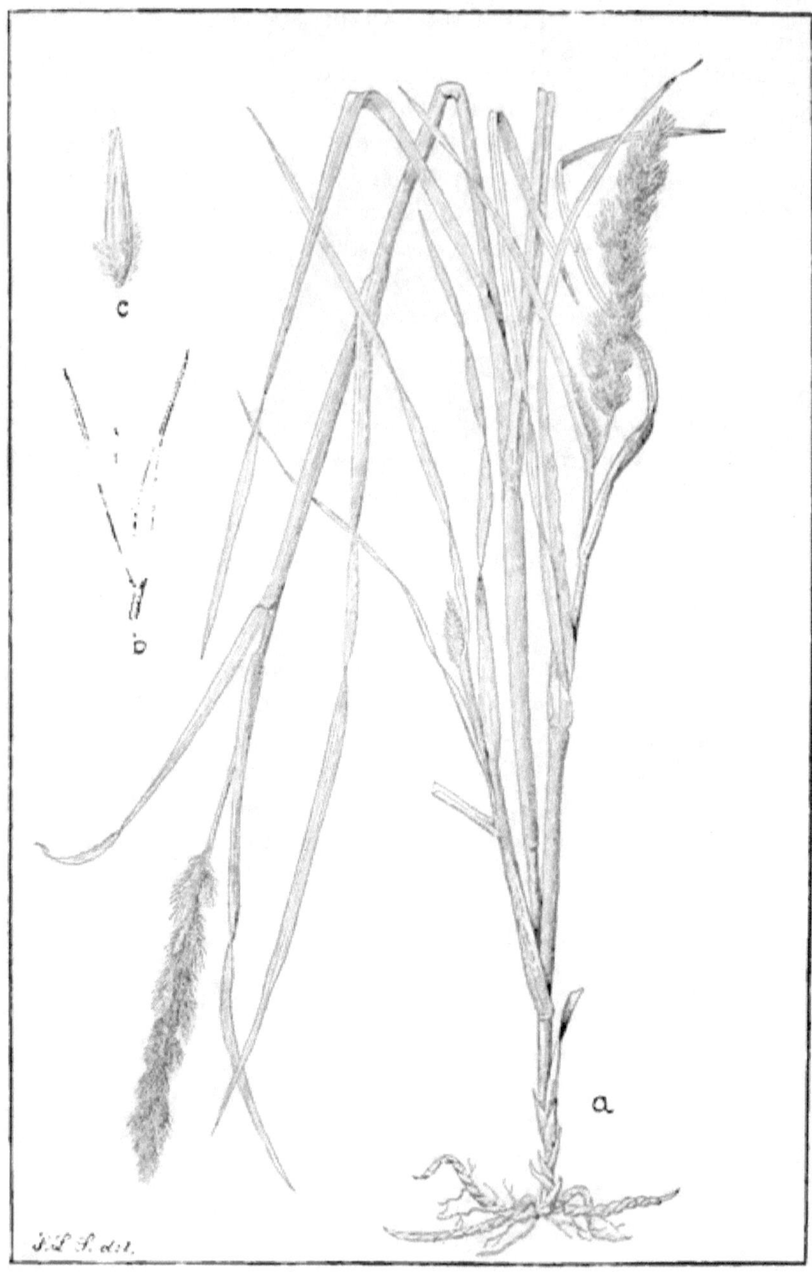

FIG. 81.

The seeds are quite small and some time is required for the grass to become well established.

Concerning this grass, Gould says: "It constitutes about one-third of the natural grasses on the beaver dam meadows of the Adirondacks. It is certain that cattle relish it very much both in its green state and when made into hay, and it is equally certain that farmers who have it on their farms believe it to be one of the best grasses in their meadows."

MUHLENBERGIA, SCHREB.

Spikelets 1-flowered, small, panicled, flowers perfect. Glumes 3, the two lower empty, persistent below the joint, membranous or hyaline, equal or oftener unequal, sometimes minute, or one of them wholly wanting, keeled, acute, mucronate, or rarely short or long awned. The floret with a minute callus or sessile, usually bearded at base. The floral glume 3-5-nerved, firm or membranous, obtuse, acute, mucronate, or very often bearing a slender awn. Palea hyaline, included, 2-keeled. Lodicules 2, very small. Stamens usually 3. Styles distinct, stigmas plumose. Caryopsis narrow, subterete, inclosed by the floral glume, free.

Grasses of various habits. Panicles terminal and axillary, narrow and slender, loose and branching, dense or spike-like, spikelets small, slender.

About 60 species, mostly North American, a few found in the Andes and Asia.

M. glomerata, Trin. Muhlenberg's Grass, Satin Grass, Wild Timothy.—Culms erect, glaucous, 1-3 ft. high, branched, or rarely simple. Panicle spike-like, dense, exserted, 2-3 in. often lead colored, glumes awned, nearly equal. Common northward in bogs, or at the west on dryer land.

FIG. 81. *Muhlenbergia glomerata*; *a*, plant; *b*, spikelet, *c*, floret; (U. S. Agricultural Department and Scribner).

The following is by Dr. C. E. Bessey, now of Lincoln, Nebraska:

"Ten or twelve years ago I had my attention first called to this wild grass as one possessing many valuable qualities, making it desirable for introduction and cultivation. I found that the liverymen of central Iowa were in the habit of cutting those parts of the prairie which lie between the sloughs and the high land. The hay obtained from these places was of fine quality, being composed of leafy, branching stems of fine length and medium hardness. It was always cut late, but even then it was not often in seed. In fact, the rarity of the seeding is so great that I have heard it averred, over and over again, that it is a seedless grass. Of course this was an error, as all grasses are seed-bearing at some stage or other of their existence. In fact, it appears to seed freer under cultivation than in the wild state.

"So much for this grass in a general way. As to common name, I find no uniformity whatever. It is known here and there under many different names. For example, in some places it is known as Nimble Will; in others as Limber Bill, names which in other regions again are entirely unknown or applied to entirely different grasses. I have heard it called Fine Slough grass, a misnomer, as it does not grow in genuine sloughs at all. Again, the name of Small Willow Top is occasionally heard, although not confined to this grass alone. In the books, all the Muhlenberg grasses are called drop-seed grasses, a name which cannot be expected to come into general use. In reports it is often spoken of as simply fine prairie grass, which is, to say the least, exceedingly vague.

"The name I have used—Muhlenberg grass—is one which I think we might well adopt, in honor of the discoverer, old Dr. Muhlenberg, a botanist of the last century, who did much to bring before the world the natural resources of this country. Now it is curious that although this grass has been known in the

West for many years as a valuable wild one, there are to be found scarcely any references to its value in published books or reports to which I have access. Flint, in his great and valuable work, 'Grasses and Forage Plants,' describes it and then remarks, 'Of no agricultural value.' Dr. Darlington, in his book, 'American Weeds and Useful Plants,' does not even mention it; but in reference to a closely-allied species he says: 'It affords an indifferent pasture in the latter part of summer; but it is not of much worth.' Dr. Killebrew does not mention it in his book, 'Grasses, Meadows and Pastures.' Dr. Vasey, in 'The Agricultural Grasses of the United States,' says, 'Specimens have been sent from Colorado and Kansas and recommended as an excellent grass for hay.'

"Now, chemical analyses show that Muhlenberg grass is highly nutritious. In the years 1878 and 1879, at my suggestion, Mr. W. K. Robbins, a graduate of the Iowa Agricultural College, made analyses of this grass, with results which showed that in nutritiousness it ranked with red top and blue grass, and, in some instances, Timothy. More recent analyses by the government chemist at Washington make a still better showing. Taking an average of the analyses I find the following results:

"Timothy contains $4\frac{1}{5}$ per cent of albuminoids.

"Orchard grass contains $6\frac{1}{2}$ per cent of albuminoids.

"Red top contains $6\frac{3}{5}$ per cent of albuminoids.

"Blue grass contains 8 per cent of albuminoids.

"Muhlenberg grass contains 17 2-5 per cent of albuminoids.

"That is, Muhlenberg grass is more than twice as nutritious, weight for weight, as blue grass. It is nearly three times as nutritious as red top and orchard grass, and about four times as nutritious as Timothy. Now I would not for a moment be understood as considering these analyses as settling the relative merits of these grasses. It is well known, however, that the analysis of a grass is one of the important factors in determining

Fig. 82.

its value, and I bring it in here as simply corroborating what the feeders of hay have been saying for a long time."

Muhlenbergia Mexicana, Trin.—Culms ascending, branching, 2–3 feet high; lateral panicle often included at base, linear, interrupted; glumes awnless, sharp-pointed, unequal.

It is quite luxuriant, thrives in the shade, and stands drought well.

Dr. Bessey also speaks well of this grass as well as of the preceding, for Iowa and Nebraska.

He writes: "When I called Prof. Budd's attention to it he said that he grew a three acre lot of it for four years, and that it yielded from 2½–3 tons per acre of hay of the highest quality. This agrees with other testimony. In fact, I have for the last ten years, from time to time, called attention to its value in the papers of this State."

If these species are as valuable as the above notes indicate, most likely several other species of the same genus are also valuable. The very small size of the seed and its slow growth when small, would make it unprofitable for alternate husbandry.

PENNISETUM, PERS.

Spikelets ovate or ovate-lanceolate, with one perfect flower, and a second male or neutral one below, solitary, or 2–3 together, closely surrounded by an involucre of bristles which are attached above the joint. Glumes 4, rarely 3, the lower small or 0, the second often equaling the spikelet, both empty, the third empty or including a palea or staminate flower; the terminal one shorter, including a perfect or pistillate flower, firmer than the palea. Stamens, 3. Styles distinct at the base or united for more or less of their length, stigmas feathery with short or long branches. Cariopsis included, free. Annuals or perennials, often branch-

FIG. 82. *Muhlenbergia sylvatica*, -*1*, Top of a plant; 2, spikelet. A grass of value in some localities.—(U. S. Agricultural Department and Scribner).

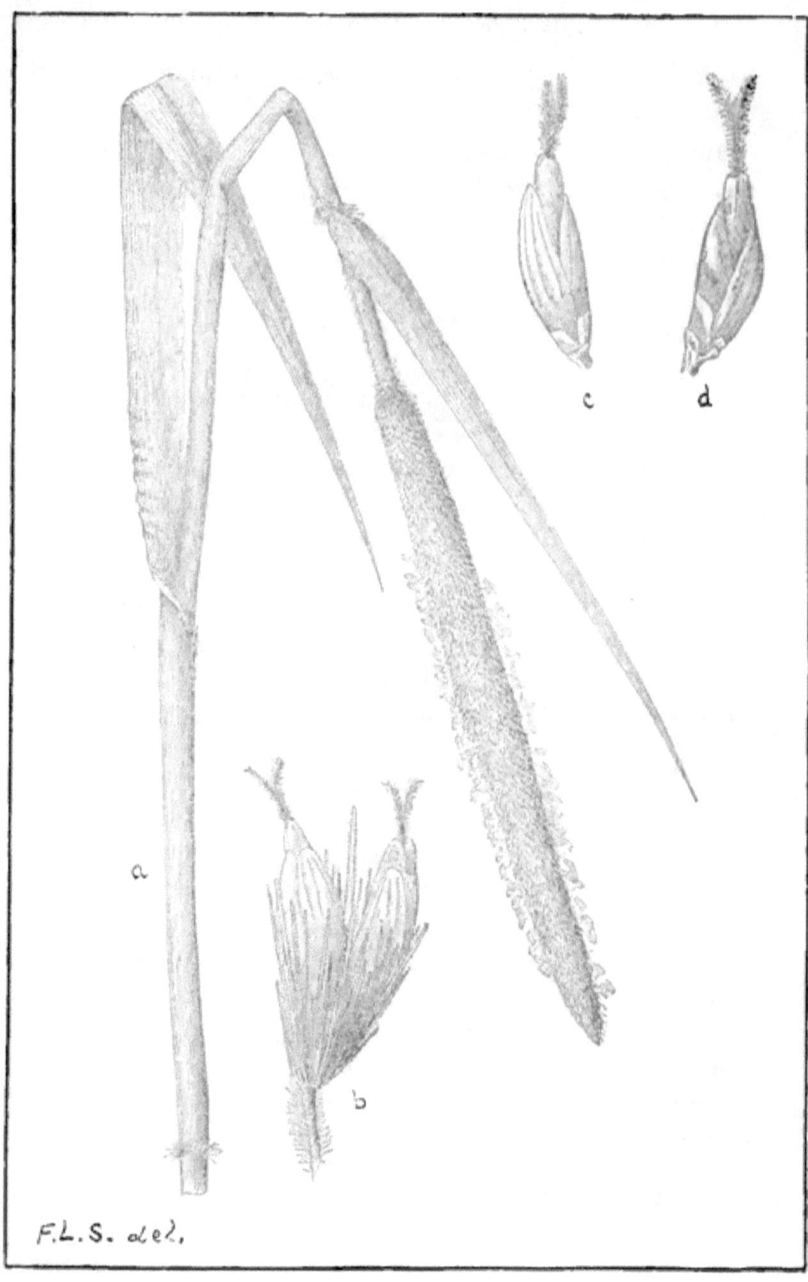
c
d
a
b
F.L.S. del.

ing. Leaves flat. Spikelets crowded in a spike-like panicle, or on spike-like branches.

Species about 40, mostly African, a few in tropical Asia and America.

P. typhoideum. Pearl, Indian, African, Cat-tails, or Horse Millet.—This grass has been spoken of very highly as a meadow grass for the South, where it has been grown for many years. It needs an abundance of heat, rich soil, and makes a rank, rapid growth six or eight feet high, each culm teminating in a stiff spike an inch in diameter and six to twelve inches in length. Branches come out in abundance near the ground, hence there will be all states of advancement in the spikes of flowers. As said of *Panicum Texanum* (Texas Millett) and *Sorghum halapense* (Johnson grass), it may be cut two or three times a year, and yield an abundant crop of rather coarse hay. It cures slowly. In central Michigan, where it has been tried, the summers are too cool for perfecting seed, and the crop does not become large till late in the season. Indian corn is certainly preferable for the North, and perhaps as suitable for the South.

PANICUM, L.

Spikelets born on a jointed pedicel, spikelets racemed or panicled, with one perfect terminal flower, and usually a second which is male or neutral. Glumes usually 4, the lowest small or minute, the second and third usually sub-equal, membranous, awnless or rarely awned, empty or the third including the rudiment of a palea or a male flower: the terminal including a perfect flower, shorter and more obtuse than the others, carioceous, as is also the included palea. Lodicules, 2, fleshy. Stamens, 3. Styles distinct or united at the base for a short distance, stigmas feathery. Caryopsis included in the firm floral glume and palea, free. Annuals or perennials of various habits.

FIG. 83. *Pennisetum typhoideum* Rich. ; *a*, top of a plant with a spike reduced one-half; *b*, a pair of spikelets on the short hairy pedicel, with bristly involucre; *c*, view of one spikelet; *d*, another view.—(Scribner).

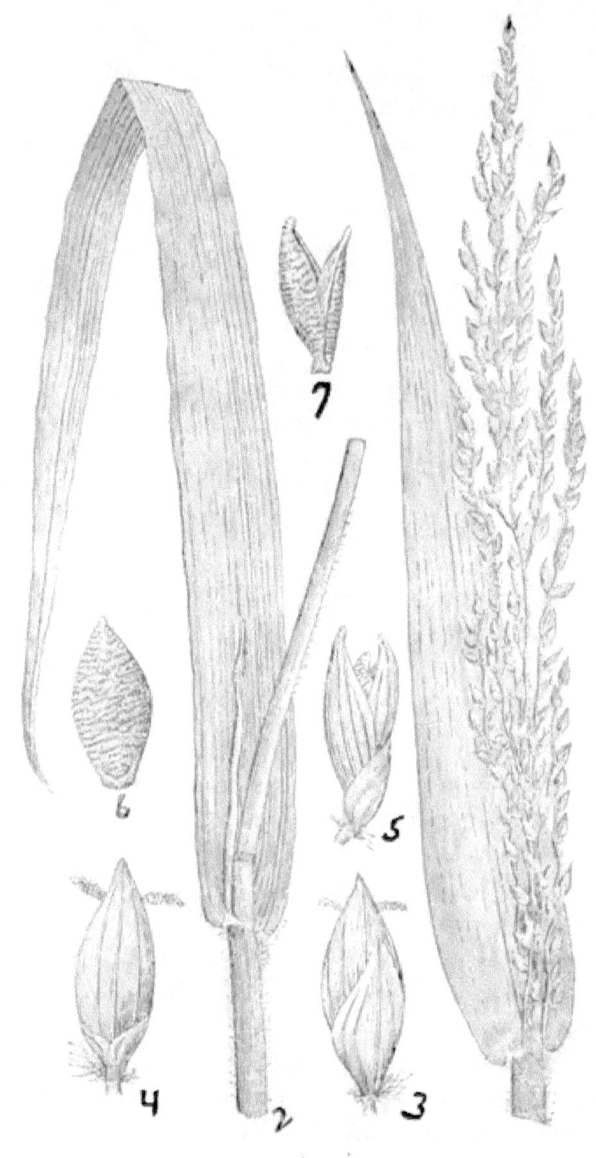
7
6
5
4
2
3

About 250-280 species, widely scattered over the earth. A large and difficult genus.

P. Texanum, Buckl. Texas Millet, Texas Panic Grass.—A leafy annual, 2-5 ft. high, sparingly branched. Leaves 6-8x ½-1 in; soft with rough margins. Panicle 6-8 in. long, narrow, erect, spikelets oblong, pointed. Lower, empty glume half as long as the second, acute, 5-nerved. The upper glume 5-7 nerved. The floral glume transversely wrinkled.

For most of the following I am indebted to Dr G. Vasey. This grass is a native of Texas. It is a grass of rapid growth, succulent, yielding a large amount of forage.

Mr. Pryor Lea, of Texas, after trying it for some years, considers it superior to any grass that he ever saw for hay. It is a much more certain crop than millet, and cultivated with less labor, and all kinds of stock prefer it. It prospers best in the warmest season of the year.

A. W. Ravenel, of S. C., has tried Texas millet for several years, and esteems it very highly.

Dr. Phares, of Mississippi, says. "In habit it is much like crab grass, which is inclined to crowd out this millet."

Prof. S. B. Buckley, of Texas, says: "It grows thick and very rapidly, one or two months being sufficient to bring it to maturity for hay. It thrives best on the Colorado bottom lands, yet I have seen it growing on poor upland soil, but it was dwarfed at least one-half. It may be cut twice or three times a year."

It need hardly be said that this grass promises nothing for the northern United States.

AVENA, L.

Spikelets 2-flowered, very rarely 1-flowered, panicled, rachilla jointed between the flowers, lower flowers, at least, perfect, the upper often male or imperfect. Empty glumes persistent below

---

FIG. 84.—*Panicum Texanum* (Texas Millet); numbers *1, 2*, top of a plant; *3*, dorsal view of spikelet; *4*, front view; *5*, side view; *6*, floral glume; *7*, side view of floral glume and palea.—(U. S. Agricultural Department, details by Scribner).

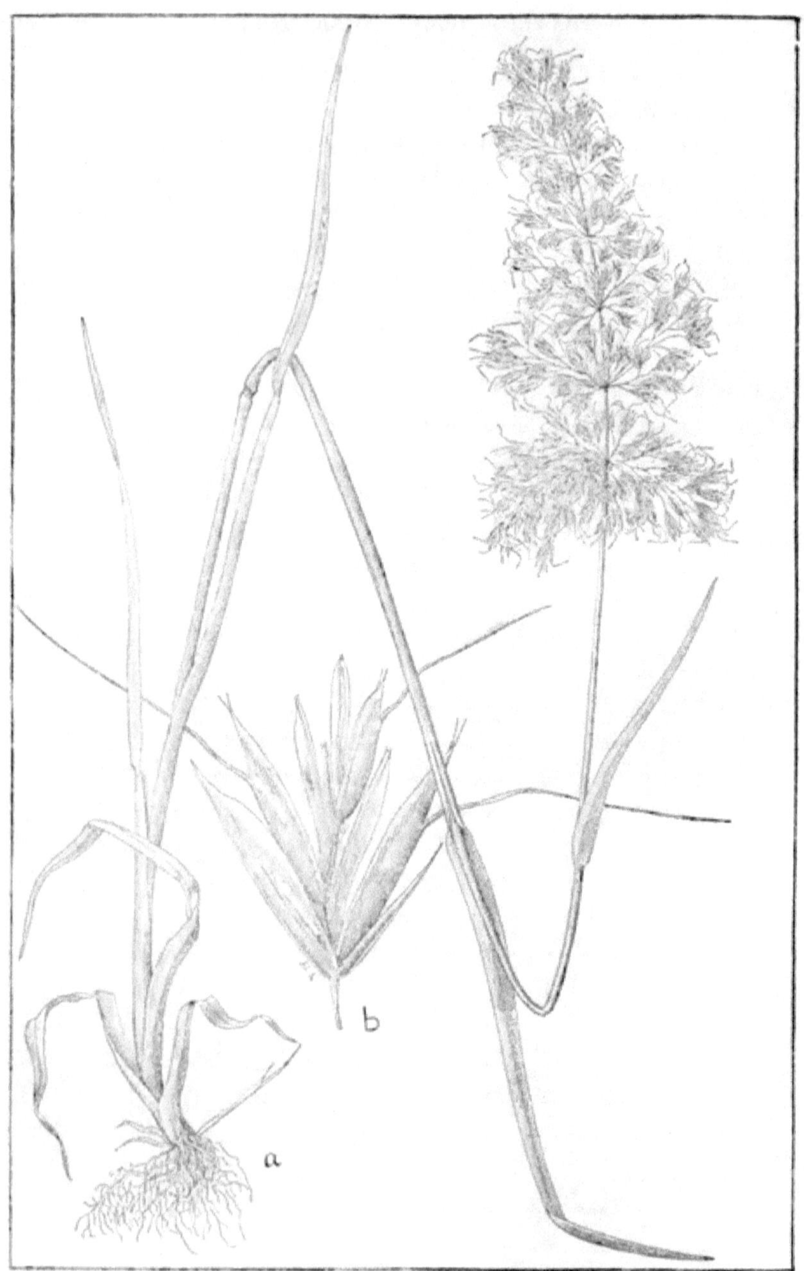

Fig. 85.

the joint, membranous, slightly unequal. Floral glumes convex on the back, acute, 5–9-nerved, often briefly 2-fid at the apex, the lower ones including a perfect flower and bearing on the back a twisted awn, the upper ones awnless, including a staminate or neutral flower. Palea narrow, 2-toothed or 2-fid. Lodicules 2-fid. Stamens 3. Styles short, distinct, stigmas hairy. Caryopsis oblong or long-fusiform, pubescent or rarely smooth, sometimes deeply grooved, included by the floral glume and palea, free or more or less adhering to the palea. Annuals or perennials.

Species about 40. Found in many temperate regions.

A. flavescens, L. Yellow Oat, or Golden Oat-Grass.—An erect, smooth, glabrous perennial, culm 1–2 ft. hi., stoloniferous. Leaves flat, sheaths hairy; ligule truncate, ciliate. Panicle open, branches in ½ whorls. Spikelets compressed ¼ in., 3–4 fld., shining, yellowish. Empty glumes ovate, acuminate. Floral glumes keeled; awns divergent.

According to Mr. Lawes, it is tufted, of rather weakly habit, the culms few and slender, producing flowers in June and July. It is found in cool, dry pastures and light soils. It is hardy and seeds early, is never sown alone, but is recommended as a minor ingredient with others for permanent pasture.

The seed is very often adulterated with seeds of *Aira flexuosa*, which is not worth raising.

I have seldom seen yellow oat grass in the pastures of the United States, and on trying it for several years in Michigan, I am compelled to say that it seems to promise little for this country.

HOLCUS, L.

Spikelets 2-fld., usually in collected dense oblong or interrupted panicles, rachilla jointed above the empty glumes, extending beyond the flowers as a small stipe; lower flower perfect, the

Fig. 85.—*Avena flavescens* (Yellow Oat Grass). *a*, A short plant.—(Sutton). *a*, spikelet. (Scribner).

Fig. 26.

upper male. Empty glumes persistent below the joint, keeled, the lower 1-nerved, acute or acuminate, the second broader, 3-nerved, acute or awned. Floral glumes shorter than the empty ones, membranous, the lower awnless, at length firm, including a perfect flower, the upper quite similar, but including a staminate or neutral flower, and bearing on its back a slender curved awn. Palea narrow, 2-keeled. Lodicules oblique, acuminate. Stamens 3. Styles distinct, stigmas feathery. Caryopsis oblong, included by the firm glume, free. Soft annuals or perennials. Leaves flat or rarely convolute.

Species 8, belonging to Europe or Africa.

H. lanatus, L., Meadow Soft Grass, Velvet Grass, Yorkshire Fog, Salem Grass, White Timothy, Velvet Mesquit Grass.—A soft perennial, culms 6–24 in., ascending, leafy. Leaves flat, upper sheaths inflated; ligule short. Panicle 2–5 in., whitish green, often pinkish, branches 2–3-nate. Spikelets $\frac{1}{6}$ in., elliptic-oblong; empty glumes acute, nerves strong. [Specific character after Hooker.]

Velvet grass is mentioned here because it is so soft, velvety, conspicuous and handsome, that every one at once becomes interested in knowing the name and value, but it is still questionable whether it is worthy of cultivation anywhere.

Holcus lanatus is very productive of seed, and somewhat resembles orchard grass. The whole plant has a grayish aspect of pale white color often tinged with red. It is very common in England, and has been introduced with other seeds into various parts of this country. During summer on the moist old pastures of New England, we have often seen bunches untouched and going to seed, while June grass, red top and white clover were kept closely cropped.

Mr. Lawes says: "This grass is not liked by cattle either

FIG. 86.—*Holcus lanatus* (Velvet Grass); *a*, A plant; *b*, spikelet; *c*, back of upper empty glume; *d*, two florets, without empty glumes. (Scribner.)

25

when green or in hay, being too soft, spongy, and insipid. It is almost a weed, tending to usurp the land, and is one of the few poor grasses which is not reduced but increased by manuring a meadow [of mixed species]. The seed should be carefully excluded."

Dr. Phares says: "It has been introduced into Texas, and constitutes nine-tenths of all the so-called mesquit grass planted in the Southern States. It grows much larger than in the Eastern States or in England; and it seems too, to be more valuable and greatly improved here. It grows two to four feet high in the South."

Holcus mollis, L., Creeping Soft Grass.—This much resembles the former grass, but is not so common. In Great Britain the creeping habit makes it very troublesome. The nodes are villous, awn inflexed, exserted. Much like *H. lanatus*, but usually more slender.

CYNOSURUS, L.

Spikelets dimorphous, fascicled in a dense one-sided spike-like panicle; the terminal fascicle 2–3 fld., flowers perfect, the lower consisting of 1–2 neutral flowers. Rachilla of the fertile spikelet usually jointed above the lower glumes. The empty glumes linear, lanceolate, acute or short awned. Floral glumes broader, membranous, 1–3-nerved, mucronate or awned at the apex or on the back. The terminal one narrower, empty, inclosing a staminate flower, or reduced to an awn. Glumes of the sterile spikelets distichous, pectinate, all empty, sub-equal, linear, subulate, 1-nerved; rachilla continuous. Palea of the fertile flower narrow, 2-toothed. Lodicules with a basal lobe. Stamens 3. Styles distinct, short, stigmas plumose. Caryopsis oblong or elliptical, included by the glume and palea and adherent. Tufted annuals or perennials with flat leaves. The sterile spikelets form an involucre to the fertile one.

Species 3 or 4, found in Europe, western Asia, and northern Africa.

C. cristatus, L., Crested Dog's Tail.—A stoloniferous perennial, 1–2 ft. hi., culms terete, erect, smooth. Leaves short, narrow, slightly hairy; ligule 2-fid. Spike 1–2 in., linear.

It has long been found in most meadows and pastures of Great Britain and the continent of Europe. It is still recommended by nearly or quite all those who sell grass seeds, especially for permanent pasture and lawns on dry light land, but we notice that some of the most observing and independent farmers in those countries seldom recommend it or use it on their lands.

Dr. Lindley said: "Its roots are long and wiry, and descend deep into the ground. It was quite early used for pastures and lawns; not very nutritious, not a favorite with stock."

Sir J. B. Lawes says: "This grass has a wide range of soils, and grows in dry, damp, and even in irrigated lands, and varies in character accordingly. It is better for pasture than hay, and was unable to maintain even a moderate degree of prominence where the conditions were favorable for the luxuriance of other graminaceous species."

The late James Buckman said: "We think it has been overmuch cultivated. It is not a favorite with deer or Southdowns. The culms soon become wiry and make poor hay, neither in quantity nor quality is it worthy a place in a good meadow.

"The culms are much used for straw-plaits, for which they are well adapted, both from their fineness and strength."

It is seldom met with in this country, and judging from our own efforts to grow this grass, we have little to expect in its favor

FIG. 87.

CHAPTER VII.

EARLY ATTEMPTS TO CULTIVATE GRASSES.

Meadows of the Romans.—It will be unsafe to enter into details in reference to the time of introduction of most of our valuable grasses. Even to the present day, there is much uncertainty and confusion of the names of grasses.

As a matter of history it may not be out of place to read a few extracts translated from *Columella*, the old Roman, who wrote about A. D. 50:

"The hay which grows naturally in a juicy soil, is reckoned better than that which is forced by constant watering. Land that shelves gently, if it is either flat or well watered, may be reduced into meadow; but such a level ground is most approved, which, having a very small gentle descent, does not suffer the showers nor the rivulets that flow into it, to abide long in it: or if any water comes upon it, it creeps off slowly; therefore, if in any part it be low and marshy, and the water stagnates upon it, it must be carried off by furrows; for either great abundance or scarcity of water, is equally pernicious to grass of all sorts."

After speaking of removing brush, briars, and weeds, he says: "It is important that we neither allow a hog to feed therein, because, with its snout, it digs up and raises the turf, nor larger cattle, unless when the ground is exceeding dry, because they sink their hoofs into it and bruise and cut the roots of the herbs. Nevertheless, the second year we will allow smaller cattle to be admitted, after the hay harvest is over, provided the dryness and condition of the place will suffer it. Then the third year, when the meadow is become more hard and solid, it may receive

FIG. 87. *Cynosurus cristatus* (Crested Dog's tail): *a*, A whole plant; *b*, *c*, two views of a spikelet.—(Plant from Sutton, spikelets by Scribner).

greater cattle also. Moreover, the leaner and pendent places must be assisted and refreshed with dung. Prudent husbandmen commonly lay more dung upon a hill than a valley, because, as I said, the rains always carry the fatter matter down to the lower grounds."

"There is a measure to be observed in drying hay, that it be put together neither over dry nor yet too green: for, in the first case, it is not a whit better than straw if it has lost its juice; and, in the other, it rots in the loft if it retains too much of it: and after it is grown hot it breeds fire, and sets all in a flame. They do not put it up in mows, before that they suffer it to heat, and concoct itself, and then grow cool, after having thrown it loosely together for a few days."

Here in a few lines we get the ancient idea of selecting lands for meadows, of drainage, of clearing out weeds, of keeping hogs and cattle off from newly seeded land, of applying manure, of storing hay, of spontaneous combustion.

The First Meadows of Great Britain.—It is not yet very long since the first efforts were made to improve pastures in Great Britain.

In his *Mystery of Husbandry Discovered and Laid Open*, J. Worlidge, in 1681, writes: "Ray grass, by which they improve any cold, sour clay weeping lands which is unfit for sainfoin, hath the precedence of all other grasses, these are lucerne, clover, tares, spurry, and trefoil." This is the first mention made of rye grass in cultivation, and for many years it was the only true grass, the seeds of which were intentionally sown. Timothy was introduced into England by the soldiers who returned from this country in 1776. Orchard grass began to be sown about the same time, and since then the number of varieties has steadily increased. Some of the above, as well as the following, is adapted from Gould:

The making of artificial meadows began to receive attention

even from the first settlement of this country. In a work written by Jared Elliott in 1749, the cultivation of Timothy and fowl meadow is strongly recommended, the latter grass is especially landed as in many respects better than any other.

Timothy and red top in the East were sown very extensively, and sea weed and fish were successfully used as manures. Although we were thus early in forming artificial meadows and pastures, our subsequent improvement has not kept pace with our early enterprise, and we are now far behind England and Scotland in this department of husbandry.

In 1824 a new and most important stimulus to their cultivation was offered by the Duke of Bedford, who published his work, giving an account of experiments made by George Sinclair. Since that time Parnell, Way, Lawes and Gilbert, Buckman and Voelcker in Great Britain have done much to advance our knowledge. Numerous prize essays and other communications have appeared, and progress has been rapid and substantial, yet even in Great Britain as late as 1882, one of the best experimenters, C. De L. F. DeLaune, says: "Unfortunately for owners and occupiers of land, the grossest ignorance prevails about grasses. To many almost every herb that is green is considered to be grass."

Progress Has Been Very Slow.—Most of the following paragraph is taken from Gould:

It will not be denied that farmers, in general, bestow much less care, or thought, upon their meadows than they do upon their grain lands. Not many can name for certain half a dozen kinds, and not one farmer in ten thousand knows the names of the grasses growing on his farm, or can discriminate between them. Grass is grass, and that is all they trouble themselves to know. Very many are not aware that they have any other varieties than Timothy, clover (which is not a grass) and red top growing on their farms, although they may have a dozen or twenty other species; much less do they understand the peculiar properties and the relative values of the different species.

"Not long ago," says Gould, "we noticed a large tract of Lyme grass, *Elymus villosus*, growing on the banks of a rivulet. We asked the owner of the land, who had lived on it over thirty years, whether his cattle relished it? He told us he did not know; he had never noticed it, and could not tell whether the cattle would eat it or not. He had seen it growing there all the time in great abundance, but never knew its name, never inquired what it was, nor what it was good for. Meadow fescue, *Festuca pratensis*, is a very common grass in the counties bordering on the Hudson river, constituting about one-fifteenth of the crop on the meadows. When it first came in flower this year we asked the first six farmers that we met with what they called it. Not one of them could name it; they were not quite sure that they had it on their farms; they had something that looked like it, but they were not sure that it was the same. Two of them thought that it was June grass. The difference between the two is so marked that an intelligent farmer should no more confound them than he should confound a horse and a cow."

Why Grasses are Not Better Known.—Improvements in agriculture have always advanced slowly, with the exception of farm implements, which have not generally been invented by farmers, but by mechanics. Probably no class of men adhere more tenaciously to old practices than the farmers. They have had great respect for fashion and the tradition of their fathers.

Grasses have often been recommended under wrong names, or from a very limited observation, or from selfish motives. Perhaps the seed was poor and failed to grow. The farmer is puzzled and returns to his old ways.

The grasses form an exceedingly natural family, and for this very reason it is difficult for a beginner to readily distinguish individual differences. A certain grass varies much in different situations and at different stages of its growth. The grasses have a great deal in common, and to a beginner all look alike.

Even for a pretty good botanist, there is no denying the fact that it is quite a task to learn to recognize our common grasses. Still, it is no more difficult than to match horses well, to judge the weight of a hog, or to pick out a good cow by her general appearance. The grasses have small flowers, and these are likely to pass unobserved, while the animals referred to, by daily association soon become familiar.

What Have Been Sown in Great Britain.—The following grasses and clovers have been recommended in various mixtures for meadows and pastures. In this connection, also, we give the number of pounds to the bushel and the number of seeds to the ounce. Most of the leading seedsmen advertise and recommend a different selection of grasses for each geological formations; one for the London clay; one for the Upper Cretaceous; one for the lower; one for the Oölite; one for the Oxford Clay; one for the Lias; others for the New Red Sandstone, Carboniferous Limestone, Coal Measures, Old Red Sandstone, Upper Silurian, Lower Silurian. Lists are made out for rich loams, poor stiff clay, light soil; for one year, two years, three years, and for permanent pasture and meadow. Many species are used over and over in different mixtures but in varying proportions. For ten to thirty or more species are usually named for each mixture.

The writer quite agrees with James Hunter, an English seedsman, who says: "Although much has been said about 'geological formations' in connection with the grasses, this has really a very unimportant bearing upon the subject, and it is more likely to lead to confusion than otherwise. Four-fifths of those desirable for permanent pasture will thrive upon all good soils. To ring the changes upon the twenty grasses and clovers through some fifty different geological formations, is nothing better than a piece of pedantry. For all practical purposes, it is quite sufficient to know the general character of the soil and the situation." To add to all this, the soils of some formations vary much in fertility and physical conditions.

The more of mystery and complication a seedsman can make out of this subject the more the farmer is likely to rely on his statements, and the more easily can he be deceived.

GRAMINEÆ.	Pounds to the Bushel.	Seeds to the Ounce.
Agrostis stolonifera, Fiorin or Marsh Bent	15	500,000
Agrostis vulgaris, Red Top	14	425,000
Aira cæspitosa, Tufted hair grass	14	132,000
Alopecurus pratensis, Meadow Foxtail	7	76,000
Anthoxanthum odoratum, Sweet Vernal	10	71,000
Arrhenatherum avenaceum, Tall Oat-grass	12	21,000
Brachypodium sylvaticum, Wood Fescue grass	10	15,500
Cynosurus cristatus, Crested Dog's Tail	26	28,000
Dactylis glomerata, Cock's Foot, Orchard grass	14	40,000
Dactylis glomerata, gigantea, Large Orchard grass	10	34,000
Elymus arenarius, Lyme grass	11	2,320
Elymus geniculatus, Lyme grass	12	2,300
Festuca duriuscula, Hard Fescue	10	39,600
Festuca elatior, Tall Fescue	15	20,500
Festuca elatior gigantea, Large Fescue	13	17,500
Festuca heterophylla, Various-leaved Fescue	12	33,000
Festuca gigantea, Giant Fescue	16	8,600
Festuca ovina, Sheep's Fescue	12	64,000
Festuca ovina tenuifolia, Slender Fescue	15	80,000
Festuca pratensis, Meadow Fescue	15	26,000
Festuca pratensis loliacea, Darnel Spiked Fescue	15	24,700
Festuca rubra, Red Fescue	10	39,000
Glyceria aquatica, Water Meadow grass	13	58,000
Glyceria fluitans, Floating Water grass	15	33,000
Holcus lanatus, Woolly Soft grass	7	95,000
Holcus mollis, Creeping Soft grass	6	85,000
Lolium Italicum, Italian Rye grass	18 to 20	27,000
Lolium perenne, perennial Rye grass	18 to 30	15,000
Milium effusum, Millet grass	25	80,000
Phalaris arundinacea, Reed Canary grass	48	42,000
Phleum pratense, Timothy	45	74,000
Poa nemoralis, Wood Meadow grass	15	173,000
Poa nemoralis sempervirens, Evergreen grass	15½	133,000
Poa pratensis, June grass	14	243,000
Poa trivialis, Rough-stalked Meadow grass	14	217,000

GRAMINEÆ—CONTINUED.	Pounds to the Bushel.	Seeds to the Ounce.
Psamma arundinacea, Sea-reed	15	10,000
Trisetum flavescens, Yellow Oat grass	5½	118,000
LEGUMINOSÆ.		
Lotus corniculatus, Bird's-foot trefoil	62	28,000
Lotus major, Large Foot trefoil	64	51,000
Medicago lupulina, Black Medick	63	16,000
Medicago sativa, Lucerne, Alfalfa	60	12,600
Onobrychis sativa, Sainfoin	26	1,280
Trifolium filiforme, Yellow Suckling clover	65	54,000
Trifolium hybridum, Alsike clover	60	45,000
Trifolium pratense, Red clover	60	18,000
Trifolium pratense perenne, Perennial clover	60	18,000
Trifolium repens, White clover	65	32,000
MISCELLANEOUS.		
Achillea millefolium, Yarrow	30	200,000
Cichorium intybus, Chicory	32	21,000
Petroselinum sativum, Parsley	41	12,800
Plantago lanceolata, Lance-leaved Plantain	52	15,600
Poterium sanguisorba, Burnet	25	3,320

From Morton's *Cyclopedia of Agriculture*, we glean the following in relation to the number of kinds selected for each use or situation:

1. For alternate husbandry, 4 grasses, 5 clovers and others.
2. For permanent pasture, 10 grasses, 4 clovers.
3. For permanent pasture, 12 grasses, 6 clovers.
4. For permanent lawn, 12 grasses, 6 clovers.
5. For permanent, another mixture, 12 grasses, 5 clovers.
6. For permanent fine lawns, 7 grasses, 2 clovers.
7. For permanent lands for irrigation, 11 grasses, 2 clovers.
8. For permanent lands in orchards, 10 grasses, 3 clovers.
9. For heathy and moory lands, 7 grasses, 4 clovers.
10. For deep mossy ground, 9 grasses, 3 clovers.
11. For marshy grounds, 8 grasses, 1 clover.
12. For sandy woods, 11 grasses, 1 clover.
13. For rocky and gravelly, 13 grasses, 4 clovers.

14. For warrens, 8 grasses, 5 clovers.

15. For drifting sands, 3 grasses, 0 clovers.

These include in all 36 species of grasses, 10 of leguminous plants and 5 of others, 51 in all. The best farmers of Great Britain in more recent times are inclined to reject quite a number of species heretofore enumerated for sowing.

What Have Been Sown in the United States.—In 1858, in a prize essay for which he received $50, S. D. Harris, of Ohio, says: "Of the grasses that may be called indigenous, and at the same time having the virtues of what are called tame grasses, there are but three kinds deserving of culture in Ohio. These are *Poa pratensis*, *Poa compressa* and *Trifolium repens*. And all worthy of cultivation from any source on arable land are Timothy, red-top, orchard grass, red clover, and, for variety of crop, occasionally German millet and common millet. We should suffer no loss were all the rest stricken from our fields at once."

In 1865, X. A. Willard reports that after making extensive inquiries of the best dairymen as to the kinds of grasses employed in old pastures, they report June grass, fowl meadow grass, meadow fescue, red-top, wire grass and sweet vernal. Timothy, orchard grass, red clover, and some other forage plants, they report, grow in pastures and meadows.

A leading farmer, in his report for the Board of Agriculture in 1868 says: "In Connecticut the almost universal practice is to sow Timothy and clover, either with rye in the fall or with oats in the spring, or in some few moist or rich meadows to use red-top."

During the same year, J. M. McMinn writes: "The pastures of Pennsylvania contain June grass (there called 'green grass'), Timothy, red-top, false red-top (*Tricuspis seslerioides*), blue grass (*Poa compressa*) and meadow fescue. In the meadows a few others were found."

As late as 1884, in the Northern States, among those who sow seeds on their lands intended for meadows, very few sow any other seeds than Timothy and clover. If left to themselves after a few years several others come in one way and another and increase the variety and quality of old meadows and pastures.

It is not quite as true in 1885, as it was when Gould wrote it in 1869, that "This Babel-like confusion of opinions demonstrates clearly enough that we have no real knowledge on this all-important subject, and that we rely only upon capricious guesses for the settlement of the problem."

Circulars or letters of inquiry in reference to the grasses for pastures or meadows seldom bring valuable or trustworthy information.

From the above it will be seen that the list of grasses now generally sown in any State can be counted on the fingers of one hand, while there are doubtless twenty or thirty which ought to find extensive sale for the various uses and the varied soils and climates of any large State. The list is growing, slowly growing larger.

We wish to impress our readers with the very important fact that little is definitely known regarding the grasses found in our pastures, and still less is known in reference to those best adapted to cultivation. As Gould says: "We must fairly grapple with the undoubted fact that the science of grass culture is yet in the early dawn of its infancy."

The Englishman selects twenty or more; not including some which are not true grasses. He selects some kinds for thin soil or upland pastures, others for stiff clays, others for rich, deep loams, others for meadows which are subject to periodic floods along the banks of rivers, and still others for irrigated meadows in which the water can be entirely controlled.

There must always be a difference of opinion as to the merits of grasses on account of the various soils, climates, seasons and uses.

In looking over a large number of agricultural reports of the Northern States we find in some of them much space is given to discussions of the grasses by the farmers at their winter meetings. There is much said about the care of meadows and pastures, with many repetitions. We cannot help being strongly impressed with the idea that we need many more careful observers—farmers who are trained students of science.

CHAPTER VIII.

TESTING SEEDS. SOME COMMON WEEDS.

Seed Stations and Their Work.—Whether a seed is liable to grow or not depends much on how it was cured and the nature of the place where it has been stored. In the following account of some experiments this subject will receive some attention.

The first station for testing seeds was established by Dr. Knobbe, of Saxony, in 1869. In Germany, in 1878, upwards of forty of the experimental stations had attached to them a seed control department, and 14 of these did nothing else.

Adulterations.—These stations discovered adulterations of seeds which were "most ingenious in character, harmful in effect, and remarkable in amount." One practice is to kill seeds by boiling or baking and mix them with some desirable seeds which they resemble. The dead seeds in that case tell no tales.

Old seeds, or seeds of another variety, are often dyed or bleached with sulphur, and used to adulterate good seeds of red clover or some other species. Old seeds are dressed with oil and sometimes rubbed by machinery to improve their appearance. Seeds of rye-grass and Italian rye-grass are often adulterated with those of chess, which they much resemble. *Holcus lanatus*, a poor grass, is also found in rye-grass. Meadow fescue is largely adulterated with that of perennial rye-grass, a cheaper seed of less value.

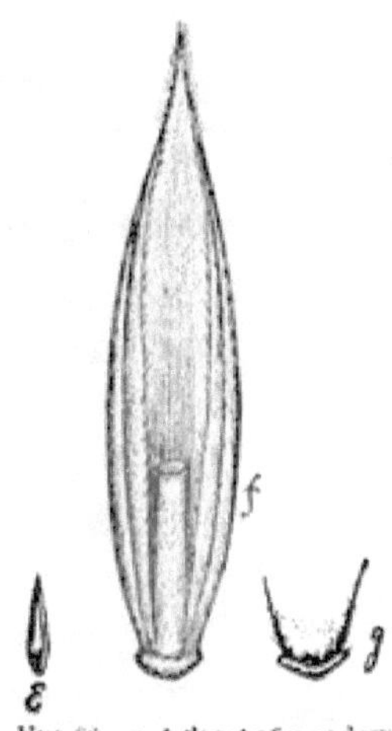

FIG. 88.—*e*, A floret of meadow fescue; *f*, the same enlarged; *g*, the other side of the base of same.

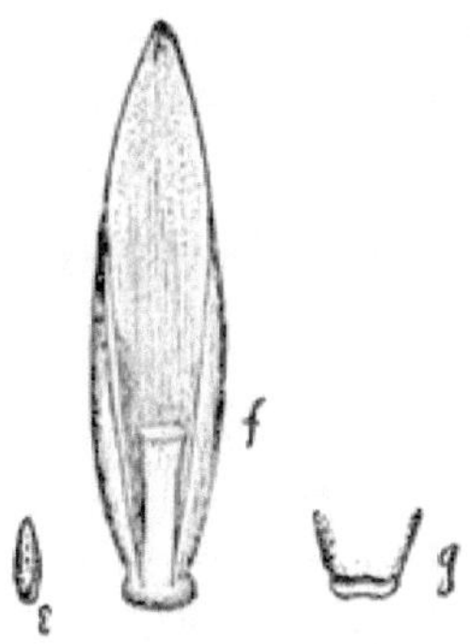

FIG. 89.—*e*, A floret of perennial rye grass; *f*, the same enlarged; *g*, the other side of the base. Observe the difference in the apexes, difference in the piece of the rachis held by each, though this is not uniformly as here shown. Observe the base of Fig. 88 is convex, while that of Fig. 89 is flat or concave.

Crested dog's-tail is largely adulterated with *Molinia cœrulea*, which is of no value.

Seeds formerly sold, even by the very best seedsmen, were more or less tampered with, and they were careful to adulterate their seeds about so much each year to prevent troublesome questions.

In Germany, the mills ground quartz; it was sifted, colored, and mixed with seeds of clover. Pure seeds are quoted as "net seed," while dead ones are quoted as *trio* or "000."

James Hunter, of England, in his seed catalogue and treatise on grasses, writes: "If it be asked how such a state of things can be possible, the only reply that can be given is, that so complete is the want of knowledge of this subject on the part of the seedsmen and agriculturists, that almost any species of adulteration of grass seeds may be practiced without fear of detection. It is probable that not one seedsman in twenty knows all the species of grasses commonly used for permanent pastures, or the seeds of the various species of grass seeds he sells."

In 1877 the writer began testing seeds sold in this country, and found many that were poor and unreliable, especially the more uncommon grass seeds, most of which are imported. Grass seeds vary much in weight, owing to the fact that they are usually sold in the chaff, which is not always well filled. For this reason it should always be bought by weight, remembering that if dry, the heavier the sample, the less empty chaff it is likely to contain.

Concerning poor seeds Professor Shelton remarks: "The difficulty experienced by farmers in securing good seed has been a serious obstacle in the way of grass culture in Kansas. We have reason to know that the complaint regarding the quality of grass seeds retailed in the State is as just as it is universal. The special cause of this trouble in Kansas seems to grow out of the fact, that, as a rule, the trade in grass seeds is not a large one as yet, anywhere; and seeds which are not sold any one season, are carried to the next. In this way, seeds which were originally good are badly damaged, or their vitality is totally destroyed by being kept year after year in damp cellars and mouldy warehouses. But more than this, seeds are often worthless in the start, from having never been properly matured, or from injury received in the field or mow before threshing."

Doctoring and adulterating and selling such seeds is worse than selling 100 yards of cotton thread for 200 yards, or deceiving in the weight or cost of tea, coffee or sugar. The sale of poor seeds affects the future crops as well as the present one.

In German seed-stations the following kind of work is done: determination of the species, the amount of impurities and their nature, the germinating power of seeds, the total weight of the seeds, their specific gravity, their weight per bushel, detection of dyeing, bleaching, oiling, etc.

The apparatus needed is very simple, consisting of a small magnifying glass, some sieves of various grades, bellows, forceps, delicate scales, thermometers, jars, test-plates, chemical tests, and a good knowledge of botany. Some genuine seeds of the common weeds and grasses are useful for comparison.

The sample should be carefully and fairly drawn from the whole, and well mixed. As a general thing for convenience, 50 or 100 seeds or multiples of these numbers are counted out. They may be placed between layers of moist flannel or thick woolen paper, and kept in a temperature of 50 to 60 degrees F. A dish of damp sand, with a paper or cloth on top to hold the seeds, over which is another cloth, is a very satisfactory arrangement.

What kind will usually Germinate and what will not.— Seeds of the commonest grasses, such as Timothy, orchard grass, June grass, red top, and the common clovers, are generally very good, containing from two to twenty per cent. of impurities, which consist mostly of dirt, straw and chaff.

The seeds of the less common grasses, such as perennial rye grass, the fescues, meadow foxtail, oat grass, crested dog's tail, sweet vernal possess a very low vitality, almost without exception. These are mostly imported from Europe.

In 1877, the writer tested grass seeds purchased of one of the less known seedsmen of New York. Four lots of 50 seeds each were tested, with the following results given in per cent. In each case what appeared to be a seed was tested. Most of these were in the chaff:

Hard fescue, 13 per cent.
Rhode Island bent, 7 per cent.
English rye-grass, 5 per cent.
Rough-stalked meadow, 2 per cent.
Schræder's Bromus, 60 per cent.

Red-top, 11 per cent.
June grass, 5 per cent.
Reed canary grass, 3 per cent.
Meadow foxtail, 1 per cent.
Sheep's fescue, 1 per cent.

Wood-meadow grass, 1 per cent.
Meadow fescue, 7 per cent.
Sweet vernal, 15 per cent.
Tall fescue, 11 per cent.
Darnel spiked fescue, 5 per cent.
Orchard grass, 27 per cent.
Hungarian grass, 51 per cent.
Yellow oat grass, 11 per cent.
Timothy, 68 per cent.
Italian rye-grass, 21 per cent.

Creeping bent, 2 per cent.
Crested dog's-tail, 8 per cent.

Large red clover, 88 per cent.
Medium red clover, 88 per cent.
Bokhara clover, 48 per cent.
Italian clover, 82 per cent.
Lucerne, 74 per cent.
White clover, 84 per cent.
Alsike clover, 64 per cent.

The writer had kept some home grown seeds at the Agricultural College for two or three years in several different rooms, one of which was a damp basement. These seeds were shelled out, as were the seeds taken from the samples purchased from the New York seedsman:

NEW YORK SEEDS.	COLLEGE SEEDS
Shroeder's Bromus, 64 per cent.	Shroeder's Bromus, 96 per cent.
Sheep's fescue, 0 per cent.	Sheep's fescue, 72 per cent.
June grass, 6 per cent.	June grass, 28 per cent.
Rye-grass, 18 per cent.	Rye-grass, 74 per cent.
Meadow fescue, 6 per cent.	Meadow fescue, 92 per cent.
Orchard grass, 66 per cent.	Orchard grass, 82 per cent.
Red clover, 94 per cent.	Red clover, 52 per cent.

Seeds taken from packages with low vitality will vary much in different tests, but good fresh seeds run high and quite uniform. Good seeds will stand the most abuse.

Will Seeds Sprout More Than Once?—It is the opinion of many that seeds once sprouted and well dried will never sprout again. "To sprout" means "to germinate," "to vegetate," "to begin to grow," "to shoot, as the seed or the root of a plant." In each of the cases considered the roots died at the end of each test, and new ones pushed out when moistened. The same *plumule* lived over, or endured all the changes. Wheat and rye and oats will start to grow after drying for several times, often for six or more times.

How to Procure Seeds that are Good and True to Name. —It has been shown that there are many difficulties in the way of making improvements in the seeding of land to grass. Our farmers usually buy two or three common sorts offered in the market. In England the seedsmen have largely prescribed the kinds to be used for meadows and pastures, and they are naturally inclined to recommend what is to them most profitable and easily obtained. Where land is to remain in grass for some years it is very important to make the right selection of seeds. The leading seedsmen keep experts, as they call themselves, for the purpose of giving information on this subject.

They take contracts at special rates for laying down a certain number of acres to grass.

The Royal Agricultural Society employs a consulting botanist to examine samples of seeds offered in the market. He has fees for performing certain work. To report on the purity, amount and nature of foreign materials, perfectness, and germinating power of a sample of seeds the fee is five shillings. The council have established a standard for the examination of seeds.

1. That the bulk be true to the species ordered.

2. That it contain not more than five per cent. of seeds other than the species ordered.

3. That the germinating power shall be, for cereals, green crops, clovers and Timothy not less than 90 per cent.; for foxtail not less than 20 per cent.; and for other grasses not less than 70 per cent.

Seedsmen in England and Germany will now guarantee seeds in accordance with this standard.

In England, in 1869, after enacting a law against "doctoring" seeds, they nearly or quite disappeared from the market. The consulting botanist had only seen two samples in five years. Killed and dyed seeds are gone, but dead seeds may still be found.

Notwithstanding the laws enacted, and care taken, it is by no means easy to secure good seeds true to name.

Mr. De Laune, in Jour. Royal Ag. Soc., in 1882, says: "However careful I was in my orders, and from whatever seed-market I ordered my seeds; the percentage of rye grass, soft woolly grass, and other bad grasses and weeds, was beyond all belief. I learned that good seed was most difficult to get. I consulted the botanist, and to my great amazement was told that my seed bought for meadow fescue was all rye grass, and the rough meadow grass was all smooth meadow grass. I have, since these experiments, never sown any seed except after the sample had been examined by the consulting botanist; and have, in consequence, obtained results most satisfactory to myself. I have found it necessary to examine seeds from different parts of every sack. I regret to say that there is no seed-merchant I would trust without the seed was examined by the consulting botanist." And yet a leading seedsman in England says: "The seedsman should be treated with much the same sort of confidence as the family doctor."

Doubtless my readers will be glad to see the following quotation from the Annual Report of the consulting botanist,—W. Caruthers, of the Royal Agricultural Society for 1884:

"During the past year I have examined 701 samples of seeds for the members of the society, besides replying to inquiries regarding the nature, habits, and names of weeds, and the best way of dealing with them: the diseases of cultivated plants; and to various matters affecting the crops of the farm. I have examined 69 samples of meadow fescue, and 46 of tall fescue, in all 115 samples, as compared with 85 in the previous year. Sixty-five per cent of the samples of meadow fescue were free from weeds and seeds of other grasses, as against 26 per cent of last year. The principal adulterant employed is rye grass; but the use of this seed is very greatly lessened. [This is on account of the work of the consulting botanist.]

"Seventy-six samples of cocksfoot [orchard grass] were on the whole pure. Six per cent had some small rye grass seeds in them, and in one case 20 per cent of Yorkshire fog (*Holcus lanatus*.) were included in the sample.

"A fair proportion of 65 samples of meadow foxtail was found good. No less than 64 per cent of the samples of *Agrostis alba* var. *stolonifera*, fiorin or creeping bent, were infested with ergot, a most dangerous fungus.

"Out of 126 samples of clover, 19 per cent of the red clover contained seeds of dodder, and 25 per cent of the alsike contained seeds of this parasite.

"Fewer samples of grass mixture have been submitted to me during the past year, but the samples examined have more firmly convinced me that it is most undesirable for growers to purchase their seed in this form. One mixture consisted entirely of rye grasses, with some trefoil and a little clover, and in addition the rye grass was infested with ergot. Another consisted of rye grass with one per cent of other grasses and clovers."

If railroad companies find it necessary to employ engineers, if trustees think it best to employ a landscape gardener to lay out a park or cemetery, if builders employ architects, why should not the farmers, at a trifling expense to each, employ a consulting botanist at an experiment station, to examine seeds before purchase?

We look forward with hope to the time when every State shall have one or more such stations.

To some extent, the following plan adopted by Professor Shelton, of Kansas, will work well:

"Our practice, which has been entirely satisfactory, has been to send to those dealers who make a specialty of grass seeds in the sections where the seeds are raised. We have always sent to the large eastern dealers for our grass seeds, and to Denver and San Francisco for our alfalfa seeds. This may not be the best plan, but it has been satisfactory as to the quality of the seed procured and as to the price, which, including the freight, we have found to be considerably less than that asked by local dealers."

Weeds in the Meadow.—A weed is now generally described as a plant out of place, or growing where it is not wanted. All the pasture grasses are weeds, if they grow in our garden or corn field. In some countries potatoes become weeds. A plant may be a troublesome weed in one country and not in another.

R. W. Emerson entertained a very hopeful view of weeds, and defined one as "a plant whose virtues have not yet been discovered. * * * * Every plant probably is yet to be of utility in the arts."

A large majority of our worst weeds are foreigners, and have come from Europe, Asia and South America. It is just so with the fields of Australia and New Zealand. Most of the weeds are introduced on to a farm by being sown with seeds of the grasses and clovers; occasionally they come from fresh manure or from waste places, or slovenly farms in the neighborhood. Most of them are following the tide of emigration and are "going west," but a few are taking the opposite course, such as *Rudbeckia hirta, L. Dysodia, Matricaria discoidea, D. C., Artemisia biennis,* Willd.

In the words of Dr. Thurber: "Weeds seem to be naturally well provided for distribution, but the careless farmer sows them broadcast by the handful, and does what nature cannot do —*he puts them in well prepared soil,* where they will be sure to grow. In the month of March hundreds of farmers will sow their clover. Next summer, or later, we shall have from some of them letters and specimens. 'A new weed has appeared in my field, or meadow, and threatens to kill out everything else. What is it, and how shall I get rid of it? Where *did* it come from?' The answer will be: 'You carefully sowed it that cool March day with your clover.'"

Plants are assisted to become weeds by producing many seeds, by ripening with the crop so the seeds are harvested with it, by ripening before the crop and scattering seeds on the ground, by producing seeds which are not easily separated by sieves, by producing very small seeds which escape notice, by having a supply of hairs that they may be carried by the wind, or some awns or hooks to hold fast to animals, by remaining a long time in the soil without losing their vitality, by producing long or thick roots not easily eradicated, by producing bulbs, or long root stocks, by being offensive to all kinds of stock so they are allowed to grow and multiply.

Clay soil is less likely to be troubled than loam or sand.

Some of the weeds which interfere with the growth of grasses and clovers in the Northern States are here enumerated, with figures of a few and remarks in reference to all.

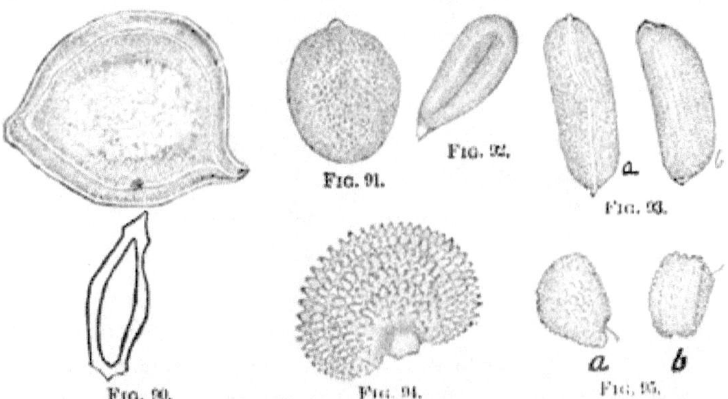

FIG. 90. An achene or fruit of *Ranunculus bulbosus*, L. (Bulbous Crowfoot.) A troublesome perennial with yellow flowers, found in the northeast. Side view and cross section, 1×10.- (Sudworth).

Ranunculus acris. L. (Tall Crowfoot.) Much resembling the last though destitute of the bulb.

FIG. 91. *Brassica nigra*, L. (Black Mustard.) 1×15.- (Sudworth.

FIG. 92. *Capsella Bursa-pastoris*, Mœnch. (Shepherd's Purse.) 1×20. (Sudworth.)

FIG. 93. *Hypericum prolificum*, L. (Shrubby St. John's Wort), two views. 1×20.- (Sudworth.)

FIG. 94. *Silene inflata*, Smith. (Bladder Champion.) 1×16.- (Sudworth.)

FIG. 95. *Cerastium vulgatum*, L. (Mouse-ear Chickweed), two views of a seed. 1×25.- (Sudworth.)

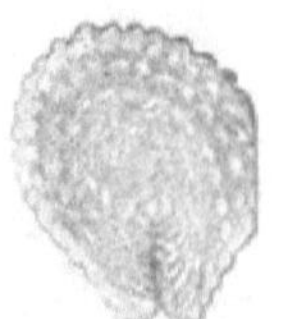

FIG. 96. FIG. 97. FIG. 98.

FIG. 96.—*Stellaria media*, Smith. (Star chickweed.) A seed lying on one of its two flat sides. 1×20.—(Sudworth.)

FIG. 97.—*Mollugo verticillata*, L. (Carpet-weed.) Two views, *a*, looking towards one edge; *b*, lying on one side. These are much like chickweeds. 1×14. (Sudworth.)

FIG. 98.—*Papaver Rhœas*, L. A seed of poppy. Two or three species are quite troublesome in some places. 1×30.—(Sudworth.)

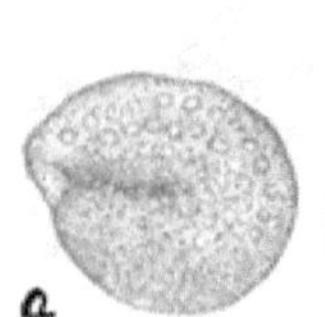

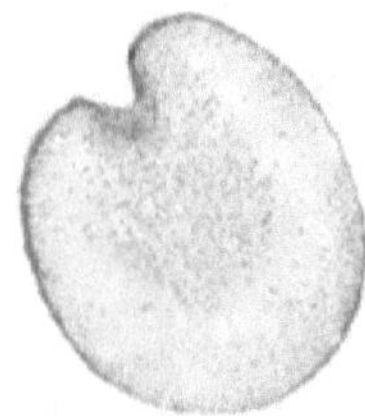

FIG. 99. FIG. 100.

FIG. 99.—*Portulaca oleracea*, L. (Purselane, "Pusley"); *a*, seed lying on one side, *b*, standing on edge showing seed scar. 1×12.—(Sudworth.)

FIG. 100.—*Malva rotundifolia*, L. (Common mallow.) This low perennial seeds freely for some months in the year, sending down a long, stout root. 1×18. (Sudworth.)

Erodium cicutarium, L'Her. Alfilaria, Pin-clover, Pin-grass. This plant belongs to the geranium family, and has become quite abundant in California. The plant makes good pasture, but the seeds get into the wool of sheep, and not unfrequently pierce the skin of the animal. The seed of this plant, by the aid of alternating drought and moisture, can penetrate the soil after the manner of *Stipa spartea* previously mentioned.

FIG. 101.—*Erodium cicutarium*, L'Her. (Alfilaria.) 1×3. (Scribner.)

Medicago lupulina, L. (Black Medick.) This is a biennial or perennial, and in habit much resembles white clover. The flowers are yellow; the plant makes good feed, though there is less

FIG. 103. *Oxytropus Lambertii*. (Loco Weed.) (U. S. Agricultural Report.)

Fig. 103.

FIG. 102. *Medicago lupulina*, L. (Black Medick Nonesuch.) Fruit or pod enlarged.—(Scribner.)

of it than would be furnished by either of the clovers in cultivation. It b longs to the same genus as Lucerne, and is here mentioned because the seeds are likely to be found mixed with the seeds of grasses and clovers. It makes a very fair pasture, especially on rich clay land. The reticulated pods adhere to the wool of sheep.

Loco Weed (*Oxytropis*) grows about a foot high, and is quite erect in habit. It is found on the dry prairies in the West. There is another plant, *Astragalus mollissimus*, which much resembles the above species, which is also called Loco Weed. From the Agricultural Report for 1884 we learn that they often cause sickness and death of cattle and other domestic animals. It causes loss of flesh, lassitude, impaired vision, and finally the brain is affected, the animal becoming crazy. The animal may linger a year or two. No antidote has been discovered.

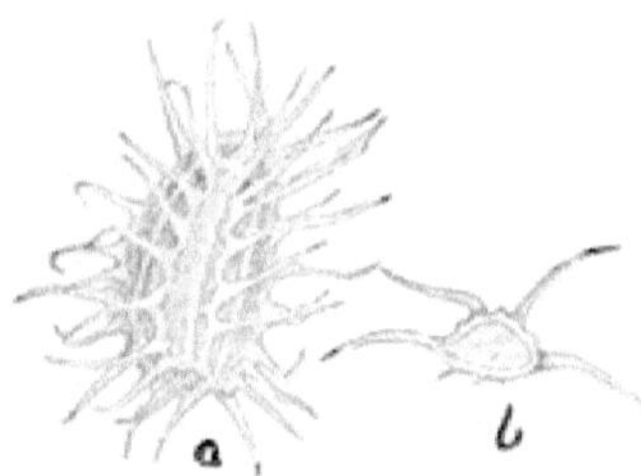

FIG. 101. *Daucus Carota*, L. (Common Carrot.) *a*, whole fruit; *b*, cross section. A troublesome weed in some places. 1×8. (Sudworth.)

Pastinaca sativa, L. Common Parsnip—has escaped from cultivation and has become a troublesome, unsightly weed, with poisonous roots.

Erigeron Canadense, L. (Flea bane, Horse-weed, Mare's tail.) This is an annual which horses and sheep will sometimes devour, though it is unsightly and of no value.

Achillea Millefolium, L. (Common Yarrow.) This plant bears small heads with white ray flowers; the leaves are twice prinnately parted. It makes an inferior pasture, though in England it is sometimes recommended to sow in mixtures for permanent pasture.

FIG. 105.—*Erigeron Canadense*, L. (Flea-bane.) 1×20.—(Sudworth.)

Erigeron annuum, Pers., *E. bellidifolium*, Muhl., *E. Philadelphicum*, L., *E. strigosum*, Muhl., are other species of similar habit to figure 105, and are quite unsightly and common in thin meadows.

Rudbeckia hirta, L. (Cone-flower.) This hairy weed has a purple cone surrounded with yellow ray-flowers. It is becoming more common, and has made its way from the West to the New England States.

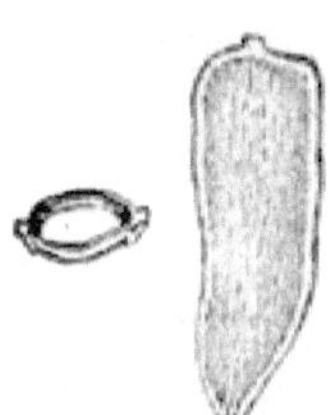

FIG. 106.

FIG. 107.

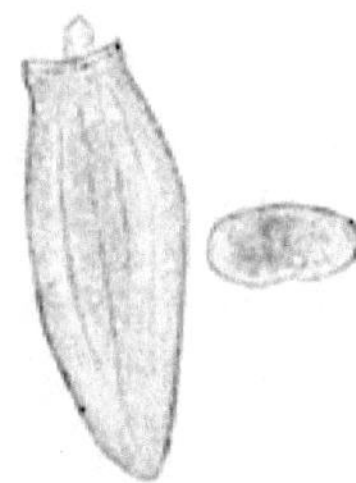

FIG. 108.

FIG. 106.—*Achillea Millefolium*, L. An achene, two views. 1×10. (Sudworth.)
FIG. 107. *Leucanthemum vulgare*, Lam. (Ox-eye.) Achene enlarged.—(Scribner.)
FIG. 108. *Cnicus arvensis*, (Canada thistle); an achene, also a cross section of the same. 1×10. (Sudworth.)

Leucanthemum vulgare, Lam. (Ox-eye, White Daisy.) This is one of the worst perennial weeds or weeds of any kind which infests the meadows and pastures of this country. The seeds are sown with grass seed.

Cnicus arvensis, Hoffm. (Canada Thistle). This is often considered the arch fiend of weeds, and is too well known in many localities. The heads are small and the scales of the involucre scarcely prickly pointed. It is a perennial rooting very deeply. Its course westward is likely to be checked by the fact that it has usually failed to produce seeds on the prairies. It is often diœcious. Some account of this pest is given in connection with the chapter on clover.

Ragweed, Hogweed, Roman Wormwood, *Ambrosia*, is very common, especially in old fields. It is a coarse, homely annual, which one, not a botanist, would scarcely suspect was a member of the aster family.

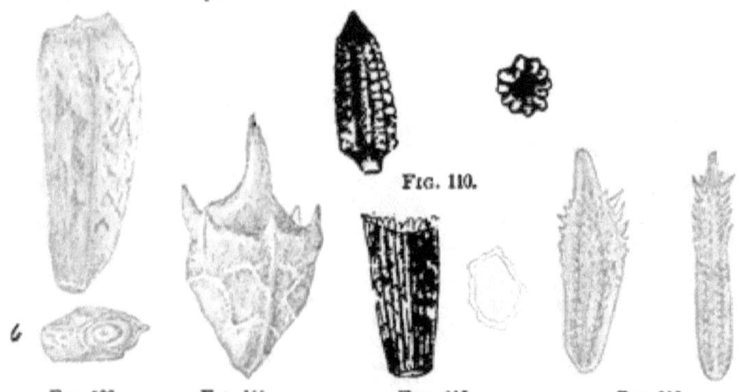

Fig. 109.—*Lappa officinalis*, var *major* Gr (Burdock.) One of our worst wayside weeds, carried on the fleeces of animals 1 × 6.—(Sudworth.)

Fig. 110.—Two views of an achene of *Anthemis Cotula*, (Mayweed) 1 × 15.—(Sudworth.)

Fig. 111.—*Ambrosia artemisiæfolia*, L. (Rag weed), an achene. 1 × 8.—(Sudworth.)

Fig. 112.—*Cichorium Intybus*, L. (Chicory, Succory), two views. 1 × 7.—(Sudworth.)

Fig. 113.—*Taraxacum Dens-leonis*, Desf. (Dandelion), two views of an achene, destitute of the long beak and pappus, which break off easily. 1 × 10.—(Sudworth.)

Chicory, Siccory, (*Cichorium Intybus*, L.) has been introduced as a substitute for coffee, and has spread in many waste places in the older portions of the Northern States.

Plantago lanceolata, L. (Ribgrass, Narrow-leaved Plantain.) This perennial herb has become extensively introduced with grass seed from the East. The flower stalk runs up quickly after cutting, and thus becomes unsightly on lawns. It has been found quite as nutritious as some of our best grasses, but it affords only a small amount of feed. Most kinds of stock eat it when young. Seeds ought to become familiar to every one who makes a lawn or a meadow.

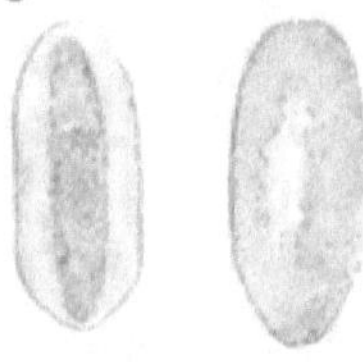

Fig. 114.—*Plantago lanceolata*, L. (Ribgrass, Narrow-leaved Plantain), 2 views. 1 × 12. (Sudworth.)

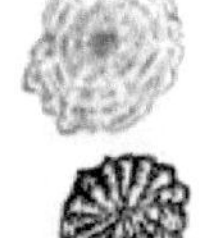

FIG. 115. *Verbascum Thapsus*, L. (Common Mullein). Three views of a seed. 1×20.—(Sudworth.)

Plantago major, L., (Common Plantain), is not very troublesome when compared with the former species. Found about door yards.

Verbascum Thapsus, L. (Common Mullein, Velvet-leaf.) The seeds of this biennial are very small and very numerous. *V. Blattaria* (Moth Mullein), is becoming common, and needs attention.

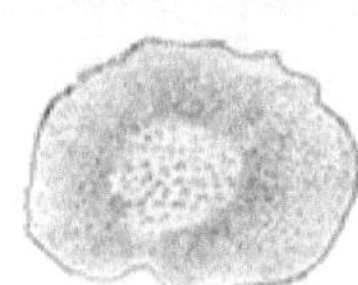

FIG. 116.—*Linaria vulgaris*, Mill. (Toad-flax.) Two views of a winged seed. 1×15.—(Sudworth.)

Linaria vulgaris, Mill. (Toad-flax, Butter and Eggs.) This is a vile nuisance in meadow or pasture. It seeds freely, and also spreads very rapidly by its root-stocks.

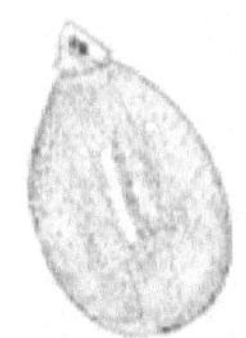

FIG. 117.—*Brunella vulgaris*, L. (Self-heal, Heal-all.) 1×12.—(Scribner.)

Brunella vulgaris, L. (Self-heal or Heal-all.) This is a small perennial herb, bearing violet flowers. In dry, thin meadows it is quite common and on the increase.

Cynoglossum officinale, L. (Hounds-tongue.) This is another plant bearing a nutlet containing hooded prickles.

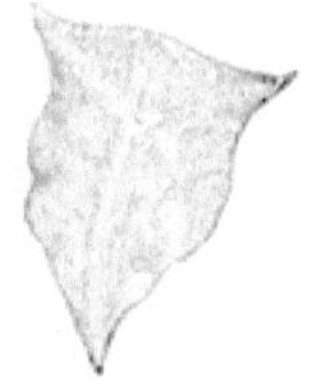

FIG. 118.—*Echium vulgare*, L. (Viper's Bugloss.) This is somewhat ornamental, but in some places has become very prolific and hard to eradicate. An angular wrinkled nut; two views, much enlarged. (Scribner.)

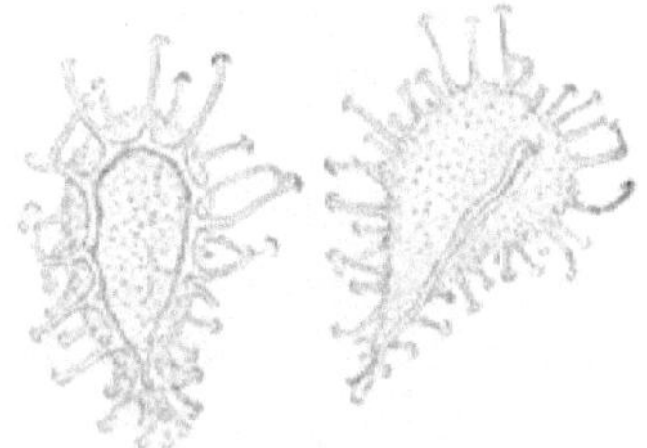

FIG. 119. *Echinospermum Lappula*, Lehm. (Stick-seed.) The small nutlets are covered with a double row of hooked prickles. It must be carefully kept out of sheep pastures; two views. 1×10.—(Sudworth.)

Cuscuta Epithymum, Murr. (Lucerne Dodder.) This is a parasitic vine which has occasionally been introduced with seed of

Lucerne. It is likely to be destroyed by the careful farmer. As soon as the slender vine from the seed gets fast to the stem of a plant, the root of the dodder perishes. It takes nourishment from the Lucerne. The seeds are very small and spherical. Sulphate of iron (green vitriol), one pound to the gallon, sprinkled on plants of dodder is said to destroy it, and will not injure the Lucerne. A sieve of the proper size will remove the seeds. The seeds will remain alive in the soil for some years. A similar cuscuta is parasitic on red clover.

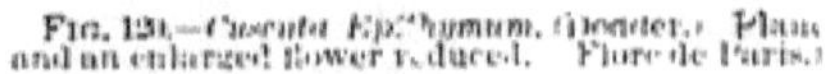

FIG. 120.—*Cuscuta Epithymum.* (Dodder.) Plant and an enlarged flower reduced. (Flore de Paris.)

FIG. 121.—*Amaranthus retroflexus.* (Amaranth, Pigweed.) Two views of a seed. 1×20.—(Sudworth.)

Asclepias Cornuti. Des. (Milkweed.) In light soils this is often a very troublesome weed. The roots run deep and spread in every direction. The seeds are carried by the wind.

Chenopodium album. L. (Pigweed, Lamb's Quarters.) This rank annual weed, with two or three other species, is quite common in waste places and in cultivated grounds. They are not troublesome in pastures and meadows, but are mentioned here because the seeds are sometimes met with in seeds of grasses and clovers.

The seeds are lenticular, black, and glossy, and are much like those of *Amaranthus.*

Another plant, or rather several plants of the genus *Amaranthus,* are often spoken of as pigweeds. They are not very likely to be annoying in meadows or pastures, but the seeds are not unfrequently found mixed with those of grasses and clovers.

FIG. 122. *Polygonum aviculare*, L., (Knot-grass. Doorweed.) 1 × 8.—(Sudworth.)

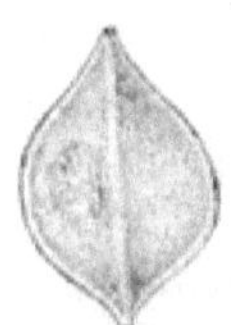

FIG. 123. *Rumex crispus*, L. (Narrow Dock.) Two views. 1 × 11.—(Sudworth.)

FIG. 124. *Euphorbia Cyparissias*, L. Two views. 1 × 10.—(Sudworth.)

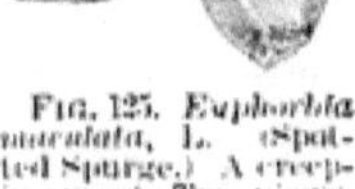

FIG. 125. *Euphorbia maculata*, L. (Spotted Spurge.) A creeping weed. Two views. 1 × 15.—(Sudworth.)

FIG. 126. *Urtica gracilis* (Great stinging nettle.) This has spread over low land meadows in some places. 1 × 15. (Sudworth.)

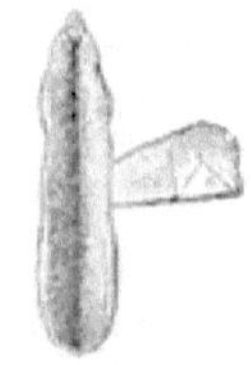

FIG. 127. *Bromus secalinus*, L. (Chess deprived of glume and palea.) 1 × 3.—(Sudworth.)

Polygonum Persicaria, L. (Lady's Thumb,) and some other species of *Polygonum* have smooth, black flat seeds.

Polygonum aviculare, L. (Knot-grass) and some others have triangular seeds, shaped much like small grains of buckwheat. The seeds of these two species, and of others, are often found among grass seeds.

Rumex crispus, L. (Curled or Narrow Dock) and *R. obtusifolius*, L. (Bitter Dock) are both common and troublesome weeds in meadows. They are perennial and have long tap roots. The seeds are three angled and like those of *Polygonum*.

Rumex Acetosella, L. (Sheep sorrel) is very common on poor, light land. The seed has the shape of the species last named

Euphorbia Cyparissias, L. (Euphorbia) has escaped from cultivation. It roots deeply, and has in some places become a great pest.

Occasionally seeds of common chess, or cheat, are met with in grass seeds. In a few instances the writer has met with seeds of chess which had been rubbed so as to deprive them of the floral glume and palea which usually adhere quite firmly. Figure 118 gives some idea of chess in this condition.

For accounts and figures of the weeds, which are themselves grasses, see the several species of grasses elsewhere described. The reader has no doubt had experience with some of them, or has heard of June grass, quack grass, *Eragrostis*, chess, barn-yard grass, crab or finger grass, the bristly fox-tails, burr-grass and others. The grasses are remarkably free from poisonous properties, there being only two or three upon which rests any serious suspicions.

For further accounts of the weeds the reader will consult the paragraphs on "*Seed Distribution*" and "*Battle in the Meadow.*"

How to Get Rid of Weeds.—There are two things to be done: 1st, prevent further seeding and the further introduction of seeds; 2d, destroy the seeds and the plants now in the soil.

Farmers cannot be too careful about the source of grass seed. Weeds, and some of the worst type, are thus freely distributed. Where possible it is better and safer to grow one's own seed, or procure it of some careful, thorough farmer near home. The older the country, as a rule, the more likely it is to furnish ox-eye daisy, yarrow, rib-grass, and other tenacious and trouble-some weeds. Foul seed is dear as a gift. It is cheaper to pay triple price for clean seed than to be perplexed with the trouble of getting rid of the weeds introduced. Some of the seeds should be spread out on a table in a very thin layer to aid in the discovery of the seeds of weeds which are liable to otherwise escape notice. Sieves and fans may remove some kinds entirely. Many sorts of seeds, especially the small ones, will pass undi-gested and unharmed through the digestive organs of horses and cattle. An ordinary compost heap does not kill all unless every portion is carefully turned in and heated.

In certain cases, one or more hoed crops may be raised on the land thoroughly summer fallowed. Pastures and meadows should always be looked over carefully and the weeds dug or pulled before the seeds are ripe, or taken off the ground if the seeds are ripe.

Sheep must be kept from pastures until stick seed, hound's tongue, burdock and the like have been removed.

The large weeds, like narrow and bitter dock, parsnip, carrot, may be left till the growing stalk has acquired some strength. Then on some day when the soil is soft and before the seeds have dropped, go over the field with a spade or a stout spud, thrusting it down perpendicularly within a couple of inches of the plant, take the stalk with one hand near the root and with the other pry it loose. In this way no roots are left below the surface to sprout and send up a new crop. Never cut off the tops of such weeds, leaving the roots in the ground.

By the following process the writer has found no trouble in killing quack grass, whether the season be wet or dry, the soil sand or clay, drained or undrained.

Plow it late in autumn, and as soon as a team can be put on the ground in the spring run over it with a cultivator every three or four days. Never allow a leaf to show itself, for then it begins to recuperate. By the middle of June every vestige has disappeared. Farther south than Central Michigan no doubt it would disappear earlier. To harrow and rake up the roots is a waste of labor. If during its growing season, the green tops are kept out of sight the plants will die. Thorough work, eternal vigilance is the only way to keep the upper hand of weeds.

For further remarks concerning the destruction of weeds, the reader will consult the paragraphs on irrigation, drainage, use of fertilizers, quack grass, care of meadows and pastures, the battle in the meadow.

CHAPTER IX.

GRASSES FOR PASTURES AND MEADOWS.

It has been repeatedly shown that a judicious mixture of several varieties will produce a larger yield than can be obtained where one variety is sown by itself. This is a rule in nature as well as in farming. Many sorts will usually occupy the ground more completely than one sort and help keep out weeds. "Each species has some special niche to fill, some separate part to play in the grand harmony of nature. Each one is better adapted for some purpose or for some soil or climate or locality than any other."—(Gould.) Sinclair found that from the beginning of spring until winter set in, there was no time when one or more species of grass was not in its most perfect state. He found that dry weather favored some, moisture that of others. He always favored a mixture for producing the best pasture. A mixed pasture is earlier, gives a better yield, and holds out better than any one species of grass.

Then there is the taste of cattle to be considered. As Plenes in her *British Grasses* states: "Sheep have strong likes and dislikes. They will hasten to a kind of grass which is a favorite with them, tramping down all the other grasses as unfit to taste. Horses again have their preference and cows theirs, and we have even seen swine exercise considerable cunning to secure a feed of a favorite grass. So the agriculturist has as much to consider as a master of ceremonies; he must consult the capabilities of situation, the qualities of his provision, and the various tastes of his company."

Because a grass is thrifty it does not necessarily follow that it is the best adapted to the farmer's use It may not be nutritious, it may be offensive to stock. A grass which thrives in one place may fail in another. Because a grass is a native to the country it is no sign that it thrives there best. On the contrary, introduced plants often thrive better than those which are indigenous.

As an instance, we have only to think of some of our worst weeds, most of which are foreigners. There is no one model grass,—a grass best suited for all purposes. This has been found to be the case with roses, with all kinds of fruits, grains, and vegetables. Some grasses are too slow in starting, or they are too sensitive to frost, or they will not endure dry weather; the stems are too woody, the leaves too thin, the tops too short, or the aftermath is of no account.

In making selections for sowing, the farmer must be guided by the climate: the soils, whether wet or dry, light or heavy. He will be guided by the uses to which he puts the field, whether for meadow, or pasture, or both, whether it is to remain seeded for some years or for one or two years. He will be guided by his location with regard to markets and supply of labor.

It is to the advantage of a grass, especially for alternate husbandry, if it seed freely, and if the seeds are comparatively large, quite certain to grow and produce good strong plants in a short time. These are doubtless some of the reasons why the rye grasses are so popular, and why Timothy and orchard grass, and the larger fescues are so much used in Europe.

For permanent meadow, there is less objection to waiting a longer time for the slower, finer grasses to become established, such as meadow foxtail, June grass, and red top.

The success of grasses depends on the supply of moisture. Liberal spring rains, with mild weather, give grasses a good start for the summer. For the dryer portions of the northern United States, a grass must endure drying for months, it must endure freezing with bare ground in a dry atmosphere, with the mercury down to 40° or 50° below zero, or endure a burning sun with the murcury up to 100° in the shade. It must be provided with means to withstand fires in dry weather. In the south it must stand great heat, much moisture and much drought. In dry climates it is often hard to start grasses. They must be perennial, or produce an abundance of seeds capable of starting quickly when the season is favorable.

From *Crops of the Farm*, I quote: "Strong, rich pastures, producing succulent grasses abundantly are well adapted for fattening large cattle, either without extra food or with the aid of a little cake [oil meal]. Second rate pastures, especially if on a cold subsoil, will generally yield a better profit from the dairy, and from the rearing of young cattle. Dry, hilly pastures are most suited to sheep. The grazing of land by mixed stock of cattle, sheep, and horses, or these in frequent succession, will keep the land more evenly grazed than where one kind only is kept. Sheep eat many weeds which cattle dislike and avoid. Horses are very uneven grazers."

In the words of the late I. A. Lapham, of Wisconsin: "It is not to any one species of grass that we should look for the support of our stock. On the native prairies we find many species intermingled, each doing its part: some preferring low, wet situations, others grow only on dry ground; some prefer the shade of forest trees, while others flourish best on the most exposed parts of the broad prairies; some grow only in the water, others along the margins of lakes and streams; some attain their maturity early in the season, others late in autumn."

Farmers who have a large quantity of meadow will often find it best to have the grasses of different sorts in different meadows that they may not all be fit to cut at the same time, thus prolonging the season for haying.

For a meadow, grasses should mature at about the same time; for pasture the time of flowering or of most rapid growth should vary and extend from early spring till late autumn, or in the South they should extend over a good portion of the year.

M. Gœtz found out what grasses were best adapted to his soils by a slow process of testing each separately, then he used a mixture of the seeds of those species which he had found did the best. Chemical analysis might tell the same story or it might not.

WHAT IS NOW SOWN IN GREAT BRITAIN.

The following notes are taken from a recent admirable essay by C. L. F. De Laune in Jour. Roy. Agr. Soc., 1882:

"Gross ignorance prevails in reference to the kinds of grasses. The use of short-lived grasses and of biennial clovers, coupled with an insufficiency of proper seed, is the main cause of the deterioration of new pastures after the first two or three years.

"The five coarse grasses most valuable for permanent pasture are the following:

"*Dactylis glomerata* (Cocksfoot).

"*Festuca pratensis* (Meadow fescue).

"*Festuca elatior* (Tall fescue).

"*Phleum pratense* (Timothy).

"*Alopecurus pratensis* (Meadow foxtail).

"These five should form the bulk of all pastures on good soil, either for sheep or cattle.

"The most valuable of the finer grasses are:

"*Cynosurus cristatus* (Crested dogstail).

"*Festuca duriuscula* (Hard fescue).

"*Poa trivialis* (Rough meadow grass).

"*Agrostis stolonifera* (Fiorin).

"*Festuca ovina* (Sheep's fescue).

"*Avena flavescens* (Golden oat-grass).

"In much smaller proportion should be used permanent red clover, cow grass (a kind of red clover), alsike, and white clover. Seeds of yarrow ought never to be omitted. These should be varied with the soil.

"The grasses most pernicious to newly formed pastures are rye grass in all its varieties and *Holcus lanatus* (soft, woolly grass). These produce an abundance of seed, are cheap, and quickly make a great show, but they soon die out and leave room for weeds to fill their places. All rye grasses, or nearly all, die after once seeding."

He insists strongly that among the best feeding grasses are some of strong and coarse habit, while among the finer kinds are many that are worthless, although many seedsmen recommend the finer grasses as of the best quality. He finds rye grass a gross feeder, and where it has been sown it is very difficult to get other grasses to grow. The flower-heads of all the best permanent grasses are much liked and greedily eaten by stock, consequently they rarely seed in a pasture, whereas the flower-heads of the worthless grasses, which are disliked by stock, are constantly seeding. In this manner the poor grasses often increase while the better grasses diminish. The best grasses are cropped closely, sheep, especially, discriminating very carefully even where the grasses are intermingled.

"Nature has provided a succession of nutritious grasses. A meadow composed of a large percentage of foxtail is certain to produce a large quantity of early keep."

For obtaining a good permanent pasture, it is as important to secure good pure seeds of the right sorts as it is for a pomologist to obtain the proper varieties of apple trees.

Mr. De Laune continues by recommending the following mixtures for permanent meadow and pasture:

For Good Medium Soils.		For Wet Soils.		For Chalky Soils.	
	Lbs.		Lbs.		Lbs.
Foxtail	10	Foxtail	4	Cock's-foot	14
Cock's-foot	7	Cock's-foot	10	Cat's-tail	3
Timothy	3	Timothy	3	Meadow fescue	2
Meadow fescue	6	Meadow fescue	3	Crested dog's-tail	5
Tall fescue	3	Tall fescue	8	Hard fescue	4
Crested dog's-tail	2	Crested dog's-tail	2	Sheep's fescue	4
Rough meadow grass	1½	Rough meadow grass	2	Yarrow	2
Hard fescue	1	Hard fescue	1	Golden oat-grass	1
Sheep's fescue	1	Florin	2	Perennial red clover	1
Florin	1½	Yarrow	1	Alsike clover	1
Yarrow	1	Perennial red clover	1	Dutch clover	1
Perennial red clover	1	Cow grass	1		
Cow grass	1	Alsike	1		
Alsike	1	Dutch clover	1		
Dutch clover	1				
Total	41	Total	40	Total	38

The reader will observe that this recent experimenter omits the rye grasses, sweet vernal and some others, which are found in nearly every list generally recommended for use in Great Britain.

Rye grass was the first true grass recommended for cultivation more than 200 years ago, and has been most extensively recommended by seedsmen and used by farmers generally for permanent grass lands. It is still much used, but some of them agree with Mr. DeLaune, that it is one of the least desirable for such purposes. This slow progress and the following of an old custom is more than matched by the following in reference to the practice of medicine.

Some years ago, Dr. O. W. Holmes, of Harvard, said: "Doctors have been using common elder as a remedy for more than 2,000 years, and have just found out that it possesses no medicinal value whatever."

The fact is, the farmers of Great Britain seem to rely largely on the recommendations of seedsmen as to what sorts and how much they shall sow. These men naturally put in a liberal allowance of seeds which are easily obtained at a cheap rate, especially if such seeds start soon and make a rapid growth which soon covers the ground. Many of their mixtures contain the Italian rye grass, which generally lasts for one year only. It makes a fine growth for a while, and after taking the cream of the soil quickly perishes, leaving vacancies on impoverished land for other grasses or, more likely, for weeds to come in and occupy.

List of Grasses for the North.—The writer approaches this subject with some want of confidence on account of the great size of our country, the diversity of soils, climate, and uses, the lack of well conducted and accurately reported experiments. He will, therefore, not puzzle the farmers with numerous long lists of mixtures, on a guess, but give a few of the best and advise experimenting for themselves. A point is gained when a farmer ventures to deviate from the long established customs of his fathers or his neighbors, many of whom have fallen into certain practices without very good reasons therefor.

The lists are recommended for climates similar to that of Michigan.

In selecting seeds for alternate husbandry, only those grasses and clovers should be sown which rapidly make a large growth, and arrive at maturity in a short time:

Grasses and Clovers for One Year.

Dactylis glomerata (Orchard grass).

Zea Mays (Indian corn).

Secale cereale (Rye).

Avena sativa (Oats).

Panicum miliaceum (Millet).

Setaria Italica (Hungarian grass, Bengal grass, sometimes called Millet).

Lolium Italicum (Italian rye-grass).

Arrhenatherum avenaceum (Tall oat-grass).

Trifolium pratense (Red clover)

Grasses and Clovers for Two Years.

Dactylis glomerata (Orchard grass).

Phleum pratense (Timothy), heavy and loamy soils, not on sand.

Arrhenatherum avenaceum (Tall oat-grass).

Festuca elatior (Taller fescue), heavy and loamy soils.

Festuca pratensis (Meadow fescue), heavy and loamy soils.

Lolium perenne (Perennial rye-grass).

Trifolium pratense (Red clover).

Trifolium medium (Mammoth clover).

Trifolium hybridum (Alsike clover).

Grasses and Clovers for Three Years.

Dactylis glomerata (Orchard grass).

Arrhenatherum avenaceum (Tall oat-grass).

Festuca elatior (Taller fescue).

Festuca pratensis (Meadow fescue).

Alopecurus pratensis (Meadow fox-tail).

Phleum pratense (Timothy), for mowing only.

Poa pratensis (June grass, Blue grass of Kentucky), for pasture only.

Agrostis alba (Red top), for rich loam or low land.

Medicago sativa (Lucerne. Alfalfa).

Trifolium pratense perenne (Perennial red clover, Cow grass of the English).

Trifolium hybridum (Alsike clover).

Trifolium repens (White clover), for pasture only.

Grasses for Marshes.

Agrostis alba (Red top).

Festuca pratensis (Meadow fescue).

Festuca elatior (Tall fescue).

Poa serotina (Fowl meadow).

Poa pratensis (June grass, Blue grass of Kentucky).

Alopecurus pratensis (Meadow foxtail).

Deyeuxia (*Calamagrastis*). *Canadensis* (Blue joint).

Concerning a selection for central Kansas, Professor Shelton remarks: "For pasture, I have no hesitation in recommending the following sorts, placing them in the order of their importance: orchard grass, alfalfa, red clover, taller fescue, Kentucky blue grass. For mowing purposes, our experience has shown, very steadily, that alfalfa, red clover, taller fescue, perhaps meadow oat-grass, and Timothy are the best. So far as the matter of withstanding the effects of drought is concerned, these sorts rank, with us, in about the following order: alfalfa, orchard grass, red clover, meadow oat-grass, Kentucky blue grass, taller fescue, and Timothy. Along the eastern borders of the State, and for thirty or more miles west of the Missouri line, Kentucky blue grass and Timothy are standard grasses which uniformly produce bountiful crops of hay and pasture, while in our experience in the central part of the State these grasses have uniformly failed."

Prof. I. P. Roberts, of New York, says: "We have tried in a small way many of the grasses and clovers in past years, and as yet we find nothing that gives as good satisfaction as medium red clover and Timothy. The seed dealers may 'boom' the tall, coarse, reedy grasses or the tender, dwarf, creeping varieties, nevertheless, in New York, clover and Timothy have come to stay."

Grasses for the South.—In many portions of the southern States, the people are still ignorant of the best grasses. They have long been wedded to cotton and have learned to believe that grasses and clovers will not thrive in their country.

In numerous places it has already been proved that many grasses and clovers grow well and produce abundantly.

The South possesses great advantages over the North in raising live stock, as the winters are so short and mild little hay need be cut and stored.

Rev. C. W. Howard of Georgia, J. B. Killebrew of Tennessee, and Dr. D. L. Phares of Mississippi, have each written valuable books concerning grasses for the South, and these books have been well received and extensively purchased.

Mr. Howard says: "It is a significant fact that the rich lands in upper Georgia, in which a mixed husbandry prevails, have rather increased in value than decreased since the war. *The depression in price has occurred only in lands devoted to exclusive cotton and rice culture, both of which require a large amount of labor.* In the South land is very very cheap, while at the North land ranges from \$50 to \$200 per acre. He looks to England, Holland, or Belgium, and finds it averaging from \$300 to \$500 per acre. Why this difference? Is the land in these countries better than ours? Not by nature: if it be better it is by the difference in treatment. Is their climate better than ours? The acknowledged superiority is on our side. Are the prices of their products any better than ours? On an average not so good. Are their taxes lighter than ours? If we were compelled to pay their tax, either at the North or in England, our land would at once be sold for taxes. Have they valuable crops which they can raise and which we cannot raise? There is not a farm product in either old or new England which we cannot raise in equal perfection at the South. Is their labor cheaper than ours? The cost of labor at the North nearly doubles the cost of labor at the South. If, then, all these things are so, why is it that their land is so valuable and ours so valueless? If we take the map of the United States and put our finger upon the State or parts of States in which land sells at the highest price, we shall find that in those States or parts of States the greatest attention is paid to the cultivation of the grasses and forage plants. If we open the map of Europe we shall find the same rule holds good. The cheapest lands in Europe are those of Spain, where little attention is paid to the grasses. Holland is almost a continuous meadow, and their land sometimer reaches $1,000 per acre.

"A Belgian gentleman, who sold his land in Belgium for $500 per acre, and bought river bottom land in Floyd county, Georgia, at $20 per acre, told the writer that he made more on the Belgium farm, valuing it at $500 per acre, under the Belgium system, than he did on the Georgia land at $20 per acre, under the Georgia system of cotton and corn. He even believed that clover and grasses would not grow in Georgia, and therefore did not attempt the Belgium system.

"If more of the land were in grass, much less labor would be required to manage it. To the wearied business man there is something charming in the thought of broad acres, a few select laborers, green grass, cool shades, running water, thrifty live stock, and all the abundance of the farm.

"A small, well manured, and well cultivated area of land in cotton and the cereals, with a large proportion of forage plants and grasses, would give to the cotton planter a pleasure in his business, and an amount of real profit which he has never before known."

Killebrew says: "These are more applicable to Tennessee than to Georgia. A stranger appears in the country desirous of investing in land, and while he would turn from the cotton plantation at ten or twelve dollars per acre, he would gladly invest in the grass farm at forty or fifty. Grasses mean less labor, less worry, fewer hands, more enjoyment, finer stock, and more charming homes, and as a consequence, happier families, more education, more taste and refinement, and a higher elevation of the moral character."

Will the cultivated grasses and forage plants grow at the South? In reply to this question, we read again from Mr. Howard: "There are some portions of the South, as is the case in all countries, where the valuable grasses will not grow, but as compared with the Northern States, the climate of the South is certainly better adapted to grass culture, if we take into consideration the whole year. At the North, during the whole winter and late in the spring, the ground is hard frozen or covered with snow. Of course, during that period the grass is useless, and this constitutes a large portion of the year. The heat and dry weather of the summer are the drawbacks to grass culture at the South. But these affect summer pasture alone. They do not affect the hay crop. Clover and hay grasses are cut before dry weather sets in. The hay crop at the South will not be injured one year in twenty by dry weather in the spring. We do not know a country more favored in this particular. In England, while the grass grows luxuriantly in the spring, it is very uncertain whether there will be enough dry weather at the proper time to save the hay. We, on the contrary, always have rain enough in the spring to mature the grass, and not enough rain to render the hay harvest at all precarious. When the hay is cut, will not the July and August sun kill the grass? There is danger of this result if live stock are turned upon the meadow as soon as the hay is hauled out, and the grass is grazed close to the ground. *A meadow at the South should never be grazed during the summer.* The aftermath will protect the roots of grasses during summer.

"After fall rains set in and cool weather begins, the meadows may be moderately grazed in dry but never in wet weather. With some grasses this grazing may be continued during all the dry weather of the winter. This winter grazing is the great advantage of the South. It more than compensates for the drought and heat of summer. It saves, to a considerable extent, the cost of cutting and curing hay, and of the construction of expensive barns. At the North, cattle and sheep are shut up in great barns for six months of the year, requiring costly feed and attention.

"At the South, in each of the plantation States, we have three different climates—that of the mountains, the interior, and the coast. For live stock, the mountains have the advantage in summer, the low country in winter, while the middle country has a share of advantages and disadvantages of both without the special excellencies or defects of either. As a general rule, a clay soil is best suited to growing grass. Any land that will bring good wheat will bring good clover, and any land that will bring good oats will bring good grass. The lands most likely to produce heavy crops of Timothy and herd's grass (red top) hay are the rice lands of the coast. They are very rich and have ample command of water. Do the rice planters know that the grass lands of Lombardy, near Milan, where irrigation is practiced, rent for from $6 to $100 per acre, while hay sells at $10 per ton? A level surface of upland, without running water, with an excess of sand, is the most unsuitable for a grass farm, and of course for stock raising.

"The writer has seen the various useful forage plants and grasses tried from the mountains to the coast of Georgia. He has been closely observing in regard to this important interest for more than twenty years. As a conclusion he does not hesitate to say if ground be made sufficiently rich and as well prepared, that if judgment be exercised in sowing and in adaptation of species to particular locality, and proper subsequent management be observed, that so far as soil and climate be concerned the South has unusual fitness for successful cultivation of the valuable grasses. The grass of the South will have some difficulties to contend with, but none so formidable as those which are incident to cotton and wheat. Broom-sedge, and crab-grass in the stubble, gives excellent summer pasture. Bermuda grass is excellent for summer. In the whole range of southern agriculture there is no crop on which manure pays as well as on winter pastures.

"On the whole, the drawbacks to successful grass culture at the South are as few and as easily removed as in any portion of Christendom. Sound political economy requires that the South should raise its own horses, mules, sheep, cattle, and hogs, and produce its own wool, butter, cheese, and hay. Grass culture is the basis of this independence."

In other portions of this work will be found quotations from Mr. Howard in reference to lucerne, sainfoin, field pea, vetch, red clover, alsike clover, white clover, millet, gamma grass, crab grass, brome grass, Bermuda grass, meadow oat-grass, orchard grass, Italian rye-grass, blue grass, Timothy, red top.

Mr. Howard believes, and rightly, too, that none of the native Texas grasses are equal, either for hay or pasturage, to some of the artificial grasses now in cultivation.

Grasses for Winter Pasture at the South.—The late C. W. Howard, of Georgia, recommends meadow oat-grass, blue grass, wild rye (species doubtful), orchard grass, red and white clover.

Dr. D. L. Phares, of Mississippi says: "The list depends on whether the pasture is for one season or for permanent pasture, as well as on location, soil, drainage, etc. For a single winter, sow wild brome grass (*Bromus unioloides*). Several other plants furnish good winter pasture, but none are so valuable as barley, though I have tried wheat, oats and rye. It does not lie on the ground like rye, but stands up."

For further remarks concerning grasses which are suited to pastures and meadows, consult what is said in regard to each of the several species usually cultivated.

CHAPTER X.

PREPARATION OF THE SOIL AND SEEDING.

Drainage.—The advantages of drainage are many, even for growing grass, though it is probably true that grass land does not require to be drained as thoroughly as that which is arable. Drainage prevents damage from flooding for long periods when not desired. It is a protection against drought: it enables the grasses to start earlier in spring and grow later in autumn; it deepens the soil and prevents baking in dry weather; it allows the plant to receive greater benefit from fertilizers applied to the soil; it allows air to penetrate the soil; it prevents frost from heaving out the plants; it makes hauling of loads easier, and renders the land less liable to injury from the treading of cattle; it improves the better grasses, which thereby encroach on those of less value, including many sedges, rushes and useless weeds. Much of our grass land, especially that in permanent pasture and meadow, would be vastly improved by thorough drainage.

If water, in a wet season, stand nearer than two feet of the surface in a small hole dug in the ground, the land needs draining.

Preparation of the Soil.—Strong, deep, calcarous soil, with a clay subsoil, is the best adapted for our most nutritious grasses.

It is most important that the land be clean, finely pulverized and of good tilth when the seeds are sown. If the land can be lightly harrowed immediately before sowing the seeds or immediately after, the seeds will be more likely to get a good start. A brush harrow is a very good substitute for a light one with slanting steel teeth.

How Much Seed to Sow.—That depends on the size and vitality of the seed, the number of seeds to the bushel, the condition of the land, whether distributed evenly, and the nature of the season that is to follow. By consulting the table which gives the number of seeds to the ounce, and a little calculation, it will be seen how thickly the grass would grow provided every seed produced a plant.

In the opinion of the writer, it would be better, in most cases, if farmers used less seed to the acre and took more pains to get the land in better condition. Suppose we sow 12 quarts of Timothy seed and 4 pounds of red clover to the acre. This will make 19,980,000 seeds of Timothy and 1,152,000 seeds of clover, a total of 21,132,000 seeds, or about 4 seeds to the square inch. Using finer seeds in mixtures, as prescribed by some of the English people, they often sow from 50,000,000 to 100,000,000 seeds to the acre, or not far from 8 to 16 seeds to each square inch. In either case there can be room for only a small portion of the plants should all the seeds grow and thrive.

Where the plants are crowded closely together, the stems of grasses and clovers are more slender and less likely to become woody. There is probably no danger of sowing too much seed, excepting in the matter of economy. If the young plants are too numerous, the stronger will soon starve and crowd out the weaker.

Under favorable circumstances one seed produces a plant which "tillers" and contains a large number of culms. Even with the best of chances, there will be much loss of seeds and young plants, what proportion no one can tell. The seeds should be well grown, well harvested, well cleaned, and true to name. Some experimenter in Great Britain found 1,100 plants (probably culms) to a square foot of good meadow land, and on water meadows the number was increased to 1,800 plants.

Sinclair found from 634 to 1,798 distinct rooted plants of various species in one square foot in nine separate localities. Where rye grass grew alone, there were only 75 plants. In a well manured water meadow, there were 1,702 grasses and 96 clovers and other plants. The smaller the number of species the smaller is the number of distinct plants to the square foot. In the words of I. A. Lapham, "Much caution must be used in applying the rules laid down in books (or given in practice) for the culture and management of grasses."

There always will be an opportunity for the farmer to experiment and use all the good judgment at his command. In connection with the account of each grass recommended for cultivation is given the amount usually sown to the acre, provided only one kind were used.

The following is by Prof. J. W. Sanborn, of Missouri, formerly of New Hampshire, and like the six following is from the *Rural New Yorker*:

"The amount of seed should vary from six quarts to about as many bushels. The poorer the farming and the more dishonest the seed dealer, the more seed will be required. Given a soil in fine tilth—that is, plowed well, harrowed by a harrow that lifts and pulverizes it, is smoothed off and fined with a harrow on the Thomas Harrow principle; if the seed is brushed in with a light brush-harrow, and if the soil is an open one, or if it's a little dry, rolled; if the seed is home raised, or not over a year old, and well kept, and the soil is fat with good available plant food—six quarts of Timothy or six pounds of clover will seed an acre. Per contra, if, instead of a fine, rich soil open to, and inviting tiny rootlets in all directions, we have a cloddy soil, plowed badly when wet, half tilled, where dry lumps repel the minute roots of the small seeds of grass, and where cavities are dry and, of course, foodless; if the seed used is poorly preserved or has been moist, and has heated, and if to it, when fresh, is added the seed of the past and of the previous year, and so on, "*ad infinitum*," and if the soil is as poor as Job's turkey, then an unlimited amount of seed will be needed, and no amount will be enough for a good crop of grass. I use twelve quarts of Timothy and ten pounds of clover seed per acre with good success, and deem this amount desirable. As my farming is rotation of crops, I seldom sow Timothy and clover together."

Gen. William G. LeDuc, of Minnesota, gives the following opinion: "As to the amount of seed per acre, an ideal meadow for me, in this soil and climate, would with present experience, start three Timothy and two clover plants to every square inch of surface. So six pounds of Timothy and eight pounds of clover, if good, sound seed, distributed evenly over an acre and fortunate in time and conditions of planting, according to my experience, give a good stand and lay the foundation for a good meadow."

Prof. E. M. Shelton, of Kansas, writes: "If for pasturage, use one bushel each to the acre of orchard grass and Kentucky blue, to which six or eight quarts of medium red clover may well be added. Liberal seeding is necessary if land is not first class either in quality or mechanical condition, to allow for lost seed. And if the land is in first rate order, liberal seeding pays well in a close, even sod.

"A late crop can be obtained from mixing red-top and Kentucky blue grass, a bushel of each, and if the land is somewhat light and moist, Alsike clover (say four quarts) may be added."

Prof. G. E. Morrow, of Illinois says: "On our prairie soils heavy seeding has not been found necessary. We aim to sow a bushel of Timothy seed to four or five acres, with a bushel of clover seed to eight or ten acres. When clover is sown alone, I should sow about one peck per acre."

Concerning the amount of seed required, the following is from Waldo F. Brown, of Ohio: "Good hay is not produced by thin seeding; for the grasses will grow coarse and rank, whereas heavy seeding will give fine, soft hay."

Prof. Wm. Brown, of Ontario: For rotation, hay and pasture,

sows fifteen pounds of grass and eight pounds of clover seed per acre.

Daniel Batchelor, of New York, recommends a bushel and a half of orchard grass and half a bushel of tall oat-grass.

"A heavy but not wet, clay loam devoted to meadow, should be sown with Timothy, red-top, fowl meadow, rough-stalked meadow, and Italian rye, at the rate of about six pounds each to the acre, in a mixture; to this may be added three pounds of medium clover.

"For a wet, peaty, black soil: Rough-stalked meadow, six pounds; red top, eight pounds; meadow foxtail, four pounds, and Alsike, six pounds, would be a good mixture per acre.

"For land much shaded the following mixture is excellent: One bushel of orchard grass, one of meadow oat-grass, and five or six pounds of wood meadow grass to the acre."

For New England, A. W. Cheever recommends the following: "If Timothy be sown alone we do not consider a bushel of seed any too much for an acre. Of red-top we would sow at least two bushels. Of orchard grass, two bushels, and a bushel of June grass with it. Rhode Island bent requires less seed by measure than red-top, as the seed is usually much less chaffy. No rule need be given for clover, so much depends upon the amount of seed contained in the land, and in the manure applied."

Professor S. A. Knapp recommends for Iowa and similar soils and climates, for early and late pasture, the following mixture:

	Lbs.		Lbs.
Blue grass	8	Orchard grass	6
Timothy	6	White clover	1

For summer pasture:

	Lbs.		Lbs.
Timothy	6	Red clover	4
Orchard grass	6		

For permanent dairy pastures on most heavy soils of the East, Sibley & Co. recommend:

	Lbs.		Lbs.
Perennial rye-grass	5	Meadow foxtail	3
Blue grass	3	Red clover	2
Orchard grass	3	White clover	1
Meadow fescue	3	Alsike clover	1

For the lighter soils of the North and East, they suggest the following:

	Lbs.		Lbs.
Timothy	5	White clover	1
Taller oat grass	10	Red clover	2
Rhode Island bent	4	Alsike clover	1
Orchard grass	3		

For wet soils in the North, they suggest:

	Lbs.		Lbs.
Blue grass	5	Rye or Ray-grass	4
Red-top	5	Alsike clover	1
Fowl meadow grass	4	White clover	1

Sowing the Seed.—The usual practice in many portions of the Northern States is to sow the seeds of the grasses in early autumn with a crop of winter wheat or rye, or to sow after these crops have been growing for a few weeks.

The clovers are sown in early spring, because the young plants are likely to winter kill if seeds are sown in autumn.

The following is from Professor E. M. Shelton, of Kansas: "The time to sow grass seed is, we believe, without any exception, in the spring: and recent experiments show that this work should not be undertaken too early in the season. In the spring of 1880, a field seeded early in April came to nothing, the violent dry winds that followed the sowing completely sweeping the seed away. Seed sowed after the spring rains have fairly set in, has never failed since 1874 to give a good stand of grass. In a few instances, and where the winter following has proved warm and open, we have had good success with Timothy and clover sowed in the fall; but the result of sowing orchard grass, alfalfa, and blue grass in the fall, has been almost invariably disastrous. Our experience with grass seeds sown in the fall has been this:

they germinate readily, even more quickly than in the spring, but, as the native vegetation fails from the action of frosts, the common grasshoppers collect upon the young grass, doing it serious damage: what remains suffers seriously, and is often quite destroyed by the action of the winter frosts and violent winds of early spring. On the other hand, when the seeding is done very late in the spring, the young and tender plants are consumed by the sun as fast as they appear above the ground. Seed sown any time during the month of April will rarely fail to germinate and make a vigorous growth. However, we cannot advise seeding, as we have before said, until the warm spring rains have set in. We have sown both alfalfa and orchard grass during the early part of May with uniformly excellent results."

Mr. Howard, of Georgia, prefers August and early September as the best time for sowing seeds. There is usually sufficient rain at that season to cause the seeds to germinate. The young plants will have time to make sufficient root to stand the severest cold of winter. Clover and lucerne, and several of the grasses, if sown without grain at this season in the South will give a cutting in the following spring. Grass seed sown late in the fall is liable to be winter killed.

If one could know the nature of the season to follow, he would much prefer to sow grass seeds in a dry day preceding mild, moist weather. It is hardly safe to give fixed rules for the sowing of grass seeds. Where several kinds of seeds are sown, it is well to sow those of equal weight and size together, going over the field again with the heavier sorts.

An experienced person on a still day will sow small seeds quite evenly by hand, but we now have several kinds of light machines, accompanied with directions for use, which will distribute the seeds more evenly than can be done by hand. If there is much to be sown, the cost of a good machine will be more than saved by sowing the seeds in a better manner. If evenly distributed, less seed will be required. For sowing by a machine, the seeds should be well cleaned and freed from leaves and straws and the machine frequently examined to see that the seed is passing through evenly.

Where the soil is loamy, sandy, or light, it is an excellent plan to roll the surface after seeding. This process brings the soil in close contact with the seed and renders it more likely to germinate.

Seeding by Inoculation.—This is rarely practiced, but has sometimes been resorted to in England. It consists in cutting ropes of turf from an old pasture, and these are chopped up into pieces about 1½ by 2 inches. They are placed by hand about nine inches apart over the ground. This is for meadow what sodding is for a lawn. The results are quick and sure, but rather expensive.

Quack grass, Bermuda grass, and Johnson grass are often seeded by scattering or planting fragments of the rootstalks, either with a hoe or by dropping in part of the furrows as the field is plowed.

Seeding Grass with Grain.—The following was prepared by John J. Thomas, of New York: "The most rapid way of obtaining a grass crop is to sow the grass seed alone without any grain. If done early in the spring, on clean, well prepared ground, we may get a cut of hay the same year, usually about two-thirds of a full crop, and a heavy one the second year. It will make a vast difference whether we sow plenty of seed or only a small quantity. We have sown a mixture of Timothy and clover at the rate of a bushel per acre, and had about twice as heavy a crop as that afforded by a scant seeding of less than a peck per acre. It is very important to have it covered with good, mellow earth, buried at a depth not greater than five or six times its largest diameter. To grow freely, one-fourth to one-half an inch is deep enough in moist soil, but clover will germinate and a grow at a depth of an inch. Much will depend on the mellowness and richness of the surface soil. A peck of seed will give a better growth on a fine, fertile surface, than a bushel on a hard crust or among clods.

"A very common cause of failure is sowing clover seed in the spring, on a heavy soil, with winter wheat, where the crust has not been broken since the previous September. Harrowing the surface with a light harrow will make a much better seed bed. But if Timothy seed has been sown in autumn a coarse harrow may tear it up.

"The objection to sowing the grass alone is that we are compelled to plow and prepare the ground for a single crop, while in seeding with grain we obtain both grain and grass at one operation, and with economy of labor. Farmers will therefore commonly prefer seeding with grain, except in certain cases where obtaining an early crop of grass is a paramount object. By seeding with winter grain, if a light top dressing of fine manure was applied in autumn to prevent a hard crust, the seed may be sown as early in spring as may be desired, without waiting for any preparation of the soil, and Timothy may be sown the previous autumn. Or if the soil is likely to settle and become hard, both Timothy and clover may be sown together in spring, after or before the surface is brushed with a light harrow which will not injure the grain. Seeding with spring grain, if properly performed, has much to recommend it. It always furnishes a freshly moved soil as a bed for the seed. But caution should be used not to cover the grass seed too deep, nor to sow a thick and shading crop of the grain.

"The best winter grain with which to sow clover seed is rye. It shades the young crop less, and if the work is properly done it rarely fails of entire success. On the other hand, the seeding rarely succeeds well with a crop of oats, and nearly the only chance for success is in sowing the oats thinly, or not to exceed a bushel of seed to the acre. Spring wheat and barley are intermediate for the purpose between rye and oats.

"It is usually more certain to rely on a slight artificial covering with soil, than the natural covering which may or may not take place by a shower of rain after the action of early spring frost, which, however, may sometimes succeed perfectly. The pressure of a common farm roller on clover or grass seed, sown on the freshly harrowed soil, covers most of it slightly, and is one of the best means for insuring germination. Another good way to cover the fine seed is to pass an evenly made brush harrow over it. This harrow is easily and cheaply constructed by placing several branches cut from a thickly set tree side by side, and stringing them together by running a stout stake through the forks at the cut ends, taking care that no large projections root into the ground in passing. Such a harrow, skillfully made, will leave the ground nearly as smooth as a floor. After the grain has been removed, it is well to look over the field and sow a few more seeds where the grass seems too thin."

The following upon this topic is by Prof. J. W. Sanborn, of Missouri:

"I have had but little of the bitter experience complained of by many in sowing grasses with other crops. Grass is, in its early stages, a slow grower, and I can get good results in grain, and under the best conditions, a crop of grass the first year. On a fine tilth and a fertile soil, I think few will fail to secure good grass with a light seeding of grain. Grass alone, sown in the spring, is out-grown by the rapid growing annual weeds, which have to be cut or the misery of their seeding is experienced.

"For several years on a large farm I grew little or nothing except grass. When I had corn and raised no other grain, I sowed the grass occasionally in the corn after the last hoeing. I have done much seeding in August, and some on the early frosts of spring, or late spring snows (the latter method is a bad one).

but now, in a rotation system of crops, I sow in the spring, and meet with good success: and if, in any degree, I have a failure, I can sow again on all thin spots after the grain is off, and still again in the spring when the frost is working the ground, or in the fall, after the spring sowing, on the fall frosts after it is too late for germinatien. These night frosts and day thawings open and close the ground and let the seed in well, and, on the whole, fall is a good time to sow for several reasons, when one is determined to sow grass seed alone, or when one is sowing over dead spots for new grass. It is an important matter to inspect all fields or sections that need re-seeding. Indeed, an annual seeding of fields is nature's way, and is often profitable."

Sowing Grass without Grain.—At the author's request, the following was prepared by A. W. Cheever, of Sheldonville, Mass.:

"Experiments repeated time and again have convinced me, here in east Massachusetts, where grain crops, compared with hay, have come to take a secondary place in the estimation of most farmers, grass sown alone is almost invariably worth more the first year than the grain and straw together would be if grain were sown with the grass, and the former made the leading crop.

"This is especially true where the land is particularly adapted to the production of hay. For the past fifteen years I have sown nearly all my grass seed alone, and in no single instance have I been sorry I did not sow grain with it to afford protection. When seeding with grass alone, I have generally cut two crops the first year. The first crop should be cut rather early, even if not fully grown. This kills or checks many of the annual weeds.

"I have often sowed the seed in spring, but this is not nature's time for sowing the grasses. Early autumn is undoubtedly the best time in the whole year for sowing most grasses. If sown then the annuals will find themselves laboring under a disadvantage and will soon give up the race.

"Grass sown in early fall will produce as full a crop the following year as it ever will, and the quality will be excellent; but if sown in connection with winter rye or wheat, it will be put back a whole year. The grain is of no advantage to the grass whatever, but rather the contrary.

"The only exception I would make in favor of sowing grass and winter grain together is when the grain is to be cut early, as soon as it heads, for feeding green or to make into hay. Cutting the grain so early in the spring gives time for the grass to make one and sometimes two full crops the first year.

"On good, moist, rich land, I have had excellent success in seeding grass in spring upon green sward turned over the previous fall and the surface thoroughly pulverized before winter and again made fine and mellow in spring. By this method, grass land may be kept producing full crops of grass every year without planting, but it will need reseeding oftener than if an occasional hoed crop is grown. Timothy is one of the poorest kinds of grass for spring seeding without grain, but if sown in August it will produce a full crop the next summer. Orchard grass is one of the best varieties for spring seeding. A great amount of grass seed is annually lost by sowing it in connection with spring or winter grain; the grain crops being harvested in the hottest part of the year, leaving the tender and previously shaded grass plants to be burned up leaf and root by the scorching sun."

Mr. Howard, of Georgia, writes: "There can be no doubt that sowing seed with grain should always be avoided. It involves the loss of a year in either hay or pasture. If sown with grain, when this is cut the young grass and clover are very tender, having been shaded by the grain. The cutting suddenly exposes them to the sun at the hottest season of the year. There is great danger that they will be burned out. In the event of sowing grass seed with grain, he must always remember that by so doing he loses a year and endangers the grass."

Here we insert the opinion of Daniel Batchelor, of New York: "Grain and grass have been sown together so long on some meadows that they will no longer bear good crops either of grain or grass; and there are people who keep their arable land in crops as long as they can get anything off, and then they seed down to grass when the soil is so sterile that it will not produce a hay crop. It is not going too far to say that over half the grass and clover seed sown with grain is smothered and utterly lost, and in many instances, after the grain is removed, especially if the weather is hot and dry, there is no hope for a grass crop without reseeding; and even where there is a tolerable catch the meadow is retarded for a whole season. I suppose some will deem it folly, but I think wisdom says: sow grass without any other crops; sow it in autumn; if you are in for rotation, put all your manure on your grass land, so that when you break up the sod or sward for corn, the land may be found in good heart for the crop. Of course, there are plenty of instances where the land is new or where the fertility has been well kept up, in which full success is obtained by growing grass and grain together."

Read what Major H. E. Alvord, of Mass., says: "We succeed well in sowing clover with oats, but prefer to cut off the oats and cure as hay while early 'in the milk.' We have not got out of the ruts sufficiently yet to prevent seeding all grasses and clovers used with wheat, rye, and oats: but I do not believe it to be good farming to try to grow two crops on the same land at the same time."

On this topic, read the experience of Professor E. M. Shelton, of Kansas:

"Oats, wheat, and rye are often recommended as excellent crops with which to sow grass seed. The argument is that the tall grain will shade and protect the young grass. But grass does not need shade when sown in proper season; it needs the sun, and, especially, it needs moisture, and this the vigorous grain is continually taking from the soil, thus robbing the young grass plants from the start. Every farmer knows how spindling, sickly, and how lacking in strength of root and stem, is a grass or clover plant growing in dense masses of grain. When this grain is harvested in June or July, just when the summer's heats are the greatest, the delicate, starved grass plants are certain to perish, unless long continued, cool, and cloudy weather prevent.

"We are aware that, upon favorable seasons, considerable success is obtained by seeding the grasses with another crop. But, even during these favorable seasons, a better and more vigorous start will be obtained without the rivalry of vigorous grains; and, upon dry seasons, a failure of the grass is almost certain, when seeded with a grain crop."

In favor of seeding in autumn without another crop, we may say: the land can be much better fitted for grass in autumn than in early spring. If weeds start in autumn frosts will kill them, while if the grass is sown in spring the weeds keep growing all summer. They may be checked, however, by mowing when the grasses are a few inches high. If there are vacant places they can be reseeded in spring. The first hay crop will be much better. Whatever plan is pursued, a failure or partial failure may sometimes occur.

All the experiments of the writer indicate that in the Northern States young grasses thrive better when they have the full benefit of all the sun and rain. The statement sometimes made, that young grasses and clovers need the shade and protection of some larger plants has no proof to sustain it. Numerous correspondents, without exception, in case they have tried both methods, speak of getting a much better catch of grass when the seed is sown without another crop. Killebrew says that in former years the farmers of Tennessee almost universally sowed grass seeds in the spring of the year on crops of grain, but since

1810 the custom has been to sow in early fall. Many sow grass alone at this time of year and get a full crop the next year. Where grass is sown with another crop they injure each other

James Sanderson, in *Transactions of the Highland Agricultural Society, 1863*, says: "If the grasses are rank and luxuriant, they greatly retard the harvesting of grain and frequently deteriorate its value. This early luxuriance is often injurious to the grass itself, as it extracts valuable ingredients from the soil. The grain denudes the grass of valuable food and renders it more susceptible of injury from extremes of weather. The plan of sowing grass seeds without a crop has recently been adopted on several farms of Great Britain with great success. Experiments have shown that the profit from the first year's pasture was more than an equivalent for the want of a crop of grain. The next year the field is fit for pasture a fortnight earlier than it would have been if sown with a crop. The grass gets a better start and makes for several years a better pasture or meadow."

He mentions the fact that many men who have tried this plan are of the same opinion. The plan of seeding without another crop has here been made prominent, because many persons have scarcely thought of any other way than that of seeding to grass with a grain crop.

Sowing Seed where Grasses already Occupy the Land.—In the Northern States where the land was more or less thickly covered with a growth of sedges and wild grasses, in numerous instances we have seen this order of things very materially changed by the introduction of other species. This was accomplished by simply sowing the seeds over the surface. In some cases a harrow passing over the land exposed the soil in small strips and patches. The change of grasses in such cases is usually rather slow and unsatisfactory, but this is not always the case.

At the Agricultural College, a good lawn on well prepared soil had for six or eight years produced only the finer grasses and one clover, consisting for a time mainly of June grass, Rhode Island bent grass, perennial rye grass, and white clover. It was mowed often and not allowed to produce flowers or seeds. After a few years the rye grass gradually disappeared, the other plants occupying its place. A year or two later some plats of grasses were established just west of this lawn. In a little while, through the help of the wind, other seeds were sown and inroads were made on this close lawn. Several of the larger fescues, orchard grass, quack grass, and a few others of less note appeared. These are rather on the increase, and with a liberal seeding I have no doubt they would soon be still more prominent. Doubtless this result would not always follow, as very much depends on the soil and climate. This suggests that by sowing seeds of better grasses improvements might often be made in our permanent grass lands.

Concerning the advisability of attempting to introduce "tame grasses" by sowing the seeds on prairie sod, Prof. E. M. Shelton, of Central Kansas, makes the following remarks: "But, whatever may be the character of the soil, prepare the land as well and thoroughly, by plowing and harrowing, as for any grain crop. This is a rule with scarcely an exception; and its violation in various ways explains a large proportion of the failures that have attended the cultivation of tame grasses in Kansas. The question is asked us many times every year: Why may I not scatter the seed upon the sod, as is often done in the East? This is often done, but the practice, so far as our observation has gone, has resulted in almost uniform failure. Where the prairie sod has been largely destroyed by the tramping of cattle, we have known blue grass to succeed partially by this method; but, even in this case, a better sod would have been obtained in less time by thoroughly subduing the land, by two or three years of cropping, before applying the grass seed. As before stated, in the eastern counties of the State, this practice is successfully followed, but even here we are satisfied that it would pay the farmer much better, and he would obtain a better sod and nearly as quickly if he should take the 'wildness' out of the land with two or three grain crops before seeding."

CHAPTER XI.

CARE OF GRASS LANDS.

Permanent Pasture vs. Alternate Husbandry.—Fifty-five correspondents in a recent report in England agree that "It is certainly unadvisable to break up any tolerably good pastures for the purpose of converting them into arable land."

With his experience and observation in mild and moist Europe, Bousingault believed that there is no system of rotation, however well conceived and carried out, which will stand comparison in point of productiveness with a natural meadow properly situated and properly attended to.

In 1881, nearly half the land occupied for agricultural purposes in Great Britain was in permanent pasture and meadow, and the proportion is on the increase. In Ireland the proportion is still greater in favor of permanent grass land. The proportion is greatest where the air contains most moisture.

The late George Geddes, in the *Country Gentleman* for 1882, reports a discussion of the Onondaga Farmers' Club. Men who had moist lands, with water under them, believed in permanent pasture. Men who cultivated dry soils, well adapted to a rotation of crops, easily plowed, and especially subject to severe droughts, were very decided in the opinion that permanent pastures are of little value as compared with grain crops, and hay and pasture in rotation. Rocky land and steep hillsides are best kept in grass. The amount of rain-fall has much importance in deciding which is the best use of the land. On dry lands subject to frequent and severe droughts, the grass soon runs out. After reseeding, they give a great crop the first year, less the next, and gradually the clover and Timothy die out.

We are informed that the best pastures in England along the banks of the Axe, the Brue, and the Parret, rent annually for five to eight pounds sterling per acre, or about twenty-five to forty dollars. In one instance, £3,000 was offered for 10 acres of such pasture land and refused. Such pastures are green in the spring when everything else is brown, and they grow on into late autumn when other pastures have ceased to support stock. They supply food for a much longer period than inferior pastures, and save a couple of month's winter keep.

In Holland an acre of permanent pasture is said to carry one cow and a sheep. In Herkimer county, New York, rich permanent pastures carry one cow to each acre and a half, while in much of New England, Professor Stockbridge says, "Eight acres are required for one cow, and then she comes home at night looking disappointed."

Secretary W. I. Chamberlain, of Ohio, in the *Country Gentleman*, says: "Our pastures are not so productive as we suppose. A fine old pasture of three years standing, when mowed in a good season, yielded less than a ton to the acre, and in one season less than half a ton per acre. The grass was short June grass, red-top, red clover, white clover, and some Timothy. The land is capable of better things. Next to it is a field, no better land, from which I have twice within 10 years taken over three tons per acre of cured hay. Not even tile draining and top dressing will restore such old pastures and meadows. A rich ten-acre field of good, newly seeded pasture will 'carry' more cows than forty acres of old pasture."

The seeding down to good, permanent pasture, even under the most favorable conditions, is a slow and costly process. As we must infer, the climate and situation have much to do in helping solve the question whether to keep land permanently in grass, or whether to include the grass in a rotation of crops.

At the present day, even in England, some of the most advanced farmers favor breaking up the dryer arable land, and believe in this way they can obtain the largest yield of animal food.

In favorable climates, land which will permanently support a good growth of grass must be naturally of the very best quality, in good heart, well prepared, and afterwards liberally manured for some years.

There are a few excellent farms in Southern Michigan, in Ohio, Kentucky, Wisconsin, and neighboring States, which contain permanent pasture of good quality, and which yield liberally. Generally the grass does not continue uniform. It dies out or becomes thin in some places, and vacancies are filled with grasses of poorer quality, or with weeds of no value.

In Johnson's *Agricultural Chemistry* we read: "It is pretty generally acknowledged that land laid down to grasses for one, two, three, or more years is in some degree rested or recruited, and that it diminishes in value again after two, three, or five years, more or less, unless some manure be given to them. The opinion is due largely to the annual production of roots (and rootstocks) on old grass land, which is equal to one-third or one-fourth of the weight of hay carried off."

The roots of grasses extend deeper than is generally supposed. These with the stubble, old leaves, and turf, make a large amount of vegetable matter. Mr. Lawes estimates that on a good pasture they will weigh from five to ten tons per acre of dry matter, containing accumulated nitrogen to the extent of one ton.

In rather dry climates, where the rootstocks and roots of an old pasture have formed a mat of vegetable materials, the yield may be much increased by plowing and harrowing the land and let the grass again occupy the soil. This plan is especially well adapted to renewing the yield of June grass, quack grass, Bermuda grass, and Johnson grass.

In reference to permanent grass lands, J. Julie, of England, in his "Gold Medal" essay, makes the following remarks in Jour. Roy. Ag. Soc. for 1882: "The cultivation of roots and cereals deprives the soil of nitrogen, whilst that of grass and leguminous plants, temporary or permanent, on the contrary causes it to accumulate in the soil. That nitrogen being the most expensive to buy, it is not economical to devote part of the land absolutely to arable and part to grass, for whilst the one uses up the nitrogen, the other accumulates it in excess. It is preferable to alternate on the same piece of land the cultivation of roots and cereals with that of grass lays. By this means cultivation can be kept up indefinitely without purchasing nitrogen, provided the land be maintained in a fit state of richness as regards the mineral elements. The occupation of land by a grass for two or three years which takes its turn in the rotation of crops is preferred to permanent occupation by grass."

The late J. J. Mechi, of England, objects to old pastures in countries which are rather dry. The crop is too light: arable land is more profitable.

The Advantages of a Rotation of Crops.—Some of them are as follows: Manure is economized, as crops do not all feed alike: the fertility of the soil is better and more economically preserved: weeds are more easily controlled: it enables a person to distribute his labor more evenly through the year: it gives a proportion of grain for feed and coarse straw for litter: crops in alternation are less liable to the attacks of fungi and insects. Where fields are occasionally cultivated, moles are less likely to become troublesome. Leguminous plants are not *specially* benefited by nitrogenous manures, but they are nitrogen "producers," and leave the land in fine condition for the grasses proper, including the cereals.

It would be better for the farmer if he looked more upon grass as a leading crop in his rotation, instead of one of minor importance.

Pasture Yields more Nourishment than Meadows.—The following experiment is reported by C. L. F. DeLaune in Jour. Roy. Ag. Soc., 1882: "After the grasses and clovers had grown one year and had become well established, one plat was mowed twice, and a similar plat was mowed six times during the year. The latter was to imitate the frequent cropping of grass by cattle.

	Total Per Acre. Tons.
Green, cut twice	17.06
Dried, " "	4.49
Green, " six times	21.26
Dried, " " "	3.602

"The following shows the amount per acre of the most valuable substances contained in the hay:

	Nitrogen.	Phosphoric Acid.	Lime.	Magnesia.	Potash.
Cut twice	229,24	82,05	208,72	41,95	279,18
Cut six times	236,36	90,06	121,30	37,43	280,96
Difference	+ 7.12	+ 8,01	−87.42	−4.49	+ 1,78

"We see that, from the second piece cut six times, deficiency in weight is made up for by superior quality. It contains 7.12 lbs. of nitrogen, 8.01 lbs. of phosphoric acid, and 1.78 lbs. of potash more than the first crop. It is lower in percentage of lime and magnesia, which, however, are but of secondary importance for feeding purposes. It is certain then that cattle grazed on the crop of the piece cut six times would have been better nourished than those to which the hay from the first piece would have been given.

"Land used for pasture yields more nourishment than that where the grass is mowed off. Young shoots are much more nitrogenous than plants in flower and young green plants are more digestible than dried ones. In pastures the droppings of animals enrich the soil.

"Aftermaths are richer in nitrogen than first cuts, and they are more nitrogenous the younger they are gathered; they are richer, also, in phosphoric acid and potash. Theoretically, aftermaths constitute a better food than first cuts, yet they sell at a lower price, probably owing to their appearance, and because when dried they are more indigestible."

In considering the above experiment in reference to the great value of grass cropped often, we should not forget to take into account that cattle and sheep, while they roam over and over the field for pasture, injure the grass more or less with their feet.

Care of Pastures.—In this country, as a rule, they can scarcely be said to receive any care. No crop gets less attention, none would respond more quickly to good care. Much attention has been given to premium crops of corn, wheat, potatoes, the improvement of horses, cattle, sheep, swine. They are encouraged by liberal premiums, but we seldom hear of a premium crop of grass. It seems practicable to double the present yield without an outlay at all corresponding to the increased value of the crop. Is there any good reason why a farmer should not bestow as much care in selecting the proper seeds and in the after treatment of meadows as he would in selecting or breeding and raising a short-horn bull calf or a merino lamb?

Grass should not be pastured in very early spring before the ground settles and the sod becomes firm. By this early pasturing the tops are kept closely cut off, the roots are much injured, from which the grass does not recover for the whole year. To gain and thrive, a grass needs some green leaves as much as a horse needs fresh air and a stomach to digest a liberal allowance of food.

Experiments show the following from the *Country Gentleman* to be true: "If cut very frequently and kept short, like the grasses of a lawn, the roots will not make the same size and extent of growth as when the stalks and leaves have free development. The roots depend as much on the leaves as the leaves and stems do on the roots."

Pastures should not be allowed to grow very long in the spring without feeding, as the culms run up and blossom and make a growth distasteful to all kinds of stock. By movable fences, or otherwise, it is a good plan to feed off a piece rather closely, let it get a start, then feed off again evenly.

In large pastures, animals are likely to pick some places closely and leave others to run to seed. A mixture of animals, or one kind of animal following another, will keep pastures more evenly fed than will one kind alone. To prevent patches from going to seed, mow them a small quantity at a time, and when the grass is wilted it will generally be eaten by the stock. In such places a fresh bite very agreeable to cattle and sheep will often start up.

Pastures and meadows are very frequently eaten close to the ground late in autumn, especially if the season be a dry one. This is a severe drain on the vitality of the plants and causes them to be a long time starting in the following spring. Joseph Harris says: "On an old Timothy meadow closely pastured last fall, this year the hay was not over half a ton to the acre. On another meadow not so pastured, the grass was as thick and heavy as it could grow."

Some fall growth is necessary to give the plant strength for a good start in the spring.

With reference to the pastures of Maine, Professor Stockbridge said, on page 70 of the Agricultural Report for 1876: "What is to be done? In my humble opinion the corner-stone in regard to the improvement of pasture land must be put in the head of the farmer himself. To improve the pasture land of Maine, you must first seek to reform the farmer. The lands were once fertile, they are now sterile. Fires burned it, floods washed it. The milk and flesh of cattle have caused an immense drain upon the land. We must use fertilizers. A mixture of sulphate of ammonia, 180 lbs.; muriate of potash, 70 lbs.; a good, nice superphosphate, 100 lbs. Mix and put on to two acres of land.

"In improving my pasture, I would like to select my stock. There must be some stock and there must be somebody to raise it. I would like to let somebody else manufacture the animal carcass and let me have it to fatten. Then the animal will only take away from my farm carbon, which I can afford to have him do. So I will reach out to New York or to the West and buy cattle from somebody whom I do not know and whose farm I shall never see."

A chief reason for the light yield of grass, or a failure to get a good "catch," in many portions of our country is due to the fact that the strength of the new land was required to produce successive crops of wheat, Indian corn, cotton, and other hoed crops on arable land.

Where thin or unproductive, harrow the surface and sow on other kinds of grasses and clovers, with a top dressing of some fertilizer. This serves, to some extent, as a rotation of crops for the soil. If the cattle are fed oil meal or some other rich food, most of it goes to fertilize the land. Bare knolls will be improved by a very thin mulch of straw put on early in winter after the ground is frozen. A light, fine-tooth harrow will work the manure out of sight and out of the way. It helps to cover the small seeds.

Scatter the droppings of cattle, that no offensive bunches of tall grass may grow around them. It is a good practice at the

North to allow a part of the pasture to grow large for late fall feeding.

For feeding late in autumn and early spring, at the North, rye is excellent to piece out or save the common pasture.

The following on the care of pastures in Iowa appeared in the *New York Tribune*, and was written by Professor S. A. Knapp:

"Many farmers do not yet understand how to manage the pasture to the best advantage. They are so anxious to receive the full benefit of every crop of grass that they are alarmed if the grass gets the start of the cattle in June, lest some of it fail to be manufactured into beef or milk.

"A little more grass on the pasture than the cattle can eat in June should not be a source of anxiety any more than an extra crock of fine June butter in the refrigerator. In the West extremes of moisture and drought are the rule, and a dry period is quite likely to occur in July and August. During periods of drought there is scarcely any growth of grass, at least entirely insufficient for the stock. Close grazing in June leaves the stock with insufficient food in case of drought, and works serious injury to the grasses.

"Last season furnishes a very good illustration. The latter part of June there was upon an average on the College farm pastures enough grass to make one ton of hay per acre: by the 1st of September, with no increase of stock, they were practically bare. Daily measurements of the grass indicated only a trifling growth during July and August. The drought began in June, and there was not sufficient rain for pastures till September, and the growth in September was not equal to the demands of the herds, as the grass appeared to start very slowly. Practically the closest grazing was in September.

"The effect of full and close pasturing upon stock was tested as follows: Four yearling Shorthorn heifers were kept in a pasture with other stock and weighed the first day of each month.

They remained in the same pasture till October 1, when they were turned into a meadow. The following gains (in pounds) were made for the season: May, 332; June, 260; July, 160; August, 172; September, 78; October, 230; November, 122; total gain, 1,354 pounds. Each animal gained—taking the average—338½ pounds from May 1 till December 1. From August 15 till October 1 they received a daily ration of wheat bran and oats—four quarts per head.

"Three of these heifers were summer calves of the year previous, making them short yearlings. The majority of our common cultivated grasses grow most rapidly when the soil is quite moist and the temperature is between 70° and 80°. Our black, prairie soil, when exposed, frequently reaches a temperature of 115° at the surface. Under such conditions evaporation goes on with great rapidity and the soil becomes dry to a considerable depth. Ten inches of dense grass afford sufficient protection to the surface of the soil to keep the temperature about that of the atmosphere in the daytime and considerably warmer than the atmosphere at night, thus preventing such sudden and wide extremes of temperature that the plants fail to adjust themselves. While it does not destroy them, it retards or prevents growth.

"Close grazing may do in a cool, moist climate with some stock, but upon our western prairies it is a mistake. Where there is an abundance of grass the cattle take regular meals, and lie down to digest; upon short range they are constantly traveling and picking, which does not afford the best conditions for vigor, growth, or the production of beef or milk. In the fall a good coat of grass protects the roots from frost, and growth continues till quite late in the season, even after the open ground is frozen two inches or more. This is the secret of our so-called winter grazing."

Concerning the management of grass lands, Sir J. B. Lawes finds it very important not to feed young grass the first year.

He opposes mowing it the second year, having found that this practice destroys the clovers and the lesser grasses by encouraging the stronger growing species. He avoids mowing for several years, feeding with cattle in preference to sheep. He sows a variety of grasses, leaving the best to hold their own.

"A pasture cannot do much above ground till after the formation of a large bulk of roots below. The working capital of nitrogen and potash in a pasture must be larger than that required in an arable soil. In a pasture there is less activity and less change than there is in an arable soil. New turf will not become permanently productive until after the underground formation of stored up material. A pasture often falls off after the first three or four years. This can be avoided by a liberal feeding on the ground of cattle eating cotton cake. For the formation of a good turf after everything else is right, nothing equals cotton cake, cotton cake, cotton cake! If he sells the hay, and thereby takes potash from his soil, he can restore the waste with kainit salts or sulphate of potash."

Care of Meadows.—Much that was said in reference to the care of pastures applies equally to the care of meadows. They are injured by being shaved too closely, by continued removal of hay without any returns in the form of fertilizers, by close feeding of cattle in addition to mowing.

If land is in excellent condition when seeded to grass little need be done for the first two years, when the sod may be broken for some other crop, or for re-seeding to grass. If clover is used a dressing of plaster is often beneficial.

In a summary of the opinions of 55 prominent farmers of England, most of them advocate mowing in the first season instead of pasturing. The majority prefer mowing early the first year and again later in the season. All admit the great value of a dressing of farmyard manure, several recommend feeding cattle and sheep with cotton cake on grass land. By no means allow sheep to pasture newly seeded grass land.

What Manures to Apply.—This is a very puzzling question—one difficult to answer. In most cases no one can tell what would be best till experiments are made, but an intelligent person of experience can usually tell approximately what is for the best.

The substances most generally needed in manures for increasing a crop are those containing available nitrogen, potash and phosphoric acid. The influence of the weather, the moisture on grasses and clovers is much more marked than that caused by the richness or barrenness of the soil.

"The tendency of modern practice in manuring is to use readily soluble and quick acting manures, but to use them sparingly at each time. Little and often is the rule. It is not good policy to bury any manure very deeply, but apply it on or near the surface."—[Crops of the Farm.]

It is better to apply nitrogenous manures in spring when plants are beginning to grow. In what follows the reader will learn from the experience of others who have made many experiments with various kinds of fertilizers applied to grass land. For these we are compelled to cross the Atlantic, as but few reliable results have been published in America.

The following is gathered from J. Julie in Jour. Roy. Ag. Soc.: "If the leguminous plants are well developed and prominent we should diminish the dose of potash and increase the nitrogen. If, on the contrary, the true grasses stifle the Leguminosæ, it is better to reduce the dose of nitrogen and increase that of potash. Farmyard manure contains a large amount of nitrogen, but very little in a soluble state, unless it is well decomposed. There is a marked advantage in using a chemical manure, as farmyard manure contains an excess of several elements which will of necessity remain unemployed. Farmyard manure is far more suitable for the cultivation of arable than grass lands, for the plow mixes it with the mass of the soil. It pays better, especially on sloping lands, to apply a little manure frequently than much manure at greater intervals." Some grasses draw much more potash and phosphoric acid than others. Here are two mixtures in which the requirments of potash and phosphoric acid greatly differ. The tables are from Mr. Julie:

FIRST MIXTURE.

	PHOSPHORIC ACID.	POTASH.
	In 1,000 lbs.	In 1,000 lbs.
Lolium perenne	6.75	36.3
Phalaris bleuatre	5.68	31.4
Avena flavescens	5.98	26.55
Anthoxanthum odoratum	6.85	25.89
Festuca pratensis	5.52	21.83
Bromus Schraderii	8.07	21.55
Mean or average	6.375	27.256

SECOND MIXTURE.

	PHOSPHORIC ACID.	POTASH.
Poa nemoralis	4.12	10.85
Bromus pratensis	3.62	13.59
Poa pratensis	4.43	15.24
Cynosurus cristatus	3.72	15.24
Festuca rubra	3.94	16.37
Phleum pratense	4.13	16.61
Mean or average	3.893	14.65

It will be seen that the average of the second mixture is only about half as exhaustive to the soil as the first in producing the same amount of hay.

In 1858 Mr. Lawes said: "The best artificial manures for grass land are Peruvian guano, and nitrate of soda and sulphate of ammonia." In 1875 he wrote: "I am disposed to think a dressing of dung once in five years and 2 cwt. of nitrate of soda the other four is about as good an application as can be used. Peruvian guano, when alone, may be used at the rate of 150 or

200 pounds per acre. A very useful top dressing for the hay crop may be made of three parts of Peruvian guano, one part nitrate of soda, and one part of ammonia, using annually 200 to 250 lbs. per acre. With this apply 10 to 12 tons per acre of rotton dung once in four or five years." This is for permanent grass land.

On this important point, let us read another excellent authority, Dr. A. Voelcker, in Jour. Roy. Ag. Soc., p. 459, 1866: "Where good farmyard manure can be obtained at a reasonable price, I have no hesitation in saying I believe it will be found the most efficacious and economical manure, both for seeds (of clover) and permanent pasture. Sometimes common salt has had no effect. In one experiment the heaviest crop of clover was produced by a mixture of superphosphate of lime and muriate of potash."

In 1874 of the same Journal, he says: "On some soils, more especially on poor, light pastures, the effect of bone-dust on the herbage is truly marvelous: whilst in other localities bones do not show any marked effect upon meadow land. I would advise making field trials on a limited scale, before heavy expense is incurred in manuring pastures. Bone meal is often wasted on cold clay soils. Sinclair's remarks may be made with regard to the application of lime to grass land. Some soils are deficient in lime and will be much improved by its use. Pasture soils vary much in composition and physical character, and hence the same manures which effect a radical improvement on pastures in one locality are often found to be of little use in another place. For this reason it is difficult and hazardous to prescribe manuring compounds for grass land. In a general way it may be stated that manures rich in nitrogen and readily available phosphoric acid produce the greatest and most beneficial effect on grass land.

"There is no pasture the productiveness of which may not be largely increased by a heavy dressing of farmyard manure or by a top dressing of guano, or by artificial manuring mixtures composed of ammonia salts or nitrate of soda and sulphate of lime. Unfortunately the application of artificial manures to permanent pasture is often disappointing in an economical point of view. As a rule, no artificial manuring gives so favorable a return as good farmyard manure, and I cannot help thinking that it would be more profitable for a farmer to apply the larger portion of his yard manure rather to his pasture land than to the arable land; for there is no difficulty in growing roots and cereal crops economically with artificial manures."

A few of our best Northern farmers, such as A. C. Glidden, of Michigan, think that a much greater benefit would be derived from manures by spreading them on the pastures or meadows that were intended for corn a year hence.

A sod is the great basis for a corn crop, and the better the sod the better the crop of corn.

In many portions of the Northern States it is the custom to use most of the manure for the corn crop, with occasionally a top dressing for wheat.

Joseph Harris, of Rochester, New York, says: "The cheapest and best manure to apply to a permanent pasture is rich, well-decomposed farmyard or stable manure, and if it is not rich apply 200 lbs. of nitrate of soda per acre in addition."

We will read from still another, J. Dixon, in Jour. Roy. Ag. Soc., p. 204, 1858: "I have no hesitation, after an extensive experience, in pronouncing bones pre-eminent above all other manures for the improvement of grass lands, when permanency as well as cost are considered. I prefer them raw and ground fine. On a high varied soil in England, within two years, the value of the land was raised more than from 30 s. to 3 l. per acre."

Here are notes from a prize essay by C. Cadle in Jour. Roy.

Ag. Soc., p. 335, 1869: "After much experience, I think manuring grass lands is one of the worst subjects to treat. I have seen bones applied and produce no good whatever; and on the other hand, I have seen them used with immense advantage. I have seen guano produce a splendid crop, while the year following the crop has been worse than before guano was applied. It is impossible to give any definite rules without knowing the kind of land to be manured, and other attendant circumstances. Still money judiciously laid out in the improvement of grass land brings in a more certain return than where expended in the growth of wheat."

In 1864 Dr. R. C. Kedzie, of Michigan Agricultural College, made some experiments in top dressing the first year after seeding to Timothy and clover with oats. The dressing was applied from the 5th to the 10th of May on a soil of sandy loam of moderate fertility.

RESULT OF FIRST MOWING, JUNE 21ST:

	Yield per acre.	Gain per acre.	Gain per cent.	Top Dressing Applied.
No. 1	2,856	....	......	None.
No. 2	3,917	1,061	37	Plaster, 2 bushels per acre.
No. 3	4,515	1,659	57	Wood ashes, 5 bushels per acre.
No. 4	4,566	1,710	59	Pulverized muck, 20 loads per acre.
No. 5	4,696	1,840	64	Pulverized muck, 20 loads per acre and 3 bushels salt.
No. 6	3,813	957	33	Common salt, 3 bush. per acre.
No. 7	3,708	842	29	Horse manure, 20 loads pr. acre.
No. 8	3,931	1,075	37½	Cow manure, 20 loads per acre.

RESULT OF SECOND MOWING ON AUGUST 11TH:

	Yield per acre.	Gain per acre.	Gain per cent.	Top Dressing Applied.
No. 1	1,742	....	....	None.
No. 2	3,056	1,314	75	Plaster.
No. 3	2,977	1,235	71	Wood ashes.
No. 4	3,306	1,564	89	Pulverized muck.
No. 5	2,975	1,233	71	Pulverized muck and salt.
No. 6	2,467	725	41⅔	Common salt.
No. 7	2,678	936	54	Horse manure.
No. 8	2,856	1,114	64	Cow manure.

A little gypsum or plaster on clover, only a half bushel to the acre, will often increase the yield in an astonishing manner, making the gypsum worth $125 per ton. In some cases it will do scarcely if any good. This is the case usually on wet land or in very wet seasons.

Mr. Lawes states the following in the *Indiana Farmer* for 1883, in reference to fertilizing pastures in the United States:

"Where pasture is constantly mown, the removal of the potash from the soil becomes in time very large. Taking into account the price obtained for hay in the states, I think it is very doubtful whether restoration of fertility by means of artificial manures, might not be too costly, and I should be disposed to think that a more economical process for such restoration, would be by feeding animals on the pasture with corn or cake.

"The quality of the pastures at Rothamsted has been wonderfully improved by giving a certain amount of cotton cake to the stock fed upon them; and it is my opinion, that if at any time the blue-grass should retire from a pasture before an invading army of weeds and inferior grasses, the manure from cotton cake will furnish the proper weapon to rout these adversaries."

"The Battle in the Meadow."—As wolves quarrel over a dead animal, or hungry swine over an ear of corn, so plants struggle with each other to secure the greatest amount of food. Whether they be diatoms in the pool, fungi on the rotting apple, weeds by the wayside, or grasses in the meadow, one rule governs them all. Each strives for all it can get. Dean Herbert was more than half right: "Plants do not grow where they like best, but where other plants will let them."

On this subject, and in this connection, we are fortunate in having access to the results of the prolonged and elaborate experiments of Sir J. B. Lawes and his associates at Rothamsted, St. Albans, England.*

For more than twenty years in succession he experimented on the agricultural, botanical, and chemical results of a mixed herbage in a permanent meadow. There were 22 plots, upon some of which were placed different kinds of fertilizers and upon others none were used.

It was a very old pasture, having been in permanent grass over a century. No fresh seed of any kind was sown during the period. The land was flat, heavy loam, with a red clay subsoil resting on chalk, naturally well drained. The first crop, for a few years, was mown; the second was eaten off by sheep. There were twelve different manures employed. The total number of species observed upon these plots was 89, belonging to 22 orders, of which 20 were grasses and 10 leguminous.

On the unmanured plot, there were slight changes from year to year, due mainly to difference in the seasons and a slight exhaustion of the soil. By weight, the grasses furnished 69 per cent, the leguminous plants 8, and the other 23 per cent was of a miscellaneous character. As Masters, the botanist who was

*See Jour. Roy. Ag. Soc., 1858 9; Philosoph. Trans. Roy. Soc., 1882.

employed, describes it: "The general appearance of the unmanured plots is one of even growth, with no special luxuriance of any particular plant. The herbage is very mixed, the crop scanty, the color yellowish green, no one kind being specially favored. *Festuca ovina* is the predominating grass; *Briza media* is more abundant than on most other plots. The miscellaneous plants are generally very abundant, such as the buttercups, *Plantago lanceolata, Leontodon, Brunella vulgaris, Achillea Millefolium, Rumex Acetosella,* and others. The contrast in early summer between the scanty yellowish-green herbage, profusion of flowers of the various weeds, and the almost total absence of flowers and rich, deep blue-green foliage of the plants in the ammonia plot is very striking."

As would be expected, almost all the plants on the experimental plots, no matter how they were treated, were perennials; very few were annuals. Few of them were left long enough to produce seeds. It would be interesting to know what would have been the result had all the plants remained without cutting.

The competition of grass for room is mainly exerted by the roots and rhizomes, which form a network more or less dense and varying in depth according to the plant and the soil. In some cases the competition is chiefly above ground, where dense tufts prevent the growth of neighboring species.

The changes of the seasons cause the proportion of plants to fluctuate.

As Darwin observes, in chapter 3, *Origin of Species:* "The struggle almost invariably will be most severe between the individuals of the same species, for they frequent the same districts, require the same food, and are exposed to the same dangers."

Very marked contrasts between species of the same genus also occur, as *Poa trivialis* and *Poa pratensis.* This, perhaps, may be partially explained by the fact that the former produces no rhizomes while the latter produces many.

Because a plant diminishes in proportion to others after being treated with a certain fertilizer, it does not follow that this plant would not also be improved if it grew alone. One species of plant often receives more benefit from a certain manure than another.

The Effect of Manures.—It was observed that those manures which are the most effective with wheat, barley, or oats on arable land were also the most effective in bringing forward the meadow grasses. Again, those manures which were the most beneficial to beans or clover benefited most other species of leguminous plants.

The *Gramineæ* and *Leguminosæ* manifest somewhat different manurial requirements. There is perhaps no crop more influenced than the grasses in its character, as well as its quantity, by the attention bestowed upon it. This applies also to the leguminous plants.

The changes were most marked where the most liberal manuring was employed; the increase was much greater in the second year than in the third as compared with the second. By means of manures the yield of dry matter, per acre, in the hay crop, was in several of the experiments considerably more than doubled. Every description of manure diminished the number of species and the frequency of weeds.

Dead leaves occurred in most places where the manuring was the heaviest and the crops were the heaviest. This is a disadvantage in manuring so highly as to cause the crop to fall and die at the bottom before the bulk is fit for cutting.

In the words of the *Agricultural Gazette* for July, 1880: "They live in harmony on the unmanured, open park, having nothing to fight for in a state of nature. Season after season the same plants appear in about the same proportions. But toss them a bone, ground fine, or any other choice bit, and their harmonious companionship terminates at once. Every act of improved cultivation occasions instant war. A grass likes the best that can be got. It will swallow soda, but not when it can get potash.

"On general principles, all manures tend to drive out the weeds by increasing the better herbage."

This is certainly very satisfactory, but not true in every particular.

Mineral Manures Alone.—The leguminous plants were largely increased at the expense of the grasses and weeds. The grasses proper scarcely increased at all, whilst the whole plat was thickly covered with perennial red clover and some other leguminous plants.

Very different was the action of ammoniacal salts which caused the exclusive increase of the grasses proper, there being scarcely a leguminous plant to be found upon the plot.

Superphosphate of lime, when used alone, slightly increased the grasses and miscellaneous plants, diminishing the leguminous. It proved to be of little or no use.

Ammonia salts alone but slighty increased the crop. The crop was moderate and but little better than the plot unmanured

Farmyard manure gave a considerable increase of chiefly graminaceous hay and some few weeds, such as *Rumex* and *Achillea, Ranunculus, Carum*, attributed chiefly to its mineral and nitrogenous constituents.

The general result is, that leguminous plants in the meadow were much increased in growth and assimilated more nitrogen from unaided sources over a given area, when they were liberally supplied with certain mixed or primarily soil constituents.

Farmyard manure greatly encouraged the growth of the good grass *Poa trivialis* and the bad one *Bromus mollis*, and when in conjunction with ammonia salts the *Dactylis glomerata*, under both conditions, *Festuca duriuscula* and *F. pratensis* were nearly excluded, and *Avena flavescens, A. pubescens, Agrostis vulgaris,*

Lolium perenne and *Arrhenatherum avenaceum* were very much reduced.

It is certainly somewhat discouraging to find that the influence of farmyard manure was not favorable under all circumstances.

Nitrate of soda alone.—This generally gave an increased proportion of grasses, a late-ripening dark green crop, rather more leafy than stemmy.

Superphosphate and ammonia.—This produced much the same effect as the ammonia added to other combinations, viz: increasing the grasses and greatly diminishing the leguminous and miscellaneous plants.

Minerals and ammonia.—Here the yield was large, the grasses much increased, the legumes and weeds not improved. The larger the amount of ammonia the more marked were the results.

All poor grasses, except *Lolium*, were discouraged by the ammonia. The *Ranunculaceæ* and *Umbelliferæ*, *Compositæ*, *Labiatæ* were nearly expelled or greatly diminished.

The grasses on the plats thus treated ran much to leaves.

Minerals and Nitrate.—The proportion of grasses was large, that of legumes small, and that of miscellaneous plants much reduced.

Sulphate of lime, (gypsum) often called "plaster," sometimes slightly increases the growth of leaves and stems of grasses, but usually exerts a very marked effect to increase the growth of leguminous crops.

Disuse of manure.—In such cases the plants soon assumed the conditions of those on the unmanured plats. A disuse of potash was followed by a decrease in the produce of grasses, a marked decrease of the legumes and an increase of miscellaneous plants.

The practical conclusions may be very shortly stated. Drainage, marling, liming, must not be neglected. The application of bones is not recommended for general adoption. They appear to be chiefly adapted to the exhausted pastures of certain localities, and not to be generally applicable to meadow land which is mown for hay. The hay crop is a great exhauster of the mineral constituents of the soil; and these owing to the high price of the salts of potash, cannot, with profit, be fully restored in artificial manures. The return of the mineral constituents is better accomplished by means of farmyard manure, night soil, and the like.

"The grasses proper appear to be the most strikingly independent of any artificial supply of carbon. The hay crop is more exhaustive of potash than wheat or barley.

"A predominance of mineral elements in the fertilizers increased the proportion of the culms of grasses, while a predominance of ammoniacal salts increased the proportion of leaves.

"Those manures which much increased the produce of hay, at the same time very much increased its proportion of graminaceous plants.

"The total miscellaneous herbage (chiefly weeds) were the most numerous in kind and nearly in the greatest proportion on the unmanured land, viz: 16 per cent., while on the manured plat they decreased to 2 per cent."

"An artificial manure containing a sufficiency of mineral and nitrogenous constituents affected some of the grasses as follows:

Lolium perenne proportionally considerably increased.

Holcus lanatus proportionally largely increased.

Arrhenatherum avenaceum proportionally largely diminished.

Anthoxanthum odoratum proportionally largely diminished.

Agrostis vulgaris proportionally very much diminished.

Briza media proportionally very much diminished.

Cynosurus cristatus proportionally very much diminished.

Dactylis glomerata proportionaliy very much increased.

Poa pratensis proportionally very much diminished.

Bromus mollis proportionally reduced.

Avena pratensis proportionally increased.

Plantago lanceolata proportionally disappeared.

In the words of Mr. Lawes: "We learn from these results that good pasture grasses can never thrive upon a poor soil; and if a soil does not contain in itself the elements of fertility they must be added from external sources. I may add that if the pasture of a rich soil deteriorates from bad treatment the good grasses do not die out, but only retire from the contest to wait for better times. Under invigorating treatment it will be found that the good grasses soon reassert their supremacy."

"The general result, comparing the produce by the different manures in one and the same season, seems to be, that the more the produce is graminaceous the more it goes to flower and seed, and the more it is ripened, the higher will be the percentage of dry substance in the hay. Under the same circumstances, the higher will be the percentage of woody fiber and the lower will be that of the nitrogenous compounds and of the mineral matter. On the other hand, in a large proportion of the non-graminaceous herbage the reverse of these things is true."

In a summary of this subject, M. T. Masters, in *Plant Life*, says: "Circumstances are never exactly twice alike; a condition of absolute equilibrium is never attained. The nearest approach to it is in the case of the unmanured plats and of the plats very highly manured, but even these were influenced by very slight climatic changes. The balance in all cases was easily disturbed."

Green Manuring.—Most of this paragraph is from a lecture by my colleague, Dr. R. C. Kedzie. A complete manure is found in fresh vegetable matter turned under the surface of the soil. It is often convenient to adopt this practice on arable fields which are remote from the barn yard where stock are fed in winter. The late George Geddes, of New York, adopted this plan quite extensively, and believed he found it as cheap as any. It is often convenient to throw in a growth of something between two other valuable crops. For example, after a crop is removed in autumn, rye is sown to plow under for a late spring or summer crop. This grass grows well in cool weather, but does not return so much to the soil as red clover or some other legume.

In green manuring the whole vegetable growth is returned to the soil, and in a condition to insure rapid decomposition. In no other way can a soil in poor condition be brought into good condition so rapidly and by so little expenditure of money. By its skillful use the light and shifting sands of Belgium have been made the most fruitful fields of Europe.

Many are prejudiced against green manuring, believing that the process gives back to the soil only what it has taken from the soil. There is in most soils a large store of reserve material for plant food, but in the insoluble and inactive form. Certain plants have a singular power of corroding these insoluble minerals and bringing them into soluble condition, using them to build up their own tissues. When such plant is plowed under the soil it may give back to the soil only what it took from the soil, yet add greatly to its fertility because it has transferred such materials from the retired to the active list. But it is not true that plants give back to the soil only what they have taken from the soil. All plants take carbon from the air, and green manuring is the easiest way to increase the store of humus in the soil. Certain kinds of plants have singular power of accumulating combined nitrogen, and when these plants, rich in nitrogen, are plowed under the soil, they give to the soil in active form something which they did not take from the soil in this form. Nitrogen is the most precious and costly element of vegetable growth.

In the Northern States red clover heads the list, and is the red-plumed commander-in-chief of the manurial forces. Where the cow pea thrives, it also acts much like red clover when plowed under the soil.

Concerning the value of red clover as a manure, the reader is referred to another chapter which treats of that plant.

Rye is our hardiest cereal and grows better than any other on the poorest sandy land. It is not as valuable for plowing under as some leguminous crop, as its roots are smaller and much less in amount and the plant returns less to the soil. As it will grow rapidly in autumn and spring, and makes a large bulk of vegetation, it is not unfrequently sown for plowing under. It often happens that such a practice may be adopted without the loss of another crop.

In plowing under any crop to fertilize the soil, the reader should not forget that animals can appropriate only a small per cent of what they eat. The rest may be saved and go back to the soil.

"In estimating the value of the manure made by animals, only the nitrogenous and ash constituents of the food are considered, as the carbonaceous elements are supplied by the atmosphere. Over 95 per cent of the nitrogen and ash constituents are voided in the excrement in the cases of sheep and oxen. This shows a very small waste of the fertilizing matter of food in fattening sheep. If 90 to 95 per cent of these fertilizing constituents of food could be actually saved by farmers and returned to the soil, then it is easy to see the effect that must be produced by judicious stock-feeding upon the depleted soils of the New England and Middle States. The farmer should also remember that considerably more than half of the fertilizing of manure is to be found in the urine, and this is much the more valuable, according to the quantity, as it is all soluble and becomes immediate and active plant food."—(Feeding Animals, by E. W. Stewart.)

Manure and Drainage Improve the Quality of Grasses.—We have nothing better to offer in this connection than the results of some excellent experiments by Dr. A. Voelcker, recorded in Jour. Roy. Ag. Soc., p. 377, 1866:

"A comparison of the composition of the improved hay with that from the unimproved pastures offers several points of interest.

"The proportion of woody fiber in the good hay is much reduced.

"The amount of flesh forming material is considerably increased.

"The total amount of albuminous compounds is increased one-fourth.

"The difference in the proportion of sugar and other soluble matters is very marked, the bad hay containing only 10 per cent, the good hay nearly 15 per cent of sugar.

"The proportion of fatty or waxy constituents likewise is larger.

"The increase in the soluble mineral matter shows that the good hay is the more succulent."

This subject is also considered in the section prepared by Professor Armsby.

Here we see, then, that arable land produces grasses of better quality than marsh land, that rich land produces richer grasses than poor land, and every farmer knows that grass grown in the open meadow is more nutritious than that grown in the shade of trees, that the short growth in a dry season is more valuable per ton than the rank growth in a wet season.

Effects of Irrigation.—The writer has had very little experience in irrigation, but briefly gives the opinions and results of some experimentors, hoping thereby to set farmers to thinking, observing, reading, and experimenting on this interesting subject. To conduct irrigation properly is quite an art, but it has often been well done with surprising results, converting a lean, hungry meadow into an oasis. Sinclair, in his famous old work on grasses, says: "Irrigation is the easiest, cheapest, and most certain mode of improving poor land, in particular if it is of a dry and gravelly nature. The land is thus put into a state of perpetual fertility, without any occasion for manure."

To the farmers of Connecticut, J. S. Gould said: "You should sow many differant varieties of grasses and by the aid of irrigation you would have seven or eight times the amount of grass you now do." To the same people, Solon Robinson said he had no doubt that if the streams of Connecticut were properly utilized in irrigating the soil, they would be more productive in value than by turning all the water-wheels of the State.

After experimenting on this subject, Mr. Pusey, in Jour. Roy. Ag. Soc. for 1849, said that the money spent in irrigating grass land yielded a profit of 30 per cent. "All water is a weak liquid manure,—the warmer the water the better. A slight film of water trickling over the surface—for it must not stagnate—rouses the sleeping grass, tinges it with living green and brings forth a luxuriant crop in early spring, just when it is most wanted, while the other meadows are still bare and brown. A water meadow is the triumph of agricultural art. The best irrigated meadows are those upon a gravelly soil, with a good drainage."

Tenacious clays are less suitable for irrigation, and then only when well drained so the water can pass off at once. Water from streams is generally preferred to that from wells and springs. In cold weather water may overflow grass, and if not frozen to the grass it may remain there for weeks or months without harm, but in warm weather the case is quite different. Some spring waters contain sulphate of iron in solution or other matters injurious. Diluted liquid manure has often been artificially applied with most excellent results. Where meadows are irrigated the grasses are cut four or five times a year yielding enormous crops. Such land is seldom used for pasture, as it becomes too soft and is more profitable for mowing.

We will next read what Prof. J. Buckman says in Jour. Roy. Ag. Soc., p. 467, 1854. "By irrigation the list of grasses change; bad grasses will nearly all die out, or greatly improve in quality, whilst many good ones, few in number before, rapidly increase. Again, such weeds as *Plantago major*, *Ranunculus bulbosus*, *Panicum sanguinale*, and many others give place to a growth of grasses.

"Take the following on the observations of a meadow which was irrigated in an inferior manner. It had a subsoil of oölitic gravel, and its product was that of a thin upland pasture. How much it has changed will be seen from the annexed table, which is designed to supply information on the following points:—

1. The names of the grasses observed.
2. The proportions of those observed in the meadow before irrigation.
3. The changes effected in two years.
4. Those affected on the fourth year.

NAMES.	PROPORTIONS.		
	Before Irrigation.	After two years' Irrigation.	After four years' Irrigation.
Alopecurus pratensis, Meadow foxtail........	1	2	4
Poa pratensis, June grass	2	3	4
Poa trivialis, Roughish meadow grass.........	1	2	1
Briza media, Quaking grass..................	2	0	0
Cynosurus cristatus, Dog's tail grass..........	2	1	0
Aira cæspitosa, Hossack grass..............	1	0	0
Agrostis stolonifera, Marsh bent..............	1	2	3
Dactylis glomerata, Orchard grass, Cock's foot.	1	2	3
Avena flavescens, Yellow oat grass............	2	3	3
Avena pubescens, soft oat grass...............	2	1	1
Hordeum pratense, Meadow barley............	1	2	2
Lolium perenne, Rye grass..................	2	4	6

"This field trebled in value in four years. The table shows us that all the better grasses have increased, if we except the *Poa trivialis* and *Hordeum pratense*, in which cases there has been an increase in grasses not possessing the best character. In the first of these there is a decline in the fourth year.

"Now if we take into consideration the same set of facts, as presented by herbs of other families, the alteration is still more striking as attested by the following table:

NAMES.	PROPORTIONS.		
	Before Irrigation.	After two years' Irrigation.	After four years' Irrigation.
Ranunculus acris, Meadow crowfoot..........	1	3	1
Ranunculus bulbosus, Bulbous crowfoot........	3	1	0
Plantago lanceolata, Narrow leaved plantain...	3	1	1
Plantago media, Broad leaved plantain........	3	0	0
Trifolium repens, Dutch clover	2	0	0
Trifolium pratense, Red clover	1	2	2
Anthriscus vulgaris, Beaked parsley...........	1	2	1

"Now this table points out the important fact that large and innutritious herbs in pastures are destroyed by irrigation, and the previous one makes it clear that their places are supplied by the grasses.

"Parsley and docks should be pulled as the latter is largely increased by irrigation."

In other words, we conclude that the best grasses are a sign of good land or good treatment by manuring or draining or irrigation. They are the most sensitive to good or bad treatment: they are hearty feeders, and are the most exhaustive to the soil. Sedges, rushes, mosses, ox-eye daisies, and most other weeds, point to land that is out of order.

CHAPTER XII.

MAKING HAY.

Cutting and Curing Hay.—Within a few years we have given up the sickle for the scythe, and the scythe for the horse-mower. The hay-tedder takes the place of several weary boys in tossing the new mown grass; the horse-rake with a spring seat for the driver takes the place of the old hand rake. In some places the hay-loader is attached to the rear of the wagon and saves much heavy work. The large hay forks with a rope, a few pulleys, a horse, a boy and a little planning by the farmer, elevate the hay in large bunches to the top of the highest stack or the hay loft. This is all easy, if you have good tools and know how to manage, but no book can tell a beginner all about it. The operator must use his own judgment in deciding between that which is valuable and that which is not worthy of his attention.

Implements are all the time improving, and enterprising manufacturers see that the farmers know the fact. Instructions are freely given in reference to their use, hence little need here be said in reference to them.

Previous to haying the business farmer will put everything in good order. He has a few extra bolts, nuts, one or more extra sickle-bars and sections, and is prepared to meet slight accidents without delay. He has a good steady team and a careful driver who has some tact with tools. Before the grass had made much growth some pains was taken to remove stones, stumps or other obstructions, or to mark them so their location could be known when the grass had become tall.

For the prospects of fair, settled weather he no longer relies wholly on the almanac, the moon's phases or the weather prophets, after the manner of his forefathers, but consults the "probabilities" of the signal service.—(Killebrew.)

If the grass be heavy, the dew should be nearly off before beginning. If the cutting bar is at one side the driver strikes out with the "off horse" next to the fence; he then turns about, driving over the swath last cut, and goes around as much as he chooses. Or if he use a Eureka or other mower where the machine follows the team immediately, he may go back and forth on one side or proceed in some other manner.

Before noon, and perhaps after noon also, the hay tedder stirs the grass once or more. Towards night it is raked and put into cocks. If there be much clover the tedder must not be used after the leaves have dried, as it crumbles and wastes the most valuable part of the hay. The leaves of clover will dry a long time before the stems.

If cut late in the afternoon, or in the evening, so it does not wilt, no harm will come if a heavy dew fall on the hay. If the day be a fair one it is not good practice to cut grass in the middle of the day and leave it partially cured exposed to dew or rain.

The finest hay is made in dry, sunless weather, with little dew, and as little handling over as possible. Burning too long in the hot sun renders the hay brittle, and some of it will be lost in handling.

As usually made, the best clover hay is only fairly wilted before it is put in the cock, where it remains from four to seven days. In the meantime the cocks are carefully opened once or more each into two or three piles for an hour or two, then put up again.

If not very well cured, hay will keep better in a close mow in the barn than in a loft or in a stack where it is much exposed to the air. The closer the barn the better for the hay.

The following was prepared by Prof. H. P. Armsby, of Wisconsin:

Effect of Drying.—"All the nutrients of dry, coarse fodder are digested and resorbed to the same extent as when it is fed green. This is only true when the fodder and the hay are otherwise of exactly the same quality, when both are cut at the same time and from the same field, and when none of the leaves or other tender and especially nutritious parts are lost during the preparation of hay. These considerations are never completely reached in practice.

"*The digestibility* of the organic constituents of a fodder is in no way altered by simply drying in the air, provided it is executed without loss of parts of the plants. The ordinary method of making hay involves a considerable loss of leaves, and the product suffers not only in its quality but in its digestibility as well.

"*Effect of Storing.*—The storing of fodder for a long time, even when all necessary preventions, such as a dry and airy location, etc., are observed, may decrease both its digestibility and palatability.

"*Period of Growth.*—Early cut forage is not only superior, other things being equal, to late cut, as regards its chemical composition, but it excels it also in digestibility. This fact is established by abundance of experimental evidence.

"Digestibility is not sensibly increased by steaming or ensilage. In practice, however, the palatability of a fodder may often be very considerably increased by suitable preparation, and the animals thus induced to eat larger quantities of fodder not perhaps agreeable to them in its natural state.

"*The Fertility of the Soil affects the Quality of Plant.*—The natural quality and fertility of a soil have a very considerable influence on the chemical composition of the crop. Still greater differences often show themselves when dark green 'rank' plants are compared with pale yellowish-green ones of the same kind, occurring in the same field, and of the same age.

It is questionable whether very high manuring really gives more nutritious fodder than can be got from soil of good fertility.

"*Method of Curing.*—All methods and appliances which diminish the amount of handling which the hay must receive, especially when it is nearly dry, tend to improve the quality of the product by avoiding mechanical losses. So, too, it is desirable to dry the grass as little as is consistent with the object of curing, sufficient to ensure the keeping of the fodder, since the dryer and more brittle it becomes the greater is the loss by handling. In the process of 'ensilage' these losses are largely avoided, but the process of fermentation causes a loss. Recent results obtained at the New York Experiment Station, and at Houghton Farm, seem to show that corn-ensilage suffers very little loss from fermentation.

"*Damage by Rain.*—Both analysis and digestion experiments confirm the common observation that hay which has been wet is diminished in value.

"*Early or Late Cutting.*—Young plants while rapidly growing contain relatively more protein and less fibre than more mature ones, consequently early cut fodder must be of better quality than that cut late. It is more digestible.

"Three elements enter into the problem of selecting the best time for cutting, viz: the quality of the fodder, its quantity, and the amount of labor expended upon it. While any grass is ripening a large part of the protein and starch passes from the leaves and stem to the seeds, which are so small that they are seldom masticated or digested. Moreover, they are easily lost in curing. The hay made from fully ripe grass is essentially straw.

"If only one crop is to be obtained, probably the best time for cutting is usually when the plants are just beginning to blossom. At this time a larger crop is obtained than if cut earlier, while the digestibility is not seriously impaired.

"If cut early there is a great advantage to the second crop, as shown by an experiment at Hohenheim:

	Percentage of Protein.	Total pounds of Protein.	Total dry matter, pounds.
One cut......................	16.3	434	2,662
Two cuts.....................	24.4	668	3,274

"The following table, taken from *Chemistry of the Farm*, shows the percentage composition of meadow grass cut at three different dates in the same field. The first cutting will represent pasture grass fed off in the green state by stock; the second cutting is good ordinary hay; the third cutting is an over-ripe hay, somewhat coarse and stemmy, but well harvested.

Date of Cutting.	Albuminoids.	Fat.	Soluble Carbo-hydrates.	Fiber.	Ash.
May 14............	17.65	3.19	40.86	22.97	15.33
June 9.............	11.16	2.74	43.27	34.88	7.95
June 26............	8.46	2.71	43.34	38.15	7.34

"These numbers speak most decidedly in favor of early cutting. When the fodder was cut twice, not only was the quality far better, as shown by the percentage of protein, but the absolute quantity both of protein and of dry matter per acre was nearly one-half greater. When we take into account the greater digestibility of the young hay, the gain becomes still greater. Experiments indicate that the richest fodder and the largest yield of digestible matters per acre may be obtained by cutting two or more crops of comparatively young grass in a season, rather than one crop of over-ripe vegetation.

"In practice, however, the fertility of the soil, the length of the season, the kind of grass, the cost of labor, etc., have to be considered. Rowen is more liable to injury from wet than coarser hay. It may often be cheaper to get one large crop of hay, even of poorer quality, and supplement it with concentrated fodders.

"The only direct feeding trials that have been made on this point, so far as I know, are those made by Professor Sanborn, of Missouri. So far as they go they indicate that the value of early-cut hay may have been over-rated.

"*Legumes.*—The legumes are characterized by the large proportion of protein contained in the plant as a whole, and in the seeds. As fodders, when properly cut and cured, they are very rich, but have the disadvantage of being rather bulky, and of being easily subject to deterioration by mechanical losses. As a general rule clover is richer in nitrogenous matters than grass. Compared with meadow hay, which is made from the true grasses, its protein is about equally digestible, its crude fibre decidedly less digestible."

In trying to decide which is the proper stage of growth for cutting grass for hay we should not forget that a late growth of the plants nearly to seeding impairs their strength. In case of red clover, it greatly interferes with the crop of seed which is obtained from the second cutting.

The following on this question is by Prof. W. H. Jordan, taken from the *Philadelphia Press:*

"What if sorghum does have more saccharose and less glucose when the seeds are formed or are ripe? Is it more nutritious? We have no reason for thinking so. Starch and the various sugars and other carbohydrates have just the same office, and, so far as we can judge, nearly the same value in animal nutrition, so how does a change from glucose to saccharose, or from starch to sugar, very materially affect the nutritive value of a plant? In the processes of digestion starch is changed to glucose, and in that form passes into the blood. Sugar in the blood requires somewhat less work for its preparation for use by the animal body, and is, undoubtedly, somewhat more completely utilized than is the case with starch. But the final form and office is the same with both starch and the sugars or foods.

"It is, therefore, difficult to see how a change from glucose to saccharose in sorghum can effect the intrinsic value. But why compare sorghum and Timothy anyway? One is a sugar-bearing plant, the other is not.

"Because sorghum, a sugar-producing plant, is worth most for making sugar when the seeds are ripe, why should it follow that Timothy, a plant containing in advanced age a very small quantity of sugar, is most nutritious when the seeds are formed? We cannot determine the effect that age has upon the nutritive value of any known fodder plant by the increase or decrease of a single compound. Plant substance is complex, is made up of many compounds, and we must measure nutritive value by the total quantity of digestible nutrients, taking into account also their form and relative quantities.

"Our knowledge of changes occurring in Timothy grass through age is, briefly, as follows:

"(1.) The nitrogenous compounds decrease and the carbohydrates (starch, sugar, etc.) increase in relative amounts.

"(2.) There is no conclusive evidence that the nitrogenous compounds assume more valuable forms in the later stages of growth than when the plant is in bloom.

"(3.) With the carbohydrates there is a change of material into the form of crude fibre. Crude fibre is in part digestible, and to that extent is as valuable as digestible starch.

"(4.) The nutrients in young grass are more largely digestible than in old.

"(5.) This decrease in percentage of digestibility may be in part or even wholly compensated by the greater acreage production in the case of mature grass. Whether this is so, undoubtedly, depends largely upon the locality and season.

"Purely chemical facts favor very strongly the idea that a pound of dry substance, as existing in Timothy when in bloom, is more valuable than a pound of dry substance at any later period, in much the same way (but in a less degree) that a pound of dry substance in young pasture grass is more valuable than the same quantity of material in the mature plant."

The following opinion, based on experiments, by Prof. J. W. Sanborn, of Missouri, differs from the above: He compares cutting grass, mostly Timothy, as soon as one-fourth part of the heads were in bloom, and other lots ten days later, when out of bloom, and after the seed had begun to mature. After repeated trials in feeding steers, and cows giving milk, he says the results indicate, not only that the amount of hay gathered from a given area are much larger when cut after bloom than when cut in bloom, but the late-cut hay was more nutritious. He believes that Timothy or clover hay, particularly the former, is worth more per pound, and for Timothy thirty-five to forty per cent. more per acre, for cutting when sufficiently out of bloom in preference to cutting in bloom or before blooming. From some experience he concludes that this is also true of corn fodder, and he is inclined to believe it is true of most vegetation.

The writer thinks it very doubtful whether it is best to cut all forage plants at the same stage of advancement.

Most farmers, as a rule, prefer to cut clover when a few of the first heads begin to turn brown.

If the grass has made a pretty good growth, and the bottom is not wet from damp weather, it is the safest plan to begin haying early. Something will very likely interrupt so that the grass last cut will be older than it should be for good hay.

Unless the weather be favorable it is difficult to cure well a thick growth of very young, succulent grass.

When the growth is thick, some of the lower leaves begin to decay, while those at the top are gaining. To save all the leaves grass must be cut when young. Very much will depend on the condition of the weather. If the sun is obscured by clouds and rain descends every few hours, the grass intended for hay must be left standing even though it be going to seed. For making hay we need dry weather, but we can fill a silo rain or shine.

Another reason for cutting early must not be overlooked. It will be noticed while reading the chapter on *Insects Injurious to Grasses and Clovers*, that in many cases early cutting is recommended as an effectual remedy.

It will be seen that it is by no means an easy matter to select the best time for cutting or the best process of curing grasses and clovers, or to tell just how much it is safe to rely on chemical analyses to help determine these questions; and when we come to the test of feeding the difficulties are still increased on account of a changing climate, differences in the animals selected, and other things only thought of by men who have carefully experimented in feeding domestic animals.

Partially cured hay may be pressed into very solid bales, and not injure by heating. It keeps much like *ensilage* in a *silo*.

If the hay in the cocks be too damp, before drawing it should be opened an hour or two. No fixed rule can be laid down to guide the farmer. Remember that dew and rain wash out much of the best portion of grass after it has been cured, or partially cured.

A few minutes of an expert will show a beginner how to put hay into neat cocks of 75 to 200 lbs. or more each. The hay at the top should spread and hang down the sides to help carry off rain, should any occur. General W. G. LeDuc, of Minnesota, has the following on this topic:

"There is an art in cocking the clover hay so that it will shed rain, and the best hay makers in this locality claim to have acquired the difficult art of thatching the clover cocks by dexterity in handling the fork and laying the hay. They insist on taking up small forkfuls of the windrow, placing one on top of another until they have a miniature cock, then taking it up on a four-tined fork and turning it skillfully so that the center of the forkful comes down, inverted upon the center of the forming cock. The cocks must be small and tall—such as will stand securely until the sunshine of the morrow."

Making Clover Hay in One Day.—By Hon. L. N. Bonham, Oxford, Ohio:

"For several years I put up clover hay as did my father and other Jersey farmers. I have long since abandoned their method and now put my clover hay in the mow the same day it is cut. The hay is far better, and the labor and risk in making it are far less. I select a bright day and start the mower as soon as the dew is off.

"By 11 o'clock I have cut as much as can be hauled in between 1 and 5 o'clock. The clover is then all turned and shaken up loose before we go to dinner. By 1 o'clock it is dry enough to rake into windrows if the day is an average hay day. No time is lost now in getting it into the mow. The hay is warm and free from external moisture. The warmer it is the less moisture is left on it. By 5 o'clock we have it all in the mow, if we can. If not all in then we prefer to leave it in the windrow until near noon the next day. After we stop hauling, at 5 P. M., the mower is started to cut what we can haul in the next day. The clover cut so late in the day is not wet with dew, and will not wilt enough to be blackened by the dew. It will be ready to shake up and spread out before 10 o'clock the next day, and by 1 o'clock we can begin to haul it into the mow.

"The clover hay thus made goes into the mow bright and with every leaf and head left on it. The secret of the whole business is, it is free from external moisture, while the warmth of the hay when it goes into the mow hastens the approach of the temperature of the mass up to 122, when the germs which cause increased fermentation are destroyed, and the hay keeps bright and sweet, and comes out fragrant clover, with all the heads and leaves of good color.

"My mow is 28x28, and as tight as good siding and strips painted can make it. There are no windows in the sides to let in air. The clover is put in as compactly as we can get it, to save room, and kept level, to have the heat uniform.

"Sometimes we sprinkle a half gallon of salt to the load when putting into the mow, but this is of doubtful value.

"'To exclude the air' from the top of my clover mow, I often cover with straw. But this does not pack closely. I find it better when hauling in wheat to fill up over the clover with wheat. This excludes air, and packs the clover so that it keeps bright to the very top.

"The old theory that the mow must be open and the clover thrown in loose, and treated to 'plenty of salt,' which may mean much or little, is exploded. Green clover will keep green in the silo if well packed and the air is excluded. Clover hay, put into the mow warm and dry, the day it is cut, will keep brighter and purer and sweeter than if cured longer in the field.

"The trouble, however, in farmers adopting the method I have successfully used, is they do not attach enough importance to the fact that the conditions named must be followed.

"It will not do to cut clover in the morning and haul it in after sun-down. It will surely mould or come out brown or fire fanged, simply because dew falls at 5 o'clock.

"Nor can we cut clover and put in the mow the same day without favorable conditions of sun and air. In neither case will the hay go in free from external moisture."

The above account was clipped from the *Farmer's Review*.

Hay caps are sometimes used, and we never heard of a farmer who threw them aside after he had once used them. They will sometimes save their cost in a single season. They are about six feet square and made of good common unbleached muslin. At each corner is an eyelet for pegs to run through into the sides of the hay cock, or stout cords may be fastened to the corners and long enough to reach to the pegs which stick in the ground. In fair weather the caps need not be used, but when rain threatens a man will sleep better with his hay covered, unless perchance, as is related in the *Country Gentleman*, some stranger wakes him up to tell him there are a lot of white cows in his meadow.

Drying by Hot Air from a Furnace or the Use of a Fan.—W. A. Gibbs, of Essex, England, has patented a contrivance for driving hot air from a furnace among the half-made hay as it is tossed by revolving forks in a long trough. This is sometimes valuable in Great Britain and Ireland, where they are subject to rains, especially for curing aftermath when the sun is low and the days short.

Morton's Crops of the Farm gives another plan which seems likely to come into more general use. It consists in providing a horizontal shaft, either under the ground on which the rick is built, or, by means of suitable boarding, within the lower layer of the rick itself, and connecting with it, one or more upright shafts into the body of the rick. A fan is used at the outer end of the main shaft, and draws the air through the whole body of the hay with sufficient rapidity at once to keep the temperature within safe limits. In this way partially cured hay can be finished before stacking.

Stacking Hay.—It is almost impossible to give rules in writing which shall be of much use. The best way for a person to learn is to become a pupil of a good stacker.

The foundation should be made on boards or some timbers to keep the hay from absorbing moisture from the ground. The middle should always be kept highest; it should be evenly trod down on all sides; the hay should be pitched onto the stack from different sides, or near the center of the stack, to prevent packing the hay on one side more than on the other. The top should be finished with long, straight, coarse grass or sedges. In the old country stacks are thatched.

Fermentation of New-Made Hay.—Concerning this point, the following is from the pen of Prof. F. H. Storer in the *Rural New Yorker:*

"There are several facts, long familiar to practical men, which show clearly that the process of hay-making is something more than a mere drying-out of moisture from the grass. New hay will 'sweat' somewhat in the mow or stack, no matter how dry it seemed to be at the moment of storing; and many horse-keepers believe it is not fit for food for horses until after this sweating fermentation has thoroughly run its course.

"Even at the ordinary temperature of the air a good deal of carbonic acid, with traces of hydrogen and hydrocarbons, are given off during fermentation.

"In the beginning of an experiment, the oxygen of the air was rapidly absorbed and changed to carbonic acid. But even after the oxygen had been completely removed in this way from the confined volume of air employed in the experiment there was still evolution of carbonic acid from the hay, the oxygen for which must have come from the grass. The atmosphere surrounding the grass had but little influence on the volume or the composition of the grasses produced. The evolution of carbonic acid took place about as rapidly in the artificial atmosphere as it did in air. It was more rapid at a temperature of 97 degrees than at 60 degrees. Where corrosive sublimate was used in the hay, or where the tube containing the hay was exposed to steam-heat for several hours and then left to itself, no gases at all were evolved; hence the conclusion that the fermentation and the evolution of gas must be dependent upon the presence in the hay or grass of low forms of organic life. In confirmation of this view the microscope always revealed numerous bacteria in the water taken from tubes in which the grass had fermented.

"It is commonly held to be quite improper to bale new-made hay, no matter how dry the hay may be. The waste of nitrogen from hay by long-continued keeping has repeatedly been noticed before by agricultural chemists. It follows that although the popular belief that the new hay is bad for animals may be true enough, old hay is not necessarily good hay."

Saving Seeds.—Instead of placing all the notes on this topic under this heading the reader will consult what is said on saving seeds of orchard grass, tall oat-grass, June grass, and red clover.

CHAPTER XIII.

LOOK THE WORLD OVER FOR BETTER GRASSES AND IMPROVE THOSE WE NOW HAVE.

Some Requisites for Success in a Grass.—J. J. Thomas, in the New York Agricultural Report for 1843, says:

"Some of the essentials to the success of grasses are—1st. They should produce seed in sufficient abundance, which may be collected without difficulty. 2d. Where used in mixtures they should not exclude others, as is the case with *Poa pratensis*. 3d. They should not be so tenacious of life as to become troublesome weeds in rotation, as *Triticum repens*. 4th. Some are valuable for close pasturage, which become too hard and wiry for meadows, as the hard fescue grass. 5th. Some are chiefly adapted to moist land, as red-top and ribbon grass; some for strong soils, as Timothy; some for growing in the shade, as *Poa nemoralis*; and in experiments these specific qualities should not be forgotten."

As Dr. Bessey, of Nebraska, puts it: "The qualities which

give value to a grass for pasture and hay are in many particulars identical, although there are many species excellent for the one use and poor for the other. Both must be nutritious, so as to have any value for feeding purposes. They must, moreover, be palatable and of inviting taste, so that they will be freely eaten by animals, for it is a fact well known to those who have made the subject one of careful study, that there are species which, although highly nutritious, are not valuable to the stock grower, because they are not relished, and therefore not eaten by stock. It goes without saying, that a grass which cattle will not eat is of no value to the farmer, be it ever so nutritious, as shown by chemical analysis.

"Then, too, any grass which is to find a place on the farm must be easily propagated, and sufficiently hardy to withstand the storms and frosts of winter, the heat and drouth of summer, the close cropping and the treading of cattle. It must be able to hold its own against the persistent efforts of the weeds of all sorts to displace it, and after all must not be persistent enough to itself become a weed upon grounds where it is not wanted. Surely these are many qualities, and it is a most difficult matter to find them combined in one species. Indeed, it may be said that for most parts of the country we have not as yet succeeded in securing an absolutely perfect grass."

The Best Soil and Climate for Pasture Grasses.—Moisture in generous quantity is indispensable for good and rapid growth of grass. An ample rainfall or artificial irrigation evenly distributed will make a good pasture, even on soils of inferior quality. An average rainfall of thirty inches or more in a temperate climate is necessary to secure favorable conditions for the growth of grass. It has been found that pastures on poor soils in Wales and Ireland will improve under treatment that would be quite insufficient on the eastern coast of England.

Soils which are naturally moist, rather flat and rich, are best adapted to the most valuable grasses. There the soil suffers less from freezing, and is less exposed on account of the absence of snow.

New Grasses for New or Old Stations.—Although the above heading may be "new" the subject is now old, for as long ago as 1843, in a prize essay for the New York Agricultural Society, J. J. Thomas said: "The great deficiency in the number and variety of our cultivated grasses has been long felt by intelligent cultivators; and a more complete order of succession, afforded by a mixture in pastures, is an important requisite. That among the number of nearly two hundred species indigenous to the Northern and Middle States, there are some which may prove equal if not superior to any we now cultivate, scarcely admits a doubt. Some of our native grasses have been tested in Great Britain, and found valuable."

The late I. A. Lapham, a sagacious botanist of Wisconsin, in the State Agricultural Report for 1853, wrote: "The importance of introducing new grasses, and efforts to improve those already cultivated, cannot be over-estimated. It is not at all certain that we have the best kinds, nor that those we have are brought to the greatest degree of perfection. Doubtless they may be improved as well as fruits and live stock."

A little later, in 1858, Dr. Thurber, in the *American Agriculturist*, forcibly expresses a similar view: "A dozen sorts, probably, cover nineteen-twentieths of all the cultivated meadow land from Maine to Texas. It can hardly be supposed that so limited a number meets, in the best manner possible, all the wants of so great a variety of soil and climate. This is one of the pressing wants of our agriculture. A single new grass, that would add but an extra yield of a hundred pounds to the acre, would add millions of dollars annually to the productive wealth of the nation."

J. R. Dodge, in the Report of the Department of Agriculture

for 1870, with regard to the plant required, says: "It must be one that will do for the coarse, open, and airy soil of the plains, which is often dry for a long time, what *Poa pratensis*, Lin., has done and is doing for the States east of the Missouri River within the same parallels; one that will not only maintain its footing, but will extend its area, and overcome competitors.

"A strong-growing, coarse perennial, with rhizomas, or underground root-stocks, would suggest itself as a suitable species for trial; or a perennial producing an abundance of radical leaves, and of early growth, that would cover the soil and prevent the growth of annuals.' Of this class he suggests: *Elymus Canadensis*, L., *Elymus Virginicus*, L., *Elymus Sibiricus*, *Elymus mollis*, Triu., *Sporobolus heterolepis*, Gray, *Ceratochloa grandiflora*, Hook."

Of foreign species he thinks the most promising is *Festuca pratensis*, Huds.

Soon after beginning to give special attention to the agricultural grasses, the writer in a lecture to the Northwestern Dairymen's Association in 1872, advised hunting up new grasses in Mexico, Europe, South America and Australia, Japan and California. Depend upon it there are treasures yet undiscovered in some of those distant lands. I suggested that, likely, grasses from a dry climate will thrive better than those from England or other moist climates. Truly we may say that very little progress has been made in this subject in forty years.

In the extensive unwooded regions west of the Mississippi the native grasses afford much pastures; but many of them start very late in spring, and stop growing early in autumn. They do not completely occupy the ground; they are easily stamped out by the hoofs of cattle and sheep. Some of the tame grasses will thrive better, and afford much more pasture. Especially is there great need of some forage plants better adapted to the Southern States, and the dryer portions of all the United States.

The sedges (*Cyperaceæ*) are mostly found on marshes, but a few grow on rather dry ground. Although extensively pastured, cut and cured for hay in new countries, they have been quite uniformly condemned as utterly unworthy of cultivation. They are nearly always much past their prime when cut for hay. They are better when cured early. The writer thinks it not unlikely that some of these sedges may prove valuable in certain localities. The majority of sedges appear in limited quantity often mixed with others which grow abundantly. Some experiments might very profitably be made on the sedges with reference to their value for pasture or hay.

On this topic I glean the following from the *Country Gentleman* of January, 1886, contributed by my colleague, Prof. L. H. Bailey: "At present there are only three species, so far as known, which possess any decided merits. One is a native of Thibet, affording fair grazing when grasses fail. Another is the sand carex of Europe (*Carex arenaria*) which is largely grown along exposed sea shores to hold the sand. The third species occurs along the Columbia River, where it furnishes a valuable hay and pasture, and is known as the hay carex. It has been received from several reliable sources. It grows rapidly in the early spring, and matures its fruit or seeds just before the annual rise of the rivers cover it. As soon as the water recedes it springs up again, but does not fruit, this time yielding an excellent hay. Hundreds of tons are cut from this species alone.

"Specialists have studied this plant quite carefully, and it has been referred to no less than five distinct species. It is probably the same as a Scandinavian species (*Carex acuta* var. *prolixa*) although that plant is not known to possess any economic value."

The following is from Dr. C. E. Bessey, of Nebraska: "For many years it has been a favorite subject of investigation with me to attempt to determine whether any of our native grasses were worthy of being brought under cultivation. In this investigation I have met with some odd experiences. I have as a rule found the opinion general that the wild grasses furnished valuable pasture and hay, and still, with few exceptions, it has been very nearly impossible to obtain exact data as to what kind of wild grasses were best, and what kinds were of most value for hay or pasture. Moreover, strange as it may seem, there are as yet scarcely any common names for these valuable wild grasses, so that it is almost impossible to speak intelligently of them without having recourse to their scientific names.

"It is not to be reasonably questioned but that there may be as valuable wild grasses which have not yet been brought under cultivation, as there are already grown on our farms. It must be remembered that every grass which we now grow was once but a wild grass in some part of the world, and that by bringing them under cultivation we have in every case increased their valuable qualities, as well as productiveness."

In *Science*, vol. 1, 1883, Prof. N. S. Shaler, referring to this subject, says: "It seems possible to improve this pasture by the introduction of other forage plants indigenous to regions having something like the same climate. The regions likely to furnish plants calculated to flourish in a region of low rainfall include a large part of the earth's surface. Those that would succeed in Dakota are not likely to do well in Texas or Arizona. For the northern region, the uplands of northern Asia or Patagonia are the most promising fields of search; while for the middle and southern fields, the valley of the La Plata, southern Africa, Australia, and the Algerian district may be looked to for suitable species." He recommends three experiment stations,—one in Nebraska, one in Texas, and one in Arizona.

In this connection, when we remember that exotic plants often thrive better than natives, we see what a vast field lies ready for experimenting with the grasses. As we have seen, private enterprise has done little. Grasses look much alike to all who have not closely studied them, so that farmers are not likely to make experiments. This is a strong reason why the state and national governments should assist agriculture in an undertaking which seems so fruitful of good results within a short time, at so trifling an expense. Expeditions are sent at great expense to explore Polar seas, with a view to slightly extend our knowledge of a barren portion of the earth's surface. Large sums are employed to fit up in magnificent style, and send to the remotest parts of the earth, expeditions to spend a few minutes in observing an eclipse or a transit of Venus. Would the sending of competent persons around the earth in search of better grasses be an undertaking less praiseworthy?

Improving by Selection.—The good effects of a change of seed is in many cases already enjoyed in the case of grasses and clovers, as most farmers occasionally purchase their seed. A change of seed means a change of soil and surroundings; and these are likely to benefit the plants.

Probably every reader believes that the following from *Master's Plant Life* is true:

"In a wheat field or bean crop no two plants are exactly alike; one is more robust than another, one tillers more than the rest, the ears of one are plumper and fuller, this one grows earlier or later in spring, is therefore hardier or more tender, as the case may be. The careful observer notes these points, and instead of passing them over endeavors to turn them to account by selecting the plant which shows a tendency to vary, taking seed from it and growing that seed another season." The best is selected, the process continued.

The shrewdest horticulturists are continually and successfully following this plan. To a limited degree the general farmer does the same thing. By this process, Major Hallett in five years caused the length of the ears of wheat to double, their contents to nearly treble, and their tillering power to increase five fold.

To improve wheat, the following plan is worth considering: Select a field where wheat will yield well, and see that everything is well done to make it prosper. When about ripe, pass through the best portion of the field and select some of the best spikes of wheat from the best stools. Plant these for the next crop, in the best land, and give them the best of care, continuing the process. This is far ahead of the common practice, which is to separate the plumpest kernels from a lot of grain by means of the fanning mill. Some of the selected kernels most likely came from short spikes of small stools.

Precisely the same method here suggested for improving wheat can be applied to the improvement of orchard grass, Timothy, June grass, meadow fox-tail, any of the fescues or the clovers. Indeed, across the Atlantic something has already been done in this direction, and with excellent results. The time will doubtless come when farmers will take some care in reference to breeding and selection of grass seeds, as they now do in reference to their domestic animals.

To procure seed corn, plant a piece by itself, give plenty of room for each stalk; enrich the soil and give excellent cultivation. Remove all poor stalks before flowering that they may not fertilize any ears. Select the best of these upper ears for seed. Florists follow the same plan by removing all poor or undesirable specimens before flowering.

Improving by Cross-Fertilization of the Flowers.—After reading the former paragraphs on fertilization, with specimens in hand, the reader will have little difficulty in understanding how to cross some of the larger grasses. In all cases, to insure a cross, the young anthers must be removed before they shed pollen. Spread apart the palet and flowering glume, and carefully remove all the anthers. At the same time, an anther a little older from another variety may be inserted in place of the three removed. The pollen of the anther inserted will keep, and is ready to fertilize the stigmas as they mature. All the flowers of a spike may be operated on, or only part of a spike, and the rest cut off. The culm will be marked so as to secure the grain when it ripens.

Professor A. E. Blount, of Colorado, is an enthusiast in crossing cereals, and has met with excellent success in obtaining good new varieties. Hear him: "All the cereals are susceptible of great improvement. They can be made to produce results, heretofore unrealized, at which some of the oldest scientific farmers are amazed. The farmer can breed up his grain as he does his stock. If it is deficient in any one element, he can supply that deficiency. Should his wheat, for instance, be too soft, too starchy, or have weak straw, he can, by crossing it upon other harder, more glutinous and stiff strawed kinds, make wheats to suit his soil, climate and his miller. If his corn does not suit him, if it is too long-lived, with too large cobs, too coarse fodder, too inferior stalks, too high, low, large or small, he can select, cross and interbreed until only quantity, form, and fineness are obtained. The experimenter must be thoroughly acquainted with the plants before he can succeed in improving them by selection. If he be a wise man, and understand his business, he does not always take the largest ear or the largest spike. The largest are by no means always the best."

Many careful experiments have been made by Darwin and others proving conclusively that the chances are largely in favor of great improvements, if the flowers are cross fertilized.

The crossing of closely related plants is generally an improvement over self-fertilization; but crossing with foreign stocks of the same variety is a far greater improvement.

The reader may ask, What is meant by the term "*crossing with foreign stock.*" The following experiment will illustrate it: Select two lots of seed corn which are essentially alike in all respects. One should have been grown, at least, for five years in one neighborhood, and the other in another neighborhood fifty or more miles distant. In alternate rows plant the kernels taken from one or two ears of each lot. Before flowering thin out all poor stalks. As soon as the tassels begin to show themselves in all the rows of one lot, pull them out, that all kernels on the ears of those rows may certainly be crossed by pollen from the other rows. Save and sow the seeds thus crossed and an increased yield may be expected the next year. The benefits of such crossing will gradually diminish and probably disappear in a few years. All species which freely intercross by the aid of insects or the wind can be crossed as follows: Procure a quantity of seed grown for some years at some distance away and mix with seed kept and raised for some time at the place where the experiment is to be tested. "The two stocks will intercross with a thorough blending of their whole organizations, and with no loss of purity to the variety; and this will yield far more favorable results than a mere exchange of seeds."—(Darwin).

In brief, mix seeds of the same variety grown in different localities to grow your seed.

The late Charles Darwin in his book on *The Effects of Cross and Self-Fertilization of Plants* records the results of experiments made on fifty-seven species of fifty-two different genera of thirty families. These experiments were continued and repeated for ten years. He generally found the plants raised from seed crossed with foreign stock were the most vigorous, the largest, the hardiest, matured the earliest, yielded the most seed, and such seeds were the most certain to germinate and germinate soonest.

In 1877 the writer began some experiments of this kind with Indian corn and with beans, and has since made others. The advantage shown by crossing corn with foreign stock was as 151 exceeds 100, and in the case of black wax-beans it was as 236 exceeds 100. Other experiments have always shown a large gain in favor of plants raised from seed obtained in the above manner.

In reviewing Darwin's book, the *Gardener's Chronicle* said: "It is certain that these practical results will be a long time filtering into the minds of those who will eventually profit most by them."

The results, so far, fully accord with the prophetic statement above quoted; the people are slow, very slow, to profit by the experiments.

CHAPTER XIV.

GRASSES FOR THE LAWN, THE GARDEN, AND FOR DECORATION.

The Lawn.—"Grass is the most lowly, the simplest, and the loveliest element to be used in the adornment of home. A smooth, closely shaven surface of grass is by far the most essential element of beauty on the grounds of a suburban home."—(F. J. Scott.)

"It would be a great gain to horticulture if ten out of every twelve 'flowerbeds' in Europe were blotted out with fresh green grass."—(Robinson's Parks of Paris.)

"A lawn is the *ground work* of a landscape-garden."—(H. W. Sargent.)

Listen to A. J. Downing: "The great elements of landscape gardening are trees and grass. For this purpose we do not look upon grass with the eyes of the farmer who raises three tons to the acre. We have no patience with the tall and gigantic *fodder*, by this name, that grows in the fertile bottoms of the West, so tall that the largest Durham is lost to view while walking through it. No, we love the soft turf which is thrown like a smooth natural carpet over the swelling outline of the smiling earth.

"Fine lawns are possible in all the northern half of the Union, although an American summer does not, like that of Britain, ever moist and humid, naturally favor the condition of fine lawns. The necessary conditions for a good lawn are *deep soil, the proper kinds of grasses, and frequent mowing.* Let the whole area to be laid down be thoroughly moved and broken up two feet deep. Let the surface be raked smooth and entirely cleared of even the smallest stone. The object of a lawn is not to obtain a heavy crop of hay, but simply to maintain perpetual verdure. *Rich* soil would defeat our object by causing a rank growth and coarse stalks, when we wish a short growth and soft herbage. Let the soil, therefore, be good, but not rich; depth, and the power of retaining moisture, are the truly needful qualities.

"Now for the sowing; and here a farmer would advise you to 'seed down with oats,' or some such established agricultural precept. Do not listen to him for a moment. Do not suppose you are going to assist a weak growing plant by sowing along with it a coarser growing one to starve it."

Owing to the difficulty of learning to recognize the seeds, the purchaser is usually at the mercy of the dealer, whose interest it is to enshroud in mystery the whole subject of grasses for the lawn.

Many of the leading seedsmen of our country are advertising extensively and appear to be selling large quantities of "mixtures" of lawn grass seeds for which there is quite a variety of attractive names.

The writer has frequently examined these mixtures and has watched the success of several of them in various portions of the Northern States. For the benefit of my readers I present the results of a careful "analysis" of some samples of seeds of mixed lawn grass.

In former years, the vitality of the rarer grass seeds has universally been found to be very low, while the germinating power of the common sorts, such as are raised in this and neighboring States, has been satisfactory.

CHICAGO PARKS MIXTURE.

Sold by —— —— Chicago, Illinois.

The table shows the relative proportion of the different kinds of seeds found:

June Grass, or Kentucky Blue Grass, *Poa pratensis*, L., in the chaff	1710
White Clover, *Trifolium repens*, L., clean	90
Sweet Vernal, *Anthoxanthum odoratum*, L., in chaff	37
Perennial Rye Grass, *Lolium perenne*, L., in chaff	35
Orchard Grass, Cock's Foot, *Dactylis glomerata*, L., in chaff	30
Red Top, Brown or Creeping Bent, *Agrostis*, in chaff	16
Timothy, *Phleum pratense*, L., clean	6
Mixed and containing traces of the following	15

Velvet Grass, *Holcus lanatus*, L., in chaff (a weed).
Sedge, *Carex* (worthless).
(Narrow?) Dock, *Rumex* (a weed).
Panic Grass, *Panicum* (worthless).
Chickweed, *Stellaria* (a weed).

This mixture is advertised as *especially adapted to the inland and western States*, and costs 25 cents per quart or $4.00 per bushel.

As will be seen, it consists mainly of June grass, which the same house offers at $1.50 per bushel; and the latter, if pure and sowed alone, is far preferable for a lawn to this mixture. Besides those marked weeds, the others which are most objectionable are orchard grass, a coarse, bunchy grass, Timothy, which is too coarse and short lived, perennial rye grass, which just takes the cream of the soil for a few years and then dies out.

FINE MIXED LAWN GRASS.

Sold in bulk by —— —— Detroit, Mich.

Table showing the proportions:

June Grass, or Kentucky Blue Grass, *Poa pratensis*, L., in chaff	627
Perennial Rye Grass, *Lolium perenne*, L., in chaff	470

Timothy, *Phleum pratense*, L., clean	340
White Clover, *Trifolium pratense*, L., clean	220
Red Top, Brown or Creeping Bent, *Agrostis*, in chaff	217
Mixed and containing traces of the following	15

Velvet Grass, *Holcus lanatus*, L., in chaff (a weed).
Orchard Grass, Cock's Foot, *Dactylis glomerata*, L., in chaff.
Chess, *Bromus*, some species (a weed).
Crowfoot, *Ranunculus bulbosus* (a weed).
Dock, *Rumex* (a weed).
Lance-leaved Plantain, *Plantago lanceolata*, L., (a weed).
Shepherd's Purse, *Capsella Bursa-pastoris*, Mœnch (a weed).

This mixture is sold at 50 cents per pound, or $4 per bushel, and is not so good as the Chicago parks mixture noticed above, because it contains a much smaller proportion of June grass and a much larger proportion of perennial rye grass and Timothy.

FLINT'S LAWN GRASS.

Sold by —— —— Detroit, Mich.

Table showing the proportions:

Perennial Rye Grass, *Lolium perenne*, L., in chaff	526
Sheep's Fescue and Hard Fescue, seeds much alike, *Festuca ovina* and *var. duriuscula*, L.	295
June Grass, or Kentucky Blue Grass, *Poa pratensis*, L., in chaff	255
White Clover, *Trifolium repens*, L., clean	227
Red or Mammoth Clover, *Trifolium pratense* or *medium*, L., clean	130
Timothy, *Phleum pratense*, L., clean	105
Meadow Foxtail, *Alopecurus pratensis*, L., in chaff	103
Italian Rye Grass, *Lolium perenne*, var. *Italicum*, in chaff	47
Sweet Vernal, *Anthoxanthum odoratum*, L., in chaff	35
Hair Grass, *Aira flexuosa*, L., in chaff (a weed)	25
Chaff	80
Mixed seeds containing traces of the following	30

Chess, *Bromus* (a weed).
Fescue (species?)
Velvet Grass (a weed).
Self Heal, *Brunella* (a weed).
Sorrel, *Rumex* (a weed).

Ribbed Grass *Plantago lanceolata* L. (a weed).
Chickweed (a weed).
Nonesuch, *Medicago lupulina* L.
A sedge, *Carex*.
Two or three others not recognized.

This mixture is sold at ....per quart or per bushel.

In addition to the objections made to the two former mixtures are the following:

Sheep's fescue and hard fescue grow in tufts or bunches and will not produce a lawn of even appearance. The red or mammoth clover will also produce a coarse patchy lawn, and the former will die out in two or three years. Italian rye grass will kill out the first winter. Hair grass is a weed substituted for crested dog's tail, which is a feeble grass of no value in this country.

FINE MIXED LAWN GRASS.

Sold by —— —— Rochester, New York.

Table showing the proportions:

June Grass, Kentucky Blue Grass, *Poa pratensis*, L., in chaff......	995
Perennial Rye Grass, *Lolium perenne*, L., in chaff..................	373
Orchard Grass, Cock's Foot, *Dactylis glomerata*, L., in chaff.......	327
Red Top, Brown or Creeping Bent, *Agrostis*, in chaff................	212
Velvet Grass, *Holcus lanatus*, L., in chaff (a weed)................	22

Mixed and containing traces of the following:
Chess, *Bromus*, Sp. (?) (a weed).
Lance-leaved Plantain, Ribbed Grass, *Plantago lanceolata*, L. (a weed).
Dock or Sorrel, *Rumex* (a weed).
White Clover, *Trifolium repens*, L.
Timothy, *Phleum pratense*, L.
Crowfoot, *Ranunculus bulbosus*, L. (?) (a weed).
Shepherd's Purse, *Capsella Bursa-pastoris*, Mœnch (a weed).

The above is sold at $4 per bushel.

For objections to some of these ingredients consult the comments inserted in connection with the former mixtures.

Sold in bulk by —— —— New York.

Table showing the proportions:

June Grass, or Kentucky Blue Grass, *Poa pratensis*, L., in chaff...	648
Red Top, Brown or Creeping Bent, *Agrostis*..........................	528
White Clover, *Trifolium repens* L., clean..........................	158
Timothy, *Phleum pratense*, L., clean..............................	38
Ergot of *Agrostis*, or Red Top, (infested with fungus).............	10

Mixed and containing traces of the following:

Eggs of insects.

Dung of insects.

Dead insects.

Panic Grass, *Panicum* (a weed).

Chickweed,

Shepherd's Purse, *Capsella Bursa-pastoris*, Mœnch (a weed).

Dock, *Rumex* (a weed).

Orchard Grass or Cock's Foot, *Dactylis glomerata*, L.

Eleocharis, a rush or grass-like plant (a weed).

Round Leaved Mallow, *Malva rotundifolia*, L. (a weed).

This is sold for $5 per bushel, and is a good mixture, omitting the seed of Timothy and the weeds. The house claims to have have sold 70,000 packages in 1885. The same house offers June grass for $2.25, and Bent grass for $4 per bushel.

THE "HENDERSON" LAWN GRASS SEED.

Sold by —— —— New York.

Table showing the proportions:

Brown or Creeping Bent or Red Top, *Agrostis*, in chaff.............	880
June Grass or Kentucky Blue Grass, *Poa pratensis*, L., in chaff	715
White Clover, *Trifolium repens*, L., clean.........................	120
Sheep's or Hard Fescue, *Festuca ovina* or var. *duriuscula*, L., in chaff	110
Perennial Rye Grass, *Lolium perenne*, L., in chaff..................	95
Sweet Vernal, *Anthoxanthum odoratum*, L., var. *Puelli*, in chaff...	17
Timothy, *Phleum pratense*, L., clean..............................	10

A few seeds of Chickweed, some *Panicum*, Mallow, *Malva rotundifolia*, L., (a weed), Ergot, some other weeds not recognized.

This is much like the Central Park lawn grass previously noticed. This one contains some seeds of small fescues apparently mixed, a little perennial rye grass, which is no benefit to it, and a very little Timothy, which would be better to omit, and a small amount of sweet vernal, which apparently is the annual variety and of no value. The three leading ingredients are the June grass, bent grass, and white clover.

It was the freest from weeds of any mixture examined. It is sold for 25 cents per quart or $5.50 per bushel. The same house sells June grass for $2.25 per bushel, bent grass for $4.00 per bushel, white clover for 40 cents per pound.

The preceding tables and the remarks below each should be studied in connection with what follows.

At the Agricultural College, numerous plats in various seasons and soils, mixed and separate, have been tried, and those grasses of most value are June grass and a small red top. White clover often thrives well with these, but it varies much with the change of seasons. Sod taken from a rich old pasture or the roadside usually makes excellent lawn as soon as laid, but it is too expensive for a large plat. The main grasses making such a turf are those last mentioned, June grass and red top, with perhaps some white clover.

In making a lawn too little stress is usually placed on thorough trenching or subsoiling and enriching the land. The surface should be harrowed and hand-raked till it is in the finest condition.

With the writer's experience, having tested for some years over two hundred kinds of grasses and clovers, both native and foreign, for Michigan and places with similar climates, he would sow about two bushels of seeds (in the chaff) of June grass, *Poa pratensis*, L., and two bushels of some small bent grass, known as Rhode Island Bent, Brown Bent, or Creeping Bent, or as red top. The latter grasses vary much and are usually much mixed, as they were in all the samples above examined.

A few ounces of white clover may be added, if the owner prefers, but it is by no means very important. Each one of these two or three kinds of plants will appear to cover the ground all over, so it will look uniform.

To the farmer who is accustomed to sow coarse seeds for a meadow or pasture the above quantity of seeds appears to be enormous. But the aim is to secure many very fine stalks instead of a few large coarse ones.

If a little sweet vernal and a little perennial rye grass are used a careful observer, at certain seasons of the year, will see that the lawn looks "patchy." Especially in early spring, or in very dry weather, some of these and others often recommended, will grow faster than the rest and assume different shades of green. For a lawn *never* use any Timothy, orchard grass, tall oat grass, red clover, meadow fescue or other large grass or clover, but only the finest perennial grasses or clovers. Sow the seeds in September or in March or April, without any "sprinkling" of oats or wheat, and as soon as the grasses get up a little and the straggling weeds get up still higher, mow them, and keep mowing every week or two all summer.

Avoid purchasing mixtures advertised in seed catalogues, as it will be much cheaper and safer to buy each sort separately, and only one or two or three sorts are desirable. The rarer grasses are mostly imported, and up to the present time, as was said, have been found to possess very low vitality; besides, bad foreign weeds are very commonly mixed with these grass seeds. There are good reasons, then, for buying common sorts, and, if possible, those raised and cleaned in a careful manner.

James Hunter, of England, in his manual of grasses, says: "Careful analysis of the mixed lawn grass seeds sold by some large seed houses at high prices prove them to consist of from 40 to 50 per cent. of rye grass, whereas not a single seed of rye grass should be included in any mixture for producing a lawn."

The Royal Agricultural Society of England employs a consulting botanist, Wm. Caruthers, who, for small fees, tests the seeds for its members. He finds it best to avoid purchasing mixtures for lawn, pasture or meadow.

The editor of the *Gardner's Monthly* echoes the sentiments of our best judges in this matter when he advises for lawn to sow June grass or red top either one alone or both mixed.

E. S. Carman, one of the editors of the *Rural New Yorker*, and manager of a fine homestead and an experimental farm, writes: "Thirteen years ago we sowed on different parts of an acre of lawn blue grass, red top, Rhode Island bent and the 'lawn mixtures' sold by seedsmen. To-day the red top presents the finest and brightest appearance, while the 'lawn mixture' portion has since been re sown with red top and blue grass."

In conclusion, if not so already, make the soil strong, drain thoroughly, deeply pulverize, harrow and hand-rake the surface carefully. In early spring, or in early autumn if not dry, sow, without any wheat or oats, three or four bushels to the acre of June grass or red top, either one or a mixture of both in any proportion.

Ornamental Grasses.—Although grasses rank among the lowest of the flowering plants, and very few have anything like gaily colored blossoms, yet no order possesses plants which surpass some of them in grace and elegance. For beauty, grasses rely mainly upon their forms and pleasing shades of green color. A few have brilliant colored anthers, or their spikelets are covered with white hairs.

From simple, rigid heads or spikes to the most graceful of delicate, drooping panicles there are all grades of pleasing forms.

We have considered the surpassing beauty of a green velvet lawn, but who can fail to admire the glory of the meadow or the pasture on the plain or the hill-side spotted with fat cattle or "bunchy" sheep?

There is much to admire when grasses are crowded together in large masses, whether they are kept closely shorn or cropped, or whether they grow to uniform height and are viewed at various stages of their growth as the clouds drift over the fields or "they wave their fairy tassels in the wind."

Occasionally, near springs and streams, the frost deposits on the panicles a covering which is indescribably beautiful.

Within a few years, florists have given considerable attention to the grasses for winter bouquets and for other decorative purposes. Our enterprising growers and dealers offer the seeds of quite a long list of the best for these purposes.

In one other respect the grasses have not yet begun to assume the prominence their merits demand. The writer has grown a large number of our native and foreign grasses, and has studied them where each kind grew by itself in isolated bunches or patches, and he is free to say that in no other place does a grass appear to better advantage. Here is an almost endless variety, as exhibited in form, texture and color of the leaves. The culms also, and the spikes, racemes or panicles reveal their peculiarities in a manner which is most varied and pleasing.

Such bunches of many kinds of grasses are well worthy of a place among the ornamental plats of our lawns and gardens.

Where so many are fine it is difficult to discriminate. Those advertised by the florists are all good, including those with striped leaves.

Mays, sugar cane, *Sorghum*, bamboo, *Arundo donax*, *Zizania aquatica*, *Phragmites communis*, and other tall species with broad leaves are valuable for the sub-tropical garden. The two latter are excellent for growing in the shallow margins of ponds.

For plumes and bouquets the following are much used, for accounts of which consult the text elsewhere: *Briza maxima*, *B. media*, *B. gracilis*, *Bromus asper*, *Lagurus ovatus*, *Polypogon monspeliensis*, *Deschampsia cæspitosa*, *Phragmites communis*, many species of *Festuca*, *Elymus arenarius*, *Agrostis elegans*, *A. nebu-*

losa, *A. scabra*, *Panicum capillare*, *P. virgatum*, *Pennisetum longistylum*, *Asprella hystrix*, *Erianthus ravennæ*, *Coix lachryma*, *Gynerium argenteum*, *Arundo conspicua*, *Chloris radiata*, *Stipa pennata*, *Hordeum jubatum*.

There is scarcely a genus of grasses of any size which does not possess one or more species of special value for ornamental purposes. To the botanist, the artist or the florist it is hardly necessary to mention the following genera, viz: *Panicum*, *Setaria*, *Spartina*, *Andropogon*, *Phalaris*, *Alopecurus*, *Phleum*, *Milium*, *Muhlenbergia*, *Holcus*, *Avena*, *Cynodon*, *Bouteloua*, *Eleusine*, *Eatonia*, *Graphephorum*, *Eragrostis*, *Melica*, *Poa*, *Glyceria*, *Festuca*, *Bromus*, *Elymus*, *Triticum*, *Lolium*, and many others. We hardly know where to stop giving names for this purpose.

With reference to collecting and the use of grasses, A. Hassard in *The Garden* for 1875 has the following: "Not even the most delicate fern will give the same airy look to a vase of flowers that a few spikes of wild grasses will impart. In cutting grasses for use they must be selected before they are old enough to fall to pieces when dried. Each variety should be tied in separate bunches, and care should be taken that they are not bruised together, for, if this is the case, when the bunch is opened each spike will be found to have dried in its crushed position, and its form will be thus quite spoilt. All grasses should be dried in an upright position, particularly those of a drooping character. Oats, while still green, are also very pretty in large arrangements. A free use of grasses and sedges enables you to dispense with many flowers. The bloom of ribbon grass is very useful for this purpose, as it has a silver-like lustre, or a rose-pink tint, which is very pretty."

CHAPTER XV.

THE LEGUMINOSÆ. PULSE FAMILY.

Herbs, shrubs, or trees. *Leaves* alternate, usually compound and stipulate. *Flowers* irregular or regular. *Calyx* mostly 5-lobed with one lobe next to the bract. *Corolla* irregular and imbricate (often papilionaceous), or regular and valvate, rarely o. One petal next to the axis. *Stamens* usually 10, rarely 5 or many, monadelphous, diadelphous or distinct, mostly perigynous. *Pistil* with a 1-celled carpel becoming a legume or an indehiscent fruit, sometimes jointed. Embryo usually destitute of endosperm.

This vast family contains, at the lowest estimate, 6,500 species, and is excelled in numbers by only one other, viz: the Compositæ, which includes asters, golden rods, sunflowers, dandelions. Plants of the pulse family are widely distributed in every climate and in all kinds of soil. They vary in size from the little pussy clover to the giant locust trees of Brazil. We comprehend only a small portion of their uses and wealth when noticing those species which are cultivated or wild in the United States. Red, white, mammoth and Alsike clovers, lucerne or alfalfa and sainfoin fill a place which could not well be supplied in our pastures and meadows, while peas and beans are scarcely of more importance than the peanut which would be missed in our groceries and on the corners of the streets, as well as by the people of Africa and the tropical islands.

The pulse family is the most wonderful of all the families of plants in the enormous number and variety of its useful products. Its wealth is fairly bewildering. It contains barks of great use for tanning, many delicious perfumes, valuable medicines, tough fibers useful for cords, ropes or coarse cloth. It abounds in durable timber and in ornamental and fragrant woods. For gums it beats the world, and supplies also many valuable coloring materials. It is well supplied with ornamental species.

PAPILIONACEÆ. PULSE FAMILY PROPER.

Leaves mostly pinnate or palmate. *Flowers* usually in axillary or terminal racemes, spikes or heads. *Calyx* of 5 sepals, united, often unequally. *Corolla* perigynous, very irregular, of 5 or rarely fewer petals, papilionaceous; upper petal called the *vexillum*, or banner, inclosing the others in the bud; 2 lateral called *alæ* or wings oblique outside and often adhering to the 2 lower, which are usually united, and called *carina*, or the keel. *Stamens* 10, very rarely 5, monadelphous or diadelphous, mostly 9 united and a free one next the banner.

This sub-family, or sub-order includes all the clovers and other leguminous forage plants which are considered in this volume.

TRIFOLIUM, L. TREFOIL, CLOVER.

Herbs, usually low. *Leaves* digitately, rarely pinnately 3-foliolate; stipules adnate to the petiole. *Flowers* capitate or spiked, rarely solitary; red, purple or white, rarely yellow; bracts small or o, sometimes forming a toothed involucre. *Calyx-teeth* 5, subequal. *Petals* persistent; wings longer than the keel, the claws of both adnate to the staminal tube. *Upper stamen* free; all the filaments, or 5 of them, dilated at the tip; anthers uniform. *Style* filiform, stigmas oblique or dorsal; ovules few. *Pod* small, indehiscent, 1–4-seeded, nearly enclosed in the calyx. Found in the north temperate and warm regions, rare in southern; species 150. The above generic description is mainly adapted from *Hooker's Flora of the British Islands.*

Pl. 128.

T. pratense, L. Red Clover, Broad leaved Clover, Common Clover, Early red clover or cl.—More or less pubescent, leaflets oblong, stipules membranous, free portion appressed to the petiole, heads terminal, sessile, globose, at length ovoid, subtented by opposite leaves with much dilated stipules, calyx-teeth slender, setaceous, erect, or spreading in fruit, the lowest longest.

Pastures, roadsides, etc., ascending to 1,900 ft. in the Highlands [of Great Britain]: flowers from May to September. Annual, biennial or perennial. *Stems* 6–24 in., solid or fistular, robust or slender. *Leaflets* ½–2 in., often marked with a white spot or lunate band, finely toothed; stipules often 1–1½ in., with long setaceous points. *Heads* ½–1½ in. diam., pink, purple or dirty white. *Calyx-tube* with a 2-lipped connection in the throat, strongly nerved; teeth not exceeding the petals, very slender, unequal. *Pod* opening by the top falling off. Found in Europe, N. Africa, Siberia, W. Asia to India; introduced in N. America certainly before the Revolution.

Early History.—Although in a general way this is a plant familiar to all farmers, there are many things in regard to its habits, variation and other peculiarities yet to learn. The expression "To live in clover" has become proverbial, and is another way of designating a good living. With a field of clover knee high, or up to the eyes, means fat cattle and swine and bunchy sheep. Some one styles the plant "The red plumed commander-in-chief of the manurial forces."

Red clover was known and prized over 2,000 years ago by the Greeks and Romans, but it can hardly be said to have been cultivated, even in the simplest way, till used in England about 1633,—263 years ago, or 44 years before the cultivation of perennial rye grass, and nearly 100 years before that of any other of the true grasses.

FIG. 128.—*Trifolium pratense, L.* (*Red Clover*), part of a plant and a flower enlarged.—(Sudworth.)

For a long time it was propagated by scattering the seed in the chaff with all the weeds and rubbish, as it accumulated at the stack or barn.

Extent of Roots.—Red clover usually has a large tap root, with numerous branches extending in all directions. Sometimes the tap root is short and soon equaled by its branches. These roots rarely ever extend less than two feet below the surface, as in moist, compact land, or where the surface is very rich. Where the subsoil is at all open and inclined to be dry, it is not unusual for the roots of clover to reach down six feet or more below the surface; however, the main bulk of the roots are usually within a foot of the surface. Various experiments and careful estimates have shown that fully one-half the weight of a clover plant is below the ground in the form of roots.

Concerning the stems and leaves I shall speak more in detail in the paragraph which treats of variations of the plant.

The Flower.—The flower is irregular, papilionaceous and its structure rather difficult to understand without considerable study of specimens or good illustrations. I take pleasure in reproducing the excellent illustrations of Hermann Mühler, as found in his *Fertilization of Flowers.*

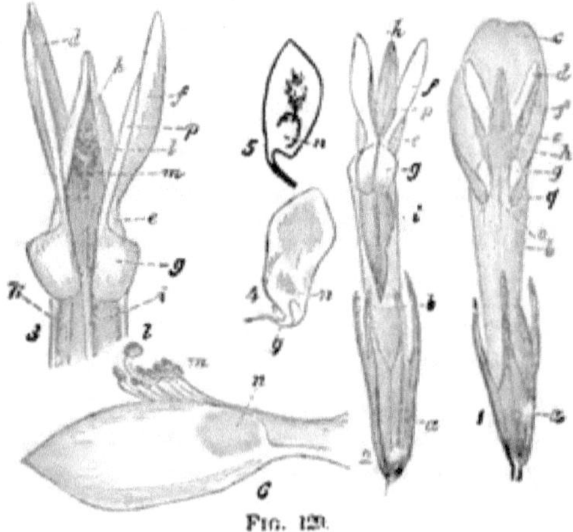

FIG. 129.

The nectar is secreted by small glands at the base, on the inside of the tube formed by the cohesion of thenine inferior filaments, and accumulates around the base of the ovary. In the center is the pistil, the style of which curves upwards, carrying the stigma a little beyond the anthers. The tenth stamen is free, and for most of its length is turned to one side, making it quite easy for the long tongue of a bee to reach the nectar.

In the words of Mül ler, "If now a bee inserts its proboscis beneath the *vexillum*, while it clings with its fore legs on to the alæ (which is coherent with the carina) resting its middle and hind legs on a lower part of the inflorescence, the carina and alæ are drawn downwards, and the stigmas and anthers are thrust up against the under side of the bee's head; the stigma, standing highest, receives the pollen brought by the bee, and instantly afterwards the anthers dust the bee with fresh pollen. Cross-fertilization is thus insured; self-fertilization may take place as the bee draws back its head, but is probably neutralized and superseded by the immediately preceding cross-fertilization.

"In order to reach the honey in this way (to the bottom of the tube) an insect must possess a proboscis at least 9 to 10 m. m. long. The pollen is accessible to all insects which can press down the carina; and such insects, whether they reach the nectar or not, will perform cross-fertilization."

Bumble Bees a Great Help in Fertilizing Red Clover.— The writer, as well as some of his students, has made many experiments which help to prove the truth of the above heading.

FIG. 129. *Trifolium pratense*, L.
1.—Flower, from below.
2.—Ditto, from above, after removing the vexillum.
3.—Anterior part of flower, twice as much enlarged; the edges of the carina have been forced apart.
4.—Right ala, from within. (The claws of 4 and 5 have been broken short off.)
5.—Right half of the carina from without.
6.—The essential organs emerging from the depressed carina.
a, calyx; *b*, tube formed by the coalescence of the nine filaments with the claws of the vexillum, alæ, and carina; *c*, vexillum; *d*, concave part of the inner side of the ala; *e*, lower border of ala, bent outwards; *f*, outward surface of ala; *g*, pouched swelling on the base of the ala; *h*, carina; *i*, style; *k*, superior free stamen; *l*, stigma; *m*, anthers; *n*, point of union between alæ and carina; *o*, point of flexure of the carina; *p*, part of the upper border of the ala, bent outwards; *q*, downward extension of vexillum.—(Mül ler.)

The following single experiment will serve as an example: Two fine bunches of the first crop of clover, apparently alike, were covered with mosquito netting. No insects were seen about either, except those mentioned below. On June 29th a bumble bee was placed inside of one netting and seen to work on the flowers.

On July 10th two more bumble bees were introduced and seen to work, and on July 12th more bees were introduced, and were seen to work on the flowers.

On July 31st 50 ripe heads were selected from each plant, and the seeds carefully shelled and counted. The 50 heads on the plant where the bumble bees were excluded yielded seeds as follows:

40 heads yielded	0
6 heads yielded one seed each	6
1 head yielded	2
1 head yielded	3
1 head yielded	5
1 head yielded	9
Total	25

The 50 heads on the plant where bumble bees were inserted and seen to work under the netting yielded seeds as follows:

25 heads yielded	0
2 heads yielded one each	2
5 heads yielded two each	10
3 heads yielded three each	9
3 heads yielded four each	12
3 heads yielded five each	15
1 head yielded seven	7
1 head yielded eight	8
1 head yielded nine	9
1 head yielded ten	10
1 head yielded twelve	12
Total	94

In the above experiment both lots of heads were covered alike with netting, that no one could say the difference in yield of seed was due to the fact that one lot was covered and the other not covered. It will be seen, that where bees were observed to work on the flowers the yield of seeds was nearly four times that where the bees were kept away. But perhaps the two plants would not have yielded the same number of seeds had they been treated in every way precisely alike.

In reply to this suggestion I can offer the following, which shows that in six examples, selected at random, only one was found in which the yield of seeds was nearly twice the number in the heads containing the fewest seeds. On September 13th, 1882, I selected of the second crop of red clover five plants within ten feet of each other, which seemed to be much alike. They had not been covered in any way. The seeds from 50 good heads of each plant were shelled out with the following results: 1, 260; 1, 275; 1, 460; 1, 485; 1, 1,820. It will be seen that 50 heads from plant number five contained only about one-third more seeds than 50 heads from plant number one.

In another place, 50 heads selected from one plant yielded 2,290 seeds, nearly twice as many as plant number one in the first lot.

Mr. C. Darwin covered one hundred flower-heads of red clover by a net and not a single seed was produced, while 100 heads growing outside yielded by careful estimate 2,720 seeds. He says: "It is at least certain that bumble bees are the chief fertilizers of the common red clover."

It may not be out of place to say here that experiments with white clover show that visits of honey bees increase the yield of seeds enormously. In one case

8 protected heads yielded	5 seeds
8 visited by bees yielded	236 seeds

This is an increase of over 47 fold in favor of the bees.

A large number of carefully conducted experiments made by

many persons on a great variety of plants show results quite as remarkable as those above cited.

Here the bees and bumble bees not only make use of a waste product, but help the plants as well. Most botanists now believe that odor and showy flowers are advertisements for attracting insects, and that nectar and surplus pollen are the wages to compensate insects for services rendered in fertilization.

If this be the case should not the farmer seek to encourage meadow mice, which make the nests sought by bumble bees in which to rear their young. The bumble bees, at least, should be encouraged. It is not improbable that the time may come when queen bumble bees will be reared, bought and sold for their benefit to the crop of clover seed.

The Sleep of Leaves.—This can in no way be compared with the sleep of animals, but refers to the fact that the leaves of clovers take different positions at night from those assumed during the day time.

FIG. 130.—*Trifolium repens*; *a*, leaf during the day; *b*, leaf asleep at night.—(Darwin.)

This difference in position is caused by turgescence in the *pulvinus*, which is the name given to a mass of small cells of a pale color found in a certain portion of the leaf stalk.

Experiments show that leaves kept open or spread apart contain more dew in the morning, and hence become cooler than those which approach each other. The leaves crowd together, or "sleep," for the same purpose that pigs crowd together in cold weather, viz: to keep warm. It has been found that the leaves which sleep do not remain quiet during the night, but continue, without exception, to move during the whole twenty-four hours. All non-sleeping leaves are also in incessant motion, circumnutating. The sleep of plants is a mere modified form of this universal circumnutation.

During a warm, dry day leaves also assume the sleeping position, which aids in checking evaporation.

There are more "sleeping" plants among the Leguminosæ than are found in all other families put together.

A Little Agricultural Chemistry.—Of the thirteen elements necessary for plant growth the farmer usually need take but little care, except in the supply of potassium, phosphorus and nitrogen, and of these three nitrogen is the most precious and costly to obtain. All agricultural plants draw much of their food from the atmosphere, and of those used by the farmer probably none are much, if any, exceeded by clover in the large proportion of nutriment thus derived. In this respect other leguminous crops are much like red clover.

"Clover seed is the best manure that a farmer can use." Clover has been called "a trap for nitrogen," as it collects and presents large quantities of combined nitrogen in a form ready to nourish growing crops.

In the words of Dr. Kedzie: "With an adequate supply of combined nitrogen all the other chemicals of agriculture become active, while a limited supply of active nitrogen correspondingly limits the action of the rest. For high farming, or the raising of exceptionally large crops, the great want is an abundant and cheap supply of ammonia and the nitrates.

"An acre of good clover will make 5,000 pounds of hay, containing 282½ pounds of mineral matter or ash. In this ash will be 97½ pounds of potash, 96 pounds of lime, 34½ pounds of magnesia, and 28 pounds of phosphoric acid. The hay will also contain 108 pounds of combined nitrogen."

The roots and stubble contain fully as much of these elements as the hay.

Sir J. B. Lawes found that in autumn, after the last crop of clover was cut, that remaining above ground, and to the depth of 72 inches was examined:

Stubble, etc., above ground contained........	2,669 pounds per acre, dry.
1st nine inches contained....................	3,017 pounds per acre, dry.
2d nine inches contained.....................	275 pounds per acre, dry.
3d nine inches contained.....................	191 pounds per acre, dry.
Total..	6,152

This was between three and four times as much dry matter as the residue of the barley.

In the words of Dr. Kedzie: "The clover hay *or sod* contains enough phosphoric acid for more than double an average crop, enough nitrogen for more than four average crops, and potash for more than six average crops of wheat! If any person were preaching the gospel of agriculture he well might hold up the triple leaf of the red clover as the symbol of trinity of blessings to the farmer, furnishing for his cereal crops, from otherwise inadequate sources, a sufficient supply of potassium, phosphorus and nitrogen. If I were designing an emblematic seal of our national agriculture I would make the central figure the clover leaf. For the farmer it is the most effective trap for nitrogen within his reach."

The late George Geddes, of New York, said: "It has been demonstrated beyond a doubt that clover and plaster are by far the cheapest manure that can be had for our lands,—so much cheaper than barnyard manure that the mere loading of and spreading costs more than the plaster and clover. Plow under the clover on the more distant fields when it is at full growth.

"A very considerable part of the cultivated land of Onondaga County has never had any other manuring than this clover and gypsum, and its fertility is not diminishing. The cost per acre is $2.32."

The Uses and Value.—The following as to the use and management of red clover is gleaned from *Harris' Talks on Manures:* "Clover is, unquestionably, the great renovating crop of American agriculture. A crop of clover, equal to two tons of hay, when plowed under, will furnish more ammonia to the soil than twenty tons of straw-made manure, fresh and wet, or twelve tons of ordinary barnyard manure.

"I prefer to make the clover into hay and feed the animals, as they seldom take out more than from five to ten per cent. of all the nitrogen furnished in the food,—and less still of mineral matter. If you plow it under you are sure of it. There is no loss. In feeding it out you may lose more or less from leaching and injurious fermentation. As things *are* on many farms, it is perhaps best to plow under the clover for manure at once. As things *ought* to be it is a most wasteful practice. Clover is good for wheat; plaster is good for clover. The roots run deep, drawing large amounts of water, and can live on very weak food. The clover takes up this food and concentrates it. The clover does not create the plant food; it merely saves it. To improve sandy land, instead of plowing the clover under or feeding it off, mow the crop just as it commences to blossom and let the clover lie. There would be no loss of fertilizing by evaporation, and the clover hay acts as a mulch. Mow the second crop about the first week in August."

The following computation of the *relative* money value of one ton of various foods for producing manure is from the experiments of Mr. Lawes:

Cotton seed meal	$27 86
Linseed cake	19 72
Beans	17 73
Wheat bran	14 59
Clover hay	9 64
Indian meal	6 65
Meadow hay	6 43
Oat straw	2 90
Potatoes	1 50
Turnips	86

All agricultural plants draw most of their food, directly or indirectly, from the atmosphere, and of those used none are exceeded by clover in the large proportion of nutriment thus derived.

If the stubble and roots contain more than half of the manurial value of red clover, and if live stock only appropriate from five to ten per cent. of the nitrogen, and the other 90 to 95 per cent. goes back to the field or dung heap, it certainly must be the best practice, as a rule. to feed red clover instead of plowing it all under.

I have not seen a more concise and valuable summary of this matter than the one by the late Dr. Voelcker, as found in the Journal of the Royal Agricultural Society of England for 1868:

1. "A good crop of clover removes from the soil more potash, phosphoric acid, lime, and other mineral matters, which enter into the composition of the ashes of our cultivated crops, than any other crop usually grown in this country."

2. "There is fully three times as much nitrogen in a crop of clover as in the average produce of the grain and straw of wheat per acre."

3. "Notwithstanding the large amount of nitrogenous matter of ash constituents of plants in the produce of an acre, clover is an excellent preparatory crop for wheat."

4. "During the growth of clover a large amount of nitrogenous matter accumulates in the soil."

5. "This accumulation, which is greatest in the surface soil, is due to decaying leaves dropped during the growth of clover, and to an abundance of roots, containing, when dry, from 1¾ to 2 per cent. of nitrogen."

6. "The clover roots are stronger and more numerous, and more leaves fall on the ground, when clover is grown for seed, than when it is mown for hay; in consequence more nitrogen is left after clover seed than after hay, which accounts for wheat yielding a better crop after clover seed than after hay."

7. "The development of roots being checked when the produce, in a green condition, is fed off by sheep, in all probability leaves still less nitrogenous matter in the soil than when clover is allowed to get riper and is mown for hay; thus, no doubt, accounting for the observation made by pastoral men that, notwithstanding the return of the produce in the sheep excrements, wheat is generally stronger and yields better after clover mown for hay than when the clover is fed off green by sheep."

8. "The nitrogenous matter in the clover-remains, on their gradual decay, are finally transformed into nitrates, thus affording a continuous source of food, on which cereal crops especially delight to grow."

9. "There is strong presumptive evidence that the nitrogen which exists in the shape of ammonia and nitric acid, and descends in these combinations with the rain which falls on the ground, satisfies, under ordinary circumstances, the requirements of the clover crop. This crop causes a large accumulation of nitrogenous matters, which are gradually changed in the soil into nitrates. The atmosphere thus furnishes nitrogenous food to the succeeding wheat indirectly, and, so to say, gratis."

10. "Clover not only provides abundance of nitrogenous food, but delivers this food in a really available power (as nitrates) more gradually and continually, and with more certainty of a good result, than such food can be applied to the land in the shape of nitrogenous spring top dressing."

The above conclusions should be posted up and read daily by every farmer till they are indelibly fixed in his mind.

Owing to the great depth to which the roots penetrate the soil, —frequently six feet or more,—they help to bring up a run-down farm; they bring the valuable ingredients from a great depth and store a large part of them in the large roots near the surface, where they are available for future plant growth.

Red Clover in Many Lands.—Red clover is well adapted to many portions of the temperate regions of the earth. It likes best a soil of clay loam, rich in lime, but will thrive better than Timothy and most other true grasses where the land is sandy or gravelly. On good grass land it is usually the custom to sow Timothy with red clover, although it blossoms some three weeks later. Many prefer to sow orchard grass with clover, as they flower and are ready to cut at the same time. Timothy is well adapted to sow with the large, late, or mammoth clover.

Red clover is not only a general favorite in the United States from Maine and New Jersey to Iowa and Illinois, but is very valuable further West and South.

For Kansas, Professor Shelton reports that it deserves a prominent place in the list of forage plants. In some very dry seasons it fails almost entirely, but during the favorable seasons it flourishes abundantly and yields more—both of hay and pasture—than is generally obtained in the East. When land is once seeded it never runs out, as is the case in the Eastern States, but thickens and spreads continually by self-seeding. We believe that nowhere are such large crops of clover seed grown as in Kansas.

In Mississippi, Professor Phares says, red clover grows most luxuriantly on all their lands with tenacious red or yellow clay subsoil, even though the soil be thin; and once set, it remains as long as the farmer desires, provided he does not mow more than twice each year, nor graze too heavily.

In Georgia, the late C. W. Howard says: "This is the most valuable herbaceous plant to the Southern farmer. It bears grazing admirably, makes excellent hay, and in large quantity, and thrives on land of moderate fertility The doubts as to whether red clover would succeed at the South have been dispelled. At the South it lasts for several years."

Red clover is valuable to enrich the land and hence to enrich the owner; it is not excelled by any forage crop as a wholesome summer pasture for swine, and some have spoken very highly of its use in winter when fed to swine in the form of hay.

For soiling, a good growth of red clover is very valuable, and it has often been packed into the silo to feed as ensilage in the winter.

Clover as a Weed-Exterminator.—We have ample testimony from a great variety of sources that red clover, with a little gypsum and perhaps a top dressing of some other fertilizer, is excellent to smother and kill out our worst weeds.

The following was furnished by special request by J. S. Woodward, now one of the editors of the *Rural New Yorker*: "Canada thistles have long roots which store up nourishment during the latter part of summer and fall to feed the spring growth. I kill the thistles without the loss of a crop as follows: Have the land rich, if possible, at least have it well seeded to clover and by top dressing with plaster, ashes, or by some means get as good growth to the clover as possible. As soon as the clover is in full bloom, and here and there a thistle shows a blossom, mow and make the crop, thistles and all, into hay. After mowing, apply a little plaster to quickly start the growth of clover. You will find this to come much quicker than the thistles. As soon as the clover has a good start, from July 20th to August 5th, plow down, being careful to plow all the land and to fully cover all growth. Then roll and harrow at once, so as to cover every thistle. But few thistles will ever show themselves after this, and they will look pale and weak. When they do show, cultivate thoroughly with a cultivator having broad, sharp teeth, so as to cut every one off under the ground. In two days go over with a sharp hoe and cut off any that may have escaped the cultivator. Watch the thistles, and keep using the hoe and cultivator until freezing weather. You will see them getting scarcer and scarcer each time and looking as though they had the consumption. By plowing this field just before freezing up you will have the land in the finest condition for a spring crop. This plan not only kills thistles but ox-eye daisies and other weeds. It is much better than a summer-fallow, and without the loss of any crop."

Putting in the Seed.—Too little care is exercised in selecting the seed, as most of it contains more or less seeds of pernicious weeds, and especially does this caution become more and more necessary as the country becomes older. The troublesome weeds of a farm can generally be directly traced to foul seeds sown with grasses and clovers for the meadows and pastures.

In the northern portion of the United States numerous experiments seem to clearly indicate that it is best to sow seeds of red clover in spring. In some sections it is sown even before freezing ceases, but many now practice sowing just in time for the young plants to begin growth with the first early vegetation. If sown in autumn, especially if late, the young plants are very likely too feeble to survive the winter. If sown in autumn the date should be early enough to give plants a good start. In the warmer portions of our Union clover is often sown in autumn, or even in winter.

Clover seed is most generally sown where wheat and some Timothy were sown the autumn previous, though it is not unfrequently sown in spring, with a thin seeding of oats or barley.

It is a common practice with our best farmers to harrow the ground very lightly before sowing the clover seed. This benefits the wheat as well as favors the growth of the clover.

Where no grass seeds have been sown, at the West, the farmer sows 6, 8 or 10 or even 12 pounds of clover seed to the acre, but at the East 25 or 30 pounds is not thought too much.

In Great Britain, which possesses a moist climate favorable to the development of grasses and clovers, it is the practice to sow much more seed than is usually sown in the United States.

There are 18,000 clover seeds to the ounce, or 288,000 to the pound. In ten pounds there are 2,888,000 seeds. In England farmers often sow seeds of grasses and clovers enough, if all grew, to produce 16,878,000 to 27,000,000 plants, which is ten to fifteen times the amount of seed thought sufficient by our western farmers.

The Englishman seeks to get large numbers of fine, small stems instead of fewer large, coarse ones.

In various portions of our country, isolated farmers have sown clover in the spring on well prepared land without the presence of another crop, and they get a crop of grass or clover the first year. This practice deserves more thought from the average farmer.

For further remarks on this last idea consult a former paragraph on seeding without a crop.

Care of the Young Clover.—It has often been shown, beyond question, that the young plants will be more certain to live and will grow faster and become stouter, if not sown with a grain crop. If the wheat is thick and large the clover is apt to suffer, if the wheat is thin and light clover is likely to become large and crowd it.

It must not be forgotten that young clover is most generally greatly benefited by even a very light dusting with gypsum, say one-fourth to one-half or even a bushel to the acre.

Sheep and swine must not be allowed to feed young clover, at least very long, because it may be much damaged, or even killed. Clover needs a little time to get its roots well established, and this cannot be done without the aid of green tops.

Clover fails "to catch" for a great variety of reasons. The soil may be very much "run down," or the seed is poor, sown too late, the ground is too rough, not harrowed nor rolled; the oats or wheat get the start and choke it out or enfeeble the plants; the weather in spring is too dry, too hot; the young plants are fed too closely. The frosts of spring may kill the young plants.

Winter Killing and Remedies.—Red clover not unfrequently "winter kills" or "heaves out," and the dead plants in spring stick up out of the ground several inches, especially in winter, when there has been little snow on the ground and frequent alternations of freezing and thawing. To prevent winter killing see that the plants are well established in autumn and that they are not fed off too closely. Thorough tile drainage is a great benefit. A moderate amount of tops left on the ground will often be of some assistance, or a very thin mulch of straw put on after the ground has first become well frozen. A mulching of straw early in autumn has sometimes done more harm than good. No attempt, at the North, should be made to save red clover over to the third year, as such efforts are not successful.

As spring approaches and the soil warms up it is rather discouraging to find the clover killed out. The proprietor often plows up the ground and puts in another crop, thus leaving the land in a still worse condition for the next seeding to clover. He very likely raises millet or Indian corn or rye or buys of his neighbors a supply of winter feed. In case of partial winter killing the writer cannot help thinking that too little attention has been given "to patching up" such meadows in spring. By this is meant to harrow, re-seed, and, if possible, top dress with some sort of manure.

The Best Time for Cutting Clover for Hay.—The following is from the pen of Prof. H. P. Armsby: "What has been shown to be true of meadow hay in this respect applies also to clover. The earlier it is cut the more concentrated and digestible the fodder, while as it grows older the crude fibre increases and it becomes coarse and less easily digestible. In regard to the best times for cutting clover the same rules apply as those given for cutting grass. In regard to the advantages of early and frequent cuttings, the experiments do not all give such striking results as those on grass."

For further notes in regard to securing clover hay the reader is referred to a former chapter of this work.

Saving Clover Seed.—The proper time to cut for seed is a difficult one to state, especially as the heads ripen unevenly. These heads should be examined, for sometimes the earliest contain most seeds, and sometimes the main bulk of the seed is found in heads which mature later in the season.

Some persons have observed that clover, when cut rather early, from the 5th to the 15th of June at the North, is more certain to seed well than that cut later. In some cases they report double the amount of seed from the clover which was cut early. Considering its high price, if there is any prospect of greatly increasing the yield of seeds more experiments are much needed. Some were suggested in the paragraph which treats of the agency of bumble bees in fertilizing the flowers.

In England Dr. A. Voelcker tried some different sort of manures for this purpose with results by no means satisfactory or conclusive. Probably the efforts were made in the wrong direction, as indicated in the preceding paragraphs.

For securing the seed, red clover is ordinarily cut with a reaper which delivers the clover in small gavels. In this way the clover is moved to one side and is not damaged by the tramping of the horses.

The clover is allowed to lie until it is well dried, and probably black and brittle. It may need turning once or more before dry and ready to thresh or draw to the stack or the barn. Clover seed during the harvesting will stand a good deal of abuse and not lose its vitality.

During a very unfavorable season for curing, when there was much rain, the writer tested samples from about sixty different farms in Michigan, and found they averaged 85 per cent. of good seeds, rarely going as low as 75 per cent., though one small lot went down to 25 per cent. Some went up to 95 per cent.

Clover is usually threshed and cleaned with a machine made for the purpose. The yield runs from less than a bushel to the acre to two bushels, a fair yield, four bushels, a fine yield, or even six bushels, an exceptionally good yield.

Relative Value of Dark and Light Colored Seeds.—Dark colored, bright looking seeds are generally considered the best. The results obtained on testing numerous samples on different seasons indicate that there is no difference in favor of the dark seeds either in vitality or the quality of the plants which they produce. It is generally the case that all the seeds, or nearly all, from one plant resemble each other in color and size. Some plants produce yellow seeds, others produce dark ones, others produce seeds of mixed colors.

Variation of Red Clover.—The late Professor James Buckman, of England, in Jour. Royal Agrl. Soc., p. 446, 1866, says the American red clover is a much larger and coarser and more hairy plant than that cultivated in England, doubtless due to a longer and warmer summer.

The wild clover, as early introduced into Europe, is usually the small hairy plant that we meet with (in England) and greatly different from that described by Sinclair, which is larger and quite smooth. The Professor goes on to say that: "Both when wild and when cultivated it is perhaps as protean in form as any plant the farmer has to deal with. Some are more perennial than others; all are more or less hardy, more or less productive, and these differences have a high significance. However, it seldom happens that any particular type can be obtained pure, though the value of the seed varies just in proportion as it is so.

"There are three desiderata with regard to clover.

"1st. A good sort or sorts.

"2d. Pure seed of the sort.

"3d. Seed from a known and suitable climate."

The Professor then describes six of the leading varieties, none of which are just like those I find in Michigan.

Not long ago our seeds of red clover came from Europe, and already we have a great change in the plants.

I have for some years past studied quite carefully in different stages of growth, at different seasons and on different soils, many hundreds of plants. I have preserved some of the plants and seeds of a few of the most striking.

There is nearly or quite a month's difference in the time of first flowering. Some plants stool out and send up many stalks; others few. On hot, dry days some plants wilt while others show no signs of wilting. Some plants are tall and large or slender; others are short, even where the soil seems to be uniform. Some are erect, even where there is nothing to crowd them; others spread out at once, even where somewhat crowded. The leaves and stems of some plants are densely pubescent; others are nearly smooth, and between these are all gradations. In this respect the same plant varies a little at different seasons. The stems vary much in length and number of branches and in the color. On some plants the leaves are dark green: on others light green. The leaflets often contain a light spot, which varies in shape, size and intensity. Some are destitute of any trace of spots. Some leaves are firm, and a quarter or more thicker than others; some are thin and flabby. Some leaflets are as broad as long; others are elliptical—lanceolate. The stipules vary in shape, color and position taken.

The heads of flowers vary in size and shape, and so far as seen were sessile, with an involucre of two leaves. The calyx tube and the lobes of the calyx vary in size and hairiness.

The petals vary in length, direction taken, and differ in color from dirty white to pink and bright scarlet. Varying with the season, and probably with the plant, the pistils contain each from none to two, three or even four seeds. There is a marked difference, as before observed, in the color of the seeds.

Of some plants observed I give the following brief description:

No. 1. Early, stems purplish, few and small, erect, quite hairy, leaflets spotted, rather narrow, leaflets of the involucre lance-elliptical.

No. 8. Late, stems few, stout, sprawling, quite smooth, purplish, leaflets rather narrow, with scarcely a trace of a spot.

No. 17. Very late, stems long, of medium size, spreading, green, quite hairy; leaves light green, spot inconspicuous.

No. 19. A seedling of dark seed, early, stems numerous, large, tall, erect, smooth, purplish, leaflets rather broad, thick, very dark green, with no trace of a spot; flowers dark colored.

The Model Plant.—I have begun a few experiments in a very small way by selecting and raising different races of red clover. This variation in our fields is a broad hint at the results which may be obtained by care and study.

For the Northern States we need a red clover which starts early, grows rapidly, has numerous erect, rather stout stems, which are not large. If too woody, the stems make coarse fodder; if they contain too little woody matter, they will not be stiff enough to stand up well. The plant should be rather hairy, as such plants usually endure hot, dry weather best. The model plant should seed freely, and to aid in this, if possible, the tube of the flower should be short enough to permit honey bees to reach the nectar.

The tongue of a honey bee when stretched out is six to seven millimeters in length, while the tube of the corolla of red clover is nine to ten millimeters. It seems by this that there is a wide breech to be gained in growth of tongue or shrinkage of corolla before the honey bee can sip all the nectar from the bottom of the tube of red clover. The tongue must elongate one-third or the tube of the flower shorten as much. The occasional visits of honey bees to the flowers of red clover may be accounted for by supposing they seek pollen, or they seek the honey which has filled a considerable portion of the floral tube. The upper portion of this honey can be reached even with the tongue of the ordinary honey bee.

Clover Sickness.—This is a term used in Great Britain to indicate a failure of the plants to thrive after they have once started. Many observations and experiments have been made and much has been written on the topic in regard to the cause and remedies. Except in a very few places in the older portions of the United States, and even these are of questionable authority, no trouble of this nature has appeared on this side of the Atlantic.

Recent investigations by Kutzleb show that clover sickness is not due to parasites, to lack of nitrogen, to lack of water, or to unfavorable physical properties of the soil, but to a deficiency of easily soluble potash, especially in the subsoil. (H. P. Armsby in Science, p. 146, 1883.)

It is not improbable as our country grows older that repeated crops of clover may so deprive the subsoil of potash that clover sickness may become common. One who suspects the presence of this trouble should look carefully for insects or some fungus before coming to a conclusion.

To my inquiry in reference to the presence of clover sickness in the State of New York, Professor Roberts replied through the *Philadelphia Press* as follows: "So far nothing like what is known in Europe as 'clover sickness' is present. The clover leaf beetle, *Phytonomus punctatus*, has injured a few fields seriously, but its ravages have been confined to very small areas, sometimes to a single acre or two in a township. The clover seed midge, *Cecidomyia leguminicola*, which prevents the clover from blossoming and destroys the seed, is found in most, if not all, of the counties of western New York. The hay crop is injured by them to only a slight extent.

"The clover root borer, *Cecidomyia trifolii*, plays terrible havoc with the clover the second year. Much has been written on this subject, yet few appear to realize that their failures, after the clover has been well established, come from the injury done by the root borer. If this beetle remains, the four-years' course must come into general practice."

Hoven.—This is a term applied to cattle which have become sick and bloated after eating too heartily of clover which was fresh and wet. At such times, till the cattle have become used to the feed so as not to be greedy, they should be turned off the clover after eating for an hour or so at a time.

TRIFOLIUM MEDIUM, L. MAMMOTH, GIANT, PEA-VINE CLOVER, OR COW GRASS (OF ENGLAND).

The following description of the typical form, as it appears in England, is mainly from *Hooker's Flora:*

Plant slightly hairy, leaflets oblong, obtuse, or acute; stipules herbaceous, free portion spreading, heads subglobose, terminal, often shortly peduncled, subtended by opposite leaves, *calyx-teeth* setaceous, spreading in fruit, lowest a little longest. June to September, perennial. *Stems* straggling, flexuous, often zigzag. *Leaflets* 1–2 in., rather rigid, almost quite entire, ciliate. *Heads* 1–1½ in. diam. *Flowers* ¾ in., rose-purple. *Calyx-throat* with a ring of hairs, tube 10-nerved, glabrous, teeth reaching half way up the petals. Pod often dehiscent longitudinally. Distributed in Europe, Siberia, Western Asia; introduced in North America.

This clover is *Trifolium medium*, and so named a long time ago by Linnæus. The common name might with propriety, be "medium red clover." I mention this fact because farmers have lately got in the notion of calling the early red clover "medium" clover.

Mammoth clover is quite similar in appearance to the early red clover, but it flowers later, with Timothy, is very often a peren-

FIG. 131.—*Trifolium medium*, L. (Mammoth Clover,) part of a plant and a lower leaf —(Sudworth.)

nial, and is adapted for permanent pasture; the stems are larger, more inclined to spread, the leaflets are narrower and often destitute of a light spot, the flowers are bright red and larger than in *Trifolium pratense*, and form a less compact head.

The samples found at the Agricultural College, and in many other places, show all grades of intermediate forms. These two species seem to be freely hybridized.

At my request, my friend, A. C. Glidden, of Paw Paw, Mich., has made numerous inquiries in reference to its value in his portion of the State, where it has been largely grown.

L. B. Lawrence, of Cass County, who owns a large prairie farm, has grown it for many years. He considers it less hardy than the other species; it is more liable to "heave" in the spring, and often grows so rank as to kill itself by the burden of stalk on the surface. He thinks the roots are smaller and that it feeds on the surface, and does not work in the subsoil like the other species. He formerly pastured this clover till the first of June, when he allowed it to flower and seed, which would often come off early enough to plow for wheat. Recently he has run over the field with a mower, clipping the tops about the first of June, and allowing them to remain as a mulch, while the new growth forms seed.

D. Woodman, of Paw Paw, once sowed a field in equal divisions of the two kinds, and the mammoth clover furnished double the amount of feed for pasture, as compared with the other half of the field. The season was a dry one. Others report that it is better than the early kind for pastures in July and August.

The notion prevails that the mammoth clover does not make as good hay as the other species; it is often coarse and woody. Another point should not be overlooked. They all agree that the mammoth clover is much the most productive of seeds. As this is the case, we may expect it will soon become more common than it is at present. Farmers will select the large kind to raise seeds to sell, and many times this seed will finally be purchased by farmers and sown, supposing it to be the early or round leaved red clover.

In managing this plant, it should be understood that if left without pasturing or mowing in spring there will only be a small crop of seed.

TRIFOLIUM HYBRIDUM, L. ALSIKE CLOVER.

Plant glabrous, perennial. *Stems*, branching, 1–2 ft. high, ascending, weak. *Petioles* long; leaflets obovate or oblong, toothed. *Stipules* rather long, nerves green. *Heads* about ¾ in. diam., globular, flowers pinkish, pedicellate, recurved after flowering; peduncles 2–4 in. *Calyx* white, teeth green; pod same as in white clover. Found in Europe, North Africa, West Asia; introduced into N. America. Its common name is derived from a parish in Sweden.

In appearance it is so nearly intermediate between red and white clover that Linnaeus supposed it was a hybrid, and hence its specific name. It is not a hybrid. Alsike likes rather moist land, containing some clay. It is smoother and more delicate than red clover, and the stems are weaker, so much so that it is quite likely to lodge. The stems remain green after seeding. It does not stand dry weather well, is apt to winter kill, the flowers continue for a long time and abound in nectar, which can be reached by honey bees.

Alsike clover has a good reputation for pasture and is a favorite with bee-keepers. It frequently yields 3–8 bushels of seed to the acre, and these are only half the size of those of red clover, hence only about half as much seed is sown to the acre. This is produced from the first crop, though it is often pastured a while early in the season. It is two or three years coming to full size, and does best for pasture when sown with some stout grasses. The aftermath is very light.

When ripe it shells more easily than red clover, and is more apt to waste, hence more care is needed in the harvesting.

Trifolium repens, L. White or Dutch Clover.—A smooth perennial; stems creeping and rooting at the joints. *Stipules* small, narrow, acuminate; *petioles* 2–4 in., *leaflets* obovate or obcordate, obcurely toothed, often with a light mark towards the base. *Heads*, or close umbels, 1 in. diam.; peduncles 3–8 in. *Flowers* white or rosy, pedicels reflexed after flowering. *Pod* 4–6 seeded. In pastures of Europe, Russian Asia. Africa, India, N. America, at the North. This is the Shamrock of the modern Irish.

The following, from Wm. Gorrie, gives a fair notion of its estimate among the farmers of England: "It has long been almost universally sown for pastures, but many consider its merits highly over-rated; for although it makes a great display on favorite soils, yet it is neither fattening nor cared for by stock when they have a sufficient choice of pasturage. No attempt has been made to secure improved varieties."

Below follows the opinion of Dr. S. A. Knapp, of Iowa, who says: "It flourishes when the true grasses wither; it appears to defy equally poverty of soil, cold, excessive moisture or extreme drought. It is perennial, which gives it a great advantage over red clover, and renders it an almost necessary substitute where close grazing is practiced. It is extremely hardy, and turns its sprightly green leaves to the lingering snows of spring and stoutly resists the sharp frosts of approaching winter. It resists drought with true clover stubbornness, and thrives in the slough or upon the knoll with almost equal vigor. It furnishes a large amount of highly nutritive material. It has more protein and more fat than red clover. In flesh-forming material it is nearly 20 per cent. richer than blue grass. The product is about eight tons of

FIG. 132. *Trifolium hybridum, L.* (Alsike Clover,) *a*, part of a plant; *c*, a flower enlarged.—(Sudworth.)

FIG. 132.

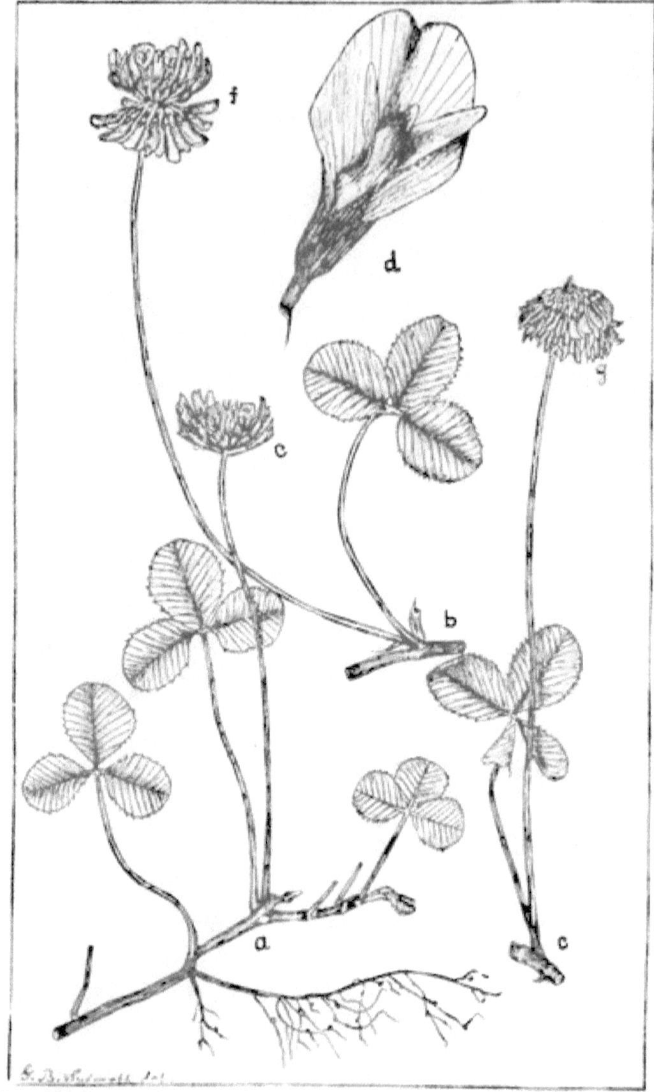

FIG. 133.—*Trifolium repens*, L. (White or Dutch Clover.) *a*, part of plant with *c*, young head; *f*, older head where part of the flowers have turned down; *g*, old head where all the flowers have turned down.—(Sudworth.)

green fodder to the acre upon rich prairie soil. The flower is excellent during most of the season, and the cattle eat it with avidity, except during the months of July and a portion of August. Almost the sole objection urged to white clover is its effect on horses during the maturing of the seed." It makes them "slobber."

Its dwarf character makes it unfit for the scythe.

If the soil is suitable it spreads so rapidly that very little seed is necessary.

White clover is a fickle plant, coming and going with the varying seasons. It often burns out in hot weather. An old hard road, once abandoned, is likely to send up white clover in advance of the grasses.

It is a well known and highly prized bee plant, although the season is often a short one, especially if hot, dry weather comes on early.

White clover is often sown with some of the finer grasses for lawns.

Trifolium incarnatum, L. Crimson or Italian Clover, French Clover.—A soft, erect, hairy annual 1–2 ft. high. *Stipules* broad, with short, broad leafy tips; *leaflets* broad, obovate, or nearly round. Heads 1–2 in., oblong or cylindrical. *Flowers* ½ in. *Calyx* soft, hairy, teeth narrow, nearly equal. *Petals* bright crimson or scarlet or a pale cream color. Found in southern Europe, and cultivated in France, Germany, Belgium.

When in flower this is a beautiful plant. As it is an annual belonging to a warm climate, it does not seem so popular at the North as red clover.

One writer, a farmer in Virginia, speaks highly of crimson clover to sow in autumn alone, or with Italian rye-grass, for cutting the next May. He says it is very productive, and is an excellent clover for one crop, or rather for one mowing, which should be taken early, as it becomes coarse and woody if allowed to mature.

After repeated trials on a small scale the writer thinks it of little or no value for Michigan. Prof. Gulley is of the same opinion in reference to Mississippi.

More recently, this clover has met with much favor in New Jersey, Delaware, Maryland and regions with like climate. It is sown in the fall and mowed or plowed under in May.

MEDICAGO, L. MEDICK.

Herbs with pinnately 3-foliolate leaves; leaflets usually toothed; stipules adhering to the peliote. *Flowers* small, in short spikes, or loose heads, violet or yellow. *Calyx*-teeth 5, nearly equal, keel obtuse, shorter than the wings. *Stamens* diadelphous, the upper one free; anthers uniform. *Pod* small, with few seeds, very much curved, or spirally twisted, indehiscent, often spiny.

Found in Europe, W. Asia, N. Africa, introduced into N. America.

M. sativa, L. Lucerne, Alfalfa, Purple Medick, Chilian Clover, French Clover, Spanish Trefoil.—An upright, deeply rooting, smooth perennial, 1–2½ ft. high. *Leaflets* obovate-oblong, toothed, tip notched. *Flowers* in a short dense raceme, blue or purple; peduncles longer than the leaves. *Pod* ¼ in. diam., spirally twisted. Origin not certainly known; now cultivated in Southern Europe and America.

The common French name is *Lucerne;* the Spanish name for the same species is *Alfalfa,* a name which followed the plant into South America and thence to Mexico and California and the dry countries this side.

It was known and prized by the Greeks and Romans 2,500 years ago, and was spoken of by Columella as the most valuable plant for fodder.

To begin with, there are a few things which the inquirer should not fail to keep constantly in mind. Lucerne is "a child of the sun;" likes a rich loam or sand with a deep porous sub-

FIG. 134.—*Medicago sativa*, L. (Lucerne, Alfalfa.) *a*, part of the top of a plant; *b*, flower enlarged; *c*, young pods.—(Sudworth.)

soil; utterly refuses to thrive on a compact clay subsoil, or in a hard bottom of any kind: while young it is a weak plant and a poor fighter; requires two or three years to become well "rooted" and established: it should be sown after settled weather has come in the spring, without another crop, on well prepared land. Sow in drills about eight inches apart, and hoe or cultivate once or more to keep the weeds and other plants in check. It is not often well worth while to use Alfalfa where the land is to be plowed up every three to five years. This plant is a perennial, and on suitable soil can be relied on to produce good crops for many years in succession. It stands dry weather admirably; is very nutritious: like other legumes, it is a collector of nitrogen. It must be mown when young and just beginning to flower, for the stems quickly become woody and rapidly deteriorate in value. This is a favorite for irrigated meadows and soiling, and is frequently cut three to eight times in the year, yielding enormous crops of valuable fodder for all kinds of live stock except in isolated places. Alfalfa or Lucerne is not a favorite north of Kentucky. Perhaps it is because clovers and the grasses thrive so well, and these can be sown broadcast and are often started with another crop. Again, the farmer looks with distrust on a plant which is so slow starting and needs weeding to keep it growing. Alfalfa endures extreme dry weather much better than the true clovers and grasses.

This is easily accounted for, when we understand that the roots become woody, as large as a pipe stem to half an inch or more in diameter, and have been known to extend ten or twelve, or even twenty feet below the surface.

An old, thick field of Lucerne is very difficult to turn over with the plow.

Those who have tried imported seed of Lucerne with seed of Alfalfa from California claim that plants of the latter will not endure the cold as well, but will stand heat and drought better.

Fifteen to twenty pounds, and even more, are usually sown to the acre.

Honey bees seem to extract the honey without any trouble.

Dr. H. P. Armsby states that "Lucerne is even richer in protein than red clover, but it is inclined to a more rapid formation of woody fiber after the flowers appear. It demands early cutting even more than clover. On account of its excess of protein it should be fed in connection with some feeding-stuff poor in protein, such as roots or straw, to realize the best effect."

Mr. Gorrie, of England, reports, that when properly managed the quantity of cattle which can be kept in good condition on an acre of Lucerne, during the whole season, exceeds belief. It is no sooner mown than it pushes out fresh shoots.

Prof. J. R. Page, of Virginia, considers it one of the most certain as well as one of the best crops the farmer can cultivate for soiling purposes. He finds no difficulty in getting a good stand and a profitable return, and recommends it very highly. It is cured in the same way as clover.

At the Agricultural College in Central Michigan, Lucerne, when hoed and properly started for the first year has not killed out during severe winters, while it beats everything to endure prolonged drought. It is not suitable to mix with clover, as the latter overtops and crowds the Lucerne. I can report no systematic attempt in Michigan to establish, mow and feed crops of lucerne. June grass in early spring and late fall crowds it out.

In 1883 Prof. E. M. Shelton, of Kansas, said: "We have no hesitation in saying that, all things considered, it is a most valuable clover, especially for the western and southwestern sections of the state. Along the Arkansas river, where irrigation is practiced, it has proved a most invaluable forage plant. More accounts come to us of failure with Alfalfa than with any other clover or grass, and this is because of the difficulty in starting the plants and in selecting and preparing the soil properly. It must not be sown with another crop, neither mowed nor pastured during the first year. The dangers which threaten it most are the common mole and pocket-gopher; the latter burrowing among and cutting the roots, has destroyed several acres on the college farm. We have cut three and even four large crops from the same ground in one season."

Early in 1885 Prof. Shelton states in the *Rural New Yorker* that Alfalfa has proved with us the most useful of all clovers for the purpose of pasturage. It endures uninjured, close cropping, all kinds of stock consume it greedily, and it has never winter killed. It requires much field room in curing, and soon spoils with light rains. For hog pastures I know of no other plant so valuable.

Prof. A. E. Blount reports for Colorado: "J. S., near the college, keeps large herds of sheep, some cattle, horses and hogs. When fed on Alfalfa cattle grow faster; cows give more and better milk; horses are more healthy and do more work with a fourth of the grain; sheep make better mutton and lose less wool; and hogs fatten, almost ready for market, without any grain. He cuts his crops three times, averaging about two tons to the cutting. By letting the first crop grow until July he raises from 5 to 10 bushels of seed per acre."

He says he has samples four feet long, grown in thirty days. It does not spread except by seeding. It is too tender to sow in the fall, but should be sown in spring after the frost has gone. Harrow it in with or without a crop.

President Ingersoll, of the same place, told me that Alfalfa was the only forage plant that would grow at their place and keep green without irrigation. It is a favorite forage crop in Colorado and its cultivation is extending very rapidly.

For Mississippi and vicinity Prof. D. L. Phares considers Lucerne very valuable. It sometimes gets two feet high by the middle of February. He knows some plots of it now in fine condition that are known to have been growing for over thirty-five years, without any marks of decay.

Prof. F. A. Gulley, of the same state, thinks it is too difficult to get it well started.

The late C. W. Howard, of Georgia, believed, as a forage plant at the South, Lucerne is very far superior to all others. For feeding it should be cut a day in advance and used in a wilted state. *It must never be pastured*, as live stock in that climate bite out the crowns of the plants and kill them. It is ready to cut a month in advance of red clover.

Medicago lupulina, L. Black Medick, Nonesuch.—A procumbent, branching, pubescent annual or biennial. *Leaflets* obovate, toothed at the apex. *Peduncles* longer than the leaves bearing ovoid heads of small yellow flowers. *Pods* small, one-seeded, black when ripe, kidney-shaped. Found in Europe, N. Africa, West Asia to India; introduced in N. America.

On rich land it often affords considerable pasture, reminding one of white clover in its habit. It is not likely worth cultivating in this country because we have something better.

Medicago maculata, Willd. Spotted Medick, Burr Clover, California Clover.—A procumbent or spreading, branching annual. *Leaflets* obovate or obcordate, often with a black central spot, minutely toothed. *Peduncles* 3–5 flowered; flowers yellow. *Pod* ¼ in. broad, making 3–5 coils, quite compact, with a double row of long, curved spines. Found in Europe, N. Africa; introduced in N. America. The pod makes something like a burr, so much so that it adheres to wool. It is too tender and short lived to be of value at the North, but has some good words from people of the South.

Prof. D. L. Phares, of Mississippi, considers it a valuable plant. He has grown it about thirty-five years, and says it furnishes good grazing from February till April or May. Cattle do not incline to eat it at first, but they learn, and finally acquire a great fondness for it. It seeds freely every year. Crab grass occupies the ground from June to October after the Medick has seeded.

Medicago denticulata, Willd. Burr-Clover.—This annual much resembles the last and is often confounded with it. The pods are loosely spiral and deeply reticulated.

A writer in the *American Agriculturist* for 1878 speaks highly of the plant.

Burr-Clover grows wild all over the plains and foot-hills, and affords much pasture. Even the burrs grow in such profusion that they afford a good supply of dry concentrated food. They collect, by force of the wind, in the hollows of the ground. It is tenacious of life and will bear close feeding.

MELILOTUS, TOURNEFORT. MELILOT.

Annual or biennial, fragrant when bruised or in drying. *Leaves* pinnately 3-foliolate, nerves ending in teeth; *stipules* slightly adhering to the petiole, often cut. *Flowers* small, yellow or white, in long, loose axillary racemes. *Calyx-teeth* 5, nearly equal. *Petals* deciduous; keel shorter than the wings, obtuse. *Anthers* uniform. Pod with one or few seeds, small, straight, thick, indehiscent. Plants abound in an etherial oil (cumarin) rendering them objectionable to stock. Warm and temperate regions of the old world.

Melilotus officinalis, Willd. Yellow Melilot, Sweet Clover.—This is an annual or biennial with yellow flowers, apparently of little importance except for bees.

Melilotus alba, Lam. White Melilot, Bokara Clover, Sweet Clover.—An erect, branching, woody, annual or biennial 2–6 or 8 ft. high. *Leaflets* truncate. Flowers small, white, in long racemes. *Pods* black when ripe.

FIG. 135.—*Medicago lupulina*, L. (Black Medick.) Portion of a plant in flower and in fruit, natural size.—(U. S. Agrl. Rept.)

Fig. 135.

Found with the last. At the North it does not seem to be eaten by live stock when green, but they will eat a little when cured with other forage plants. Bees find it valuable for the nectar which is abundant during the heat of summer. Prof. Phares speaks of this as cultivated for forage. As it is a leguminous plant and a near relative of red clover, and very large, it has been mentioned as quite suitable for green manuring.

Lupinus, Tourn. Lupine.—Of this genus there are several species, some of which have proved valuable in the old world, but I cannot learn that they are as valuable as some other forage plants in any portion of the United States.

Dr. H. P. Armsby says: "The yellow lupine, when cut just at the end of flowering, is the most highly nitrogenous of all coarse fodders."

Of domestic animals sheep only eat lupines well on account of their bitter taste.

In Central Michigan, at any rate, several varieties, after numerous trials, have uniformly made a slow, sickly growth. Similar trials have been reported from Georgia, Mississippi and other states.

Ulex, L. Furze, Ulim, Gorse.—Much-branched, thorny shrubs. Leaves prickly. Flowers yellow. Found in regions all about the Mediterranean Sea. Often abundant in England, Wales and Ireland. When bruised or wilted it is eaten by stock. Perhaps we might do the same with thistles. Like lupines, above mentioned, it is apparently of no value in the United States.

Onobrychis, Tourn. Sainfoin.—To this genus belongs one species which is much cultivated in parts of the continent of Europe and in Great Britain. In France this leguminous plant is much grown to improve poor, hungry land, and will last 4 to

Fig. 136.—*Melilotus alba*, *Lam.* Portion of a plant in flower and fruit, natural size.—(U. S. Agrl. Dept.)

Fig. 136.

7 years in succession. It is employed for soiling, for pasture, or made into hay after the manner of red clover. The seed is sold in two forms, that covered with the short, wrinkled pod, and that which has been separated from the pod.

In England it is considered a very suitable forage plant for calcareous soils.

Although an old plant, and so well and favorably known in Europe, I cannot learn that it has met with even moderate success in any portion of this country. The seed has been widely distributed by the Department of Agriculture. In Central Michigan we have not been able, even with the best of care, to raise respectable samples.

Vicia, L. Vetch, Tare.—Of this genus of legumes there are quite a number of species native to this country, and many in Europe. Like the last mentioned, sainfoin, it has had repeated trials in various portions of the United States, and yet we do not know that it has really succeeded anywhere. To those not familiar with vetches, it may be enough to say in this connection that they are much like peas, with slender leaflets and small stalks, flowers and seeds.

A writer in Morton's Cyclopedia says: "Of the artificial grasses it is next to clover in value. Sheep fatten faster upon this than any other herbage; horses improve more rapidly upon it than on clovers or the grasses; horned cattle thrive surprisingly upon this fodder; cows yield more butter from the tare than from any other provender; pigs voraciously consume and prosper upon it. They may be cut twice a year, and are much used for soiling. In quality they much resemble lucerne. At Lansing, Michigan, they make a weak growth, and will not endure the hot, dry weather. It is not improbable that some of our native vetches could be improved and adapted to cultivation.

Pisum, L., Pea.—To this small genus of two species, belong the numerous races of cultivated field and garden peas. They thrive in cool, moist, temperate regions. Like other legumes, they draw much from the air and subsoil, and are most excellent crops to alternate with wheat and the true forage grasses. They like moist loamy soil, but this should not be in the highest condition, else the plants "run too much to vines," at the expense of a good crop of seeds.

From $1\frac{1}{2}$ to 4 bushels of seed to the acre is sown broadcast or in drills, yielding 15 to 25 bushels of seed, which is a very nutritious food for swine and sheep.

The greatest enemies to this crop are the pea weevil or "bug" and mildew.

Of "buggy" peas only about twenty-five per cent will usually grow, and these produce feeble plants. Seed can be obtained from the North, where the bugs are not troublesome, and the young crop can be fed out bugs and all.

The weevil can be killed when young, by putting the peas as soon as threshed in a tight box with some bisulphide of carbon. There is some difference in varieties, but hot, dry weather is quite sure to favor the development of mildew, which weakens and often prevents the growth of the plants or the production of a good crop of seeds. In favorable localities enough attention is not paid to this crop, both for feeding and to precede a crop of wheat.

VIGNA, L. COW PEA.

Calyx campanulate, lobes or teeth short, often obtuse, the two upper more or less united. The *banner* rounded, with inflexed appendages at the base; the *wings* falcate-obovate adhering to the *keel*, which is incurved and often beaked, but not spiral. The odd stamen free from the banner. *Anthers* uniform. *Ovary* subsessile, many ovuled; *style* curved, barbed, or with a pencil of hairs below the terminal stigma. *Pod* shaped like a scymetar, falcate, or linear, compressed, 2-valved, often thickened at the sutures; valves flat or convex. *Seeds* thick or compressed. hilum short or long, covered or naked. Plants herbaceous or shrubby, climbing. erect, or prostrate. *Leaves* pinnate, 3-foliolate, stipellate. *Stipules* small. *Flowers* violet, flesh colored, yellow or white. Solitary or clustered in the axiles. About 20 species, found in the cooler parts of Africa, in Asia, Australia, and America.

Vigna Cotjang Walp, L. Cow Pea, Bush Pea, Chinese Pea. Leaflets vary much in shape, and are oval, broadly ovoid, or rhomboid. Flowers few at the end of the peduncle. *Pods* 3-8 in. long, mostly straight, 2, 3, or 4 to a stalk. *Seeds* black, white, red, cream colored, purple, or spotted. The style of foliage, absence of tendrils, shape of seed, and the raising of the seed leaves above the ground in germination, all indicate that it is more nearly related to the bean than the common pea.

It has been cultivated in China from remote antiquity, and is a favorite forage crop in the Southern States, where it takes the place of red clover at the North.

There are many varieties in cultivation which differ much in foliage, size of plant, size, color and shape and yield of seeds.

Some are quite bushy and spread into a tangled mass. Even in Central Michigan some of these peas make a rank growth, completely covering the ground two feet and a half high.

At the North, horses refuse to eat it, but at the South, probably from "education," all grazing domestic animals are very fond of cow peas, either fresh or dried.

The following notes are mainly gleaned from an article by P. J. Berckmans, of Georgia, as found in the *American Agriculturist* for 1876:

Almost any land will grow the cow pea, though the "Clay," "Red" and "Black" succeed better on poor land than the "Lady" or "Crowder" varieties.

Spring crops are sown in April, and fall crops after taking off wheat or oats. From four to six pecks per acre are sown broadcast, the larger amount on poor soil. On good soils two crops of forage are often cut from one sowing, provided the season be favorable. The crop is sometimes plowed under. As with young clover, so plaster is sown on cow peas.

All the plain or semi-colored varieties are of a spreading nature and are best suited for forage. The "Red," "Clay" and "Black," of the plain kinds, and the "Whippoorwill," of the semi-colored, are most esteemed. The "Red Ripper," or "Tory" may be sown in fall if preferred.

The speckled varieties are usually bushy in growth, and unfit for forage. They are raised for market and the table.

The "Lady Pea" and "White Table" are used for culinary purposes, sometimes for snaps, or shelled in the green state; when dry they are very desirable for soup, or they may be baked the same as the white bean.

The vines are fit to cut for fodder when the pods begin to turn yellow. The vines often lodge badly, and are usually cut with a scythe. A few grains of corn mixed in with the seed gives some stalks for support.

The main difficulty in curing pea hay is to retain the leaves on the stalks; to ensure which they must be handled very little. The wilted vines may be loosely piled and remain so for two or three weeks till cured and ready for storing.

On good land, and good culture, two tons of forage per acre may be expected, and sometimes two cuttings in a year, with a yield of two tons at each cutting. The yield of seed varies from 30 to 40 bushels per acre, or more commonly 10 bushels. The latter is likely to be the yield when sown in rows in corn fields.

For feeding stock, well cured cow pea hay is more nutritious than any hay produced from grasses, millet, or other plants.

When the pods are left until they are filled the value of the food is much increased. When fed upon such fodder, horses and mules should receive less corn or oats than when fed on any other provender. In some parts of the country the peas are often troubled with pea weevil or "bug." To prevent this Prof. Phares lets them remain in the pod till ready to use, or when dry, then thresh them and mix with road dust.

With reference to the cow pea for Mississippi, Professor Gulley reports as follows: "For hay and for plowing in to fertilize the land, we sow broadcast a bushel to a bushel and a half to the acre, harrow in and cut with a mower as we would clover. Black and red peas make more vines and will stand wet weather without rotting, when speckled peas will be entirely spoiled. I sow the black and red exclusively, cut for hay, feed off with stock or leave them to rot on the ground for manure. For seed we sow in drills and cultivate once or twice. Peas are a slow crop to gather, as they do not ripen evenly. I consider this crop one of the most valuable for hay or ensilage or for restoring the fertility of the soil. It stands first."

Lespedeza striata, Japan Clover.—This is a low annual herb, with small trifoliolate leaves and very small flowers, producing a small, flattish, indehiscent one-seeded pod. The seeds to this were accidentally brought to South Carolina about 1849, probably in connection with importations of tea from China. It has spread continually and quite rapidly over the South, and has quite tenaciously held its own, even crowding Bermuda grass.

The writer knows little of this plant, and ventures to quote some very conflicting opinions as to its value. Several writers speak of it as very suitable for poor soils for grazing in dry, hot weather. The stems spread close to the ground, seldom growing over a foot high. It is quite firm and hard, and at first not a favorite with stock. They learn to eat and thrive on it because of its nutritive qualities, which chemical analysis makes

Fig. 137.—*Lespedeza striata.* (Japan Clover.) Part of a plant.—(U. S. Agrl. Rept.)

Fig. 137.

quite remarkable. The plant freely produces small seeds, and it is hard to exterminate.

Henry Stewart, in the *Country Gentleman* for January, 1886, says: "I assert emphatically that unless cattle and pigs are starved to it, they will not eat the Japan clover, or any kind of *Lespedeza*. A good deal has been written in favor of this plant. In a few places it may be of some service. This statement is given to prevent your readers from being fooled into buying the seed and trying to grow it in any place north of Virginia."

Prof. F. A. Gulley, of Mississippi, says: "For the South, Japan clover is, *without exception, the most valuable plant* that grows. After once started it grows spontaneously, except on lime land. It keeps hills from washing, even coming in to fill the 'washes.'

"It can be killed by plowing for one year. On good land it grows from 12 to 24 inches high, cuts a good crop of hay, equal to first-class Timothy. For pasture from May 15th to the first frost it is as good as anything we have except Bermuda grass, and equal to the best pastures at the North. It will grow when blue grass and the clovers fail entirely. It stands dry weather admirably, and on some soils will even choke out Bermuda. It is our principal pasture during summer."

Prickly or common Comfrey, Borage, and numerous other plants, have been occasionally highly recommended as excellent forage plants, but they have not won very general favor.

It is very probable, especially for the South and the dryer portions of our country, that we shall yet find some new forage plants which will in some respects surpass any that we now cultivate.

CHAPTER XVI.

THE ENEMIES OF GRASSES AND CLOVERS.

Mice and Shrews.—These small animals often damage meadows by eating some of the stems and larger roots of grasses and clovers, especially the thickened portions stored with starch near the surface of the ground. They could be trapped if too troublesome, or caught by cats and dogs, but it has been shown that they are not an unmixed evil, as they build nests of old stems and leaves, which, when deserted, are the favorite abodes of bumble bees; and these should be encouraged, because they help fertilize the flowers of red clover, and thus increase the yield of seeds, which are very valuable.

Moles.—In permanent pastures or meadows where the land is dry and sandy, moles sometimes become very troublesome, raising large numbers of unsightly mounds, which are a great annoyance to the mowers. No doubt the moles eat some worms, large numbers of white grubs and other insects, some of which feed on the roots of grasses and clovers, but we know from experiments that moles will eat vegetation in considerable quantities. We should rather run the risk of dispensing with the services of the moles, but the writer is sorry to say that he thinks this kind of game is not often easily caught. Where fields are plowed, and a rotation of crops is followed, moles are seldom troublesome.

Pocket Gophers.—With these diggers the writer has had no experience, but from all accounts, they will often do a good deal of damage. Their burrows are a great nuisance, to say nothing of the grass and clover which they devour or tread under foot.

Woodchucks.—These large rodents are often very troublesome to the farmer who owns dry, sandy, or gravelly land. They dig long holes, raise piles of dirt, devour and tramp down large patches of meadow. They can usually be caught quite easily in steel traps; they can be shot if one has the patience to watch for them. Where the ground is not too high and dry and a good supply of water handy, by taking advantage of a wet time when the soil is full of moisture, they can often be drowned out and made to come to the surface, where they make sport for the dog. To help make the job a success, before beginning, draw several barrels of water and pour them in quick succession down the hole. **Saturate rags with bisulphide of carbon and put in the holes.**

Insects.—The rest of this chapter is prepared for this volume by my colleague, Prof. A. J. Cook.

It is generally supposed, even by those best informed and most interested, that our forage plants, including clovers and grasses, are comparatively free from the devastation of insect pests. While our fruits, vegetables and grains are known to be tunneled or devoured at the root, girdled or fed upon at stem and foliage, and blasted in the fruit, the same is not generally supposed to be as true of the plants which give value to our pastures and meadows. While Harris and Fitch give account of many insects which prey upon nearly all others of our cultivated plants, very few are mentioned that attack our grasses and clovers, even by these great scientists and wonderful observers. Mr. J. Stanton Gould, in his Forage Crops, knows only four insects which attack the clovers, while at that time over seventy were known to attack the apple. This is not because such enemies do not exist, but rather because the plants fed upon are so abundant that even great damage is either not noticed or else is supposed to be due to drought or other climatic disturbance, or forsooth to the "running out" of the crop. The very nature of our grasses and clovers conceals insect ravages, and thus the harm must become very patent or it will generally be all unobserved.

At present over seventy different species of insects are known to attack the clover, and nearly or quite as many draw their sustenance from our grasses. It is not probable that all this increase is due to more close study and observation. Insects are constantly leaving our wild plants, either from choice or necessity, and adopting our cultivated plants as a more acceptable diet. We are constantly introducing species from the old world, some of which are of recent importation, and may well cause solicitude because of the serious damage they do. Both of these causes, change of food, habits and importation will continue to increase these pests, so that constant study and experiment will be necessary to ward off the threatening danger.

Insects Injurious to Clover.—In the report of the New York Agricultural Society for 1881–82, p. 190, Prof. J. A. Lintner gives a list of insects which infest clover in Europe. There are 71 species. Nearly every genus represented by these foreign species is included in our insect fauna. The list is of much interest to us, the more so as the noxious insects of Europe are constantly being introduced into America. The following is the list:

Sitones flavescens	*Marsham.*	Apion craccæ	
Sitones lineatus	*Linn.*	Coccinella impuncta	
Phytonomus nigrorostris	*Fabr.*	Hylesinus trifolii	*Müll.*
P. meles var. trifolii	*Herbst.*	Labidostomis longimana	*Linn.*
Tychius polylineatus	*Germ.*	Lycæna Amyntas	*Schiffm.*
Tychius pisirostris	*Fabr.*	Lycæna Alexis	*Treits.*
Ceutorhynchus lineatus	*Payk.*	Lycæna Ægon	*Schiffm.*
Apion seniculum	*Kirby.*	Lycæna Cyllarus	*Fabr.*
Apion virens	*Herbst.*	Lycæna Dolus	*Hübn.*
Apion flavipes	*Fabr.*	Lycæna Adonis	*Hübn.*
Apion fagi	*Linn.*	Lycæna Argus	*Hübn.*
Apion assimile	*Kirby.*	Melitæa Athalia	*Esp.*
Apion trifolii	*Linn.*	Melitæa Cinxia	
Apion gracilipes	*Dietrich.*	Colias Hyale	*Linn.*
Apion varipes	*Germ.*	Leucophasis sinapis	*Linn.*
Apion apricans	*Germ.*	Zygæna Minos	*Hübn.*

Zygæna meliloti............*Ochs.*
Zygæna lonicerae............*Hübn.*
Zygæna trifolii............*Esp.*
Zygæna filipendulæ............*Hübn.*
Zygæna peucedani............*Esp.*
Zygæna scabiosæ............*Hübn.*
Zygæna Achilleæ............*Hübn.*
Zygæna angelicæ............*Ochs.*
Orgyia fascelina............*Hübn.*
Gastropacha rubi............*Hübn.*
Gastropacha trifolii............*Hübn.*
Callimorpha hera............*Linn.*
Lasiocampa trifolii............*Linn.*
Orthosia litura............*Hübn.*
Orthosia gracilis............*Hübn.*
Plusia gamma............*Hübn.*
Mamestra pisi............*Hübn.*
Mamestra chenopodii............*Hübn.*
Mamestra suasa............*Esp.*
Agrotis comes............*Hübn.*
Episema graminis............*Linn.*
Acontia solaris............*Hübn.*
Euclidia glyphica............*Hübn.*
Euclidia mi............*Hübn.*
Herminia crinalis............*Treits.*
Boarmia selenaria............*Hübn.*
Fidonia clathrata............*Linn.*
Ortholitha bipunctaria............*S. V.*
Ortholitha palumbaria............*Hübn.*
Ypsolophus deflectivellus............*H. S.*
Phoxopteryx badiana............*S. V.*
Gelechia tæniolella............*Treits.*
Gelechia anthyllidella............*Hübn.*
Lithocolletis Bremiella............*Frey.*
Lithocolletis insignitiella............*Zell.*
Coleophora deauratella............*Lienig.*
Acipitilus pentadactylus............*Linn.*
Agromyza trifolii............*Kaltenb.*
Agromyza nigripes............
Lopus roseus............*Fall.*
Aphis pisi............*Kaltenb.*
Epilachna globosa............
Goniostena sexpunctata............

In the same volume, p. 192, Prof. Lintner gives the following list of insects attacking the clover in the United States, nearly all of which are widely distributed:

LEPIDOPTERA.

Callidryas Cuvale............*Linn.*
Colias Cæsonia............*Stoll.*
Colias Eurytheme............*Boisd.*
Colias Philodice............*Godt.*
Terias Nicippe............*Cram.*
Terias Lisa............*Boisd-Lec.*
Terias Delia............*Cram.*
Melitæa Editha............*Boisd.*
Chrysophanus Americana............*D'Urb.*
Lycæna Comyntas............*Godt.*
Eudamus Pylades............*Scudd.*
Hyphantria textor............*Harr.*
Arctia Phalerata............*Harr.*
Arctia Achaia............*Gr.-Rob.*
Pyrrharctia Isabella............*Sm.-Abb.*
Hyperchiria Io............*Sm.-Abb.*
Agrotis saucia............*Hübn.*
Mamestra trifolii............*Esp.*
Mamestra renigera............*Steph.*
Mamestra (Ceramica) picta............*Harr.*
Leucania unipuncta............*Haw.*
Prodenia commelinæ............*Sm.-Abb.*
Drasteria erechtea............*Cram.*
Hypena scabra............*Fabr.*
Aspilates dissimilaria............*Hübn.*
Asopia costalis............*Fabr.*

LEPIDOPTERA—CONTINUED.

Asopia olinalis .... *Guen.*
Asopia farinalis .... *Linn.*
Tetralopha N. sp ....
Phoxopteris angulifasciana .... *Zell.*
Tortrix incertana .... *Clem.*
Anaphora agrotipennella ..... *Clem.*
Gelechia roseosuffusella ...... *Clem.*

COLEOPTERA.

Hylastes trifolii .... *Müller.*
(Larva in roots
Languria Mozardi .... *Fabr.*
(Larva in stem.)
Graphorrhinus vadosus .... *Say.*
(Imago on leaves.)

DIPTERA.

Cecidomyia leguminicola .... *Lintn.*
(Larva on seeds.)
Cecidomyia trifolii .... *Loew.*
(Larva on leaves.)
Oscinis trifolii .... *Burgess.*
(Larva on stem.)

ORTHOPTERA.—(ALL ON LEAVES.)

Caloptenus femur-rubrum .. *De Geer.*
Caloptenus spretus .... *Thomas.*
Caloptenus differentialis ... *Thomas.*
Caloptenus bivittatus .... *Say.*
Caloptenus atlanis .... *Riley.*

HOMOPTERA.

Pemphigus lepidii ...... *Riley, M. S.*
(On roots.)
Thrips sp. .... *Welsh.*
(On blossoms.)

The following additional species are added on page 206:

LEPIDOPTERA.

Nephelodes violans, Guenée, *Riley*, 1st Report, N. Y. St. Ent., 1882, p. 103.

Plusia brassicæ, *Riley*, Id Gen. Ind. Suppl. Mo. Repts., p. 78.

Heliothis armiger, *Hüb.*, Barret Ent. Month Mag, XIV., p. 151.

Eupithecia interruptofasciata, *Pack.*, Coquillet, Papilio I., p. 57.

Ephestia interpunctella (*Hüb.*,) Clemens, Proc. A. N. S. Ph. 1860, p. 206.

Dichelia sulfureana, (*Clem.*) Comstock Report Comm. Agr. 1880, p. 255.

Amphisa discopunctana, (*Clem.*) Comstock Report Comm. Agr. 1880, p. 258.

Platynota flavedana, (*Clem.*) Comstock Report Comm. Agr. 1880, p. 257.

Sericoris instrutana, (*Clem.*) Comstock Report Comm. Agr. 1880, p. 258.

Grapholitha interstinctana, (*Clem.*) Comstock Report Comm. Agr. 1880, p. 251.

COLEOPTERA.

Lachnosterna serricornis, *Le Conte.*, Webster Am. Nat., XVI., p. 746.
Colaspis brunnea, *Fab.*, Webster Am. Nat., XVI., p. 746.
Diabrotica longicornis, (*Say*) Forbes' 12th Rept. Ins. Ill., pp. 21, 23.
Diabrotica 12-punctata, (*Oliv.*) Forbes' 12th Rept. Ins. Ill., p. 104.
Tenebrio Molitor, *Fitch*, Fitch Trans. N. Y. St. Ag. Soc., XIII., p. 376.
Macrobasis unicolor, (*Kirby*) Webster Am. Nat., XVI., p. 746.
Epicaerus imbricatus, (*Say*) Webster Am. Nat., p. 746.
Sitones lineellus, (*Gyllenhal*) European authors.
Sitones flavescens, (*Marsh*) European authors. (Kalt et al.).
Phytonomus punctatus, (*Fab.*) Riley Am. Nat. XV., p. 750.

HEMIPTERA.

Poecilocapsus lineatus, (*Fab.*) Lintner 1st Rep., N. Y. St. Ent., p. 277.
Limothrips tritici, (*Fitch*) Lintner 1st and 2d Repts. Ins. N. Y., p. 304.

NEUROPTERA.

Smynthurus hortensis, (*Fitch*) Fitch 6th-9th Repts. Ins. N. Y., p. 189.
Smynthurus arvalis, (*Fitch*) Fitch 6th-9th Repts. Ins. N. Y., p. 191.

The following species are mentioned by Prof. S. A. Forbes in Entomological Report of Illinois, Vol. 14, pp. 72-74:

LEPIDOPTERA.

Cymatophora crepuscularia......*Tr.* | Tortrix pallorana..............*Robs.*
Haematopis grataria ......*Fab.* | Hypena scabra................*Fabr.*
Cacoecia rosaceana..........*Harr.*

HOMOPTERA.

Coccus trifolii..............*Forbes.*

ACARINA—MITES.

Bryobia pratensis........*Garman.* | Bryobia pallida...........*Garman.*

To these I would add two other coccids which I have observed on clover.

Pulvinaria innumerabilis.....*Rath.* | Lecanium tiliae...............*Fitch.*

Strecker in his catalogue of N. A. Mac. Lipidoptera gives Meganastoma caesonia, Stroll as feeding on clover. While in Rep. Comm. Ag. 1863, p. 573, and in 1865, p. 40, Epicae fallax is mentioned as a clover enemy.

It is more than probable that others of the butterflies, especially of the genera Colias, Melitaea and Lycaena, will be found upon further investigation to feed upon our clover; while it is not at all probable that the fourteen species of beetles named in the list comprises all the enemies of the clover belonging to that order. There is but little doubt that the list will be doubled. Of the Orthoptera (Locusts) but five are named, and they all of the genus Caloptenus. It is quite certain that all of our many species of that genus, and nearly all others, may be justly included in the list. Only one unnamed species of Thrips is mentioned. The past season I have found three species, one black, one light yellow, and one bright red, all to be very abundant on the clover blossoms, yet I could not see that they were greatly injurious.

Many of the insects named in the above list feed more generally on other plants. Mamestra picta prefers the cabbage. Heliothis armiger feeds on the corn and cotton, Leucania unipuncta—the army worm—on oats and the grasses, so that for the most part they are not serious enemies to our most valuable forage plant. One of the insects named in the list, Asopia costalis, feeds on the dry clover, either in the stack or mow, where it often does very great injury.

As our space will not permit a detailed description of all of the above only those whose mischief is so considerable as to create concern for the future of one of our most valued farm crops will be described. These work on the roots, foliage, and seed, and will be described in that order.

Hylastes trifolii, Clover-root Borer.

Order Coleoptera. Family Scolytidae.

Müller, Mem. Soc. Dep. Mt. Tonerre I., pp. 47-64, 1807.
Schmitt, Stett Ent. Zeit. V., pp. 389 397, 1844.
Lintner, Ann. Rep. N. Y. St. Agr. Soc., 1879, pp. 41-42, Ill.
Lintner, Rep. N. Y. Ag. Soc., 1882, p. 193, Ill.
Riley, Ann. Rep. Comm. Agr. 1878, pp. 248-250, Ill.
Riley, Am. Entomol., Vol. III., p. 180, 1880, Ill.
Saunders, Ont. En. Rep., Vol. XII., p. 43, 1881, Ill.

This insect has long been known as a not very common insect of Germany in Europe. Müller, as shown by the name, regarded it as an enemy of the clover, while Schmitt thought that it attacked such plants as were already enfeebled, and was not a serious injury. In 1878 the beetle attacked this valuable plant in northwestern New York, and the fact that it injured very seriously the clover of that region proves that Müller was correct and Schmitt wrong.

Prof. Riley investigated the habits of the insect, which he found very destructive to the clover in Yates, Ontario and Seneca counties. He described it under the name Hylesinus trifolii, or Clover-root Borer, and pointed out the fact that it is much like one of our common bark beetles, Hylesinus opaculus Lec., which is often found just under the bark of ash and elm trees. While much like the elm bark beetle, it is not only a different species, but is placed in a different genus,—Hylastes by Leconte and Horn, and by European Coleopterists.

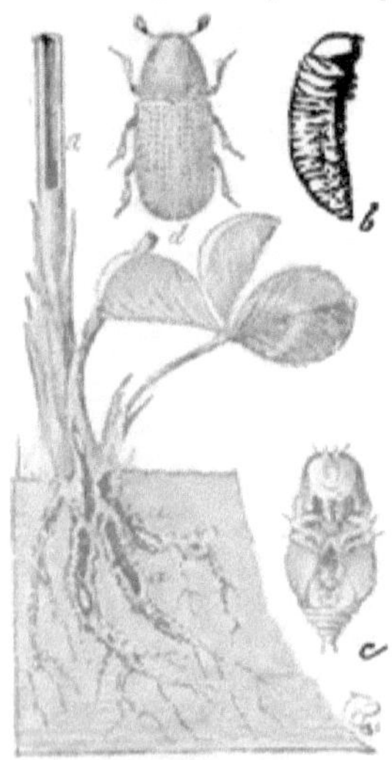

FIG. 138.

The family to which it belongs, Scolytidæ, is represented by numerous species, usually called bark beetles, as they tunnel and sculpture various evergreen and deciduous trees just beneath the bark. It is often stated that they attack enfeebled trees, yet I have often found them industrious and thriving on trees which were in full strength and vigor.

The insect is well represented in Fig. 138, *a* showing the affected plant, *b* the grub or larva, *c* the pupa, and *d* the beetle or imago. The eggs are whitish oval, the larva white, with yellow head. The length of larva is 3 m m (.12 of an inch) in length. The pupa is 2.2 m m.long and has two spinous projections on the top of the head, and two smaller anal projections. The imago is black, with brown punctured elytra. It is 2 m m (.08 of an inch) long.

The beetle hibernates, usually as an imago, but also as a pupa or larva. Mating occurs in early spring, when the female bores into the crown of the plant and deposits five or six eggs. When these hatch the larva feeds at first in the opening formed by the imago for her eggs, but soon works downward forming tunnels lengthwise of the main roots, which entirely destroys the plant.

In September many pupæ will be found in the upper part of the galleries.

The beetle has become the worst enemy to the clover that we have. It has already shown its destructive work in nearly every clover producing section between the Atlantic and Pacific. Surely such enterprise as would carry the insect from Europe to America can hardly be expected to permit it to remain stationary on this continent. The fact that it has no parasites, as yet discovered, to weaken its efforts or reduce its numbers, not only accounts for its exceeding numbers in this country as compared with Europe, but also gives prophecy of wide extension and serious ravages in the future.

It is difficult to suggest satisfactory remedies for insects which are so numerous and scattered as are these beetles. It is probable, in fact the experience in New York already confirms the suggestion, that they will not be equally destructive every year; that while they may ruin whole fields one season the very next year they may be quite rare and far less injurious. The only remedy thus far suggested is to plow the clover under when the insects are discovered to be at work, and not grow clover for a time. This green manuring would certainly be very excellent for the land. Yet it is to be feared that the insects would take to other herbage, possibly other leguminous plants, rather than perish. This supposition seems more probable in that this species has varied so far in its habits from those of its near congeners, which are all bark or wood eaters so far as I know. It is quite probable that summer plowing, followed by thorough harrowing, might destroy the insects at work in the clover. If such were the case it certainly would be a wise proceeding.

Prof. I. P. Roberts says: "In Central New York, of late years, we mow the seeded land but once and pasture in the fall the abundant second growth. Since 1878 the clover-root beetle has worked upon the clover to such an extent that it invariably fails the second year. This has caused us to change from a five to a four year rotation, viz.: hay, corn, oats and wheat."

Languria Mozardi, Fabr., Clover-stem Borer.

Order Coleoptera: Family Erotylidæ.

Latreille, Gen. Crust. et Ins. III., p. 66, 1807.

Say, Am. Entomology, III., 1828, Ill.

Lamarck, An. sans vert., deux. edit. IV., p. 486, 1835.

Melsheimer, Cat. Coleop. U. S., p. 47, 1853.

Le Conte, Proc. Acad. Nat. Sci. Phil. VII., p. 161, 1854.

Oliver, Entomol. V. p. 464, Ill.

Crotch, Trans. Am. Ent. Soc. IV. p. 350, 1873.

Le Baron, 4th Ann. Rept. Ins. Ill. p. 181, 1874.

Comstock, Ann. Rept. Comm. Ag. 1879, p. 199, Ill. 1880.

Saunders, Ont. En. Rep. Vol. XII. p. 44, 1881, Ill.

Lintner, Ann. Rep. N. Y. Ag. Soc., 1882, p. 196, Ill.

The clover-stem borer, though not very common, is widely distributed throughout the country. It is found in Michigan, Canada, New York, and south to Washington and west to Kansas. Indeed it is mentioned as far south as Louisiana. Prof. J. H. Comstock was the first to discover and describe its full life history. Though not as yet known to be a serious pest, from its wide distribution we may suggest that it does more harm than is suspected, as a great many plants in a clover field could be destroyed and yet not be missed. Even though not as yet alarmingly injurious we cannot tell when it may become so. In its distribution the seeds of mischief are wide scattered, no knowing when they may germinate.

Fig. 139.

Fig. 139 shows the eggs, larva, pupa and imago of the insect as well as the natural size of eggs and the larva—the latter as it appears in the hollowed stem of the clover. The eggs are yellow, curved, and 1.7 m m (about 1-16 of an inch) long. The larvæ, like wire worms and many other grubs, are slim, with the three pair of jointed legs well developed, and a pair of anal pro-legs. When full grown the yellow larva is 8 m m (a little more than .3 of an inch) long. Like the pupa it has two plainly marked anal spines. The pupa is also yellow and slender, and 6 m m long. I find Say's description of the imago, as usual, very exact: "It is slender, cylindrical; the dark red antennæ gradually form a club of five joints. The palpi are thread-like; the mandibles bifid at tip; the maxillae have horny teeth. The thorax is yellowish-red, smooth and unspotted. The elytra are bluish-black, with a green tinge, marked with deeply impressed punctures, arranged in regular series, but without impressed striæ. The thighs are pale rufous at base; the tibiæ have a slight rufous tinge; the tarsi are dotted with dense hairs beneath the three basal joints, the 3d being bilobate." The venter has the three posterior joints black. The length of the beetle is about 7 m m (¼ of an inch).

The female lays the eggs in June, piercing the stem with her jaws, and pushing her eggs clear in to the pith, often, says Prof. Comstock, to a depth of 6 m m. The larvæ feed upon the pith downward, forming a burrow 15 c m (6 inches) long. This greatly injures if it does not kill the plant outright. The pupa is formed at the bottom of the burrow in August, and shortly after the fully developed beetles begin to appear. They are seen to emerge from the hollowed stems from August to October. There is only one brood a year. Like many of our noxious beetles, the imago hibernates and waits for the vigorous plants of genial June before dropping her precious burden of eggs.

If this pest promises to do any serious harm we have only to cut the clover early in July, when we shall save the crop, and probably destroy the insects. This would give chance for a second crop of hay or fine pasture or crop of seed from the same plants. It is a welcome fact that Prof Comstock found two parasites working on these beetles, in such abundance that we understand why the latter are no more numerous and destructive. One a Chalcid and the other an Ichneumon fly.

Phytonomus punctatus, Fabr., Clover Leaf Beetle.

Order Coleoptera. Family Curculionidae.

Le Conte, Rhyncophora, p. 124, 1853.

Riley, Am. Naturalist, Vol. XV., p. 912, Nov., 1881, Ill.

Riley, Rep. Comm. Ag. 1881-82, p. 171, Ill.

Kilman, 15th Rep. Ont. En. Soc., 1884, p. 32.

This, like many of our most destructive insects, is an imported species. It is a common insect in Germany, and has probably been in this country for years, as Dr. LeConte received it from Canada in 1853, when he described it as Phy. opimus. As it does not exist in collections of American Coleopterists, it is possible that the insect described by Dr. LeConte by mistake was reported as Canadian, it really being itself foreign. In 1881 a serious invasion of Western New York, Yates county, was experienced, when Dr. Riley, of the Agricultural Department, investigated and gave a detailed description of the species, including its work and habits. It is worthy of remark that Phytonomus nigrirostris, also imported, exists in the United States, and doubtless works as a larva on the clover, as it is known to do in Europe. I have taken this species in considerable numbers along on our Western Michigan lake shore.

Dr. Riley records this insect as very destructive in New York in 1881, and again in 1882. Mr. A. H. Kilman, of Ontario, reports this same weevil at Ridgeway, in that province. He says they were wafted across the lake by a strong August wind. He says that Eastern New York was desolated by the insect in 1883, but that the insect in that year proceeded no further west than Rochester. August 10, 1884, they were so abundant in Buffalo that they could be gathered by the quart, and thousands were crushed by persons walking on the pavement.

As entomologists know, these weevils are armor proof against water, we can easily see how this destructive insect can be easily and quickly distributed along the shores of the northern lakes, and thus soon become a widely known and greatly dreaded pest.

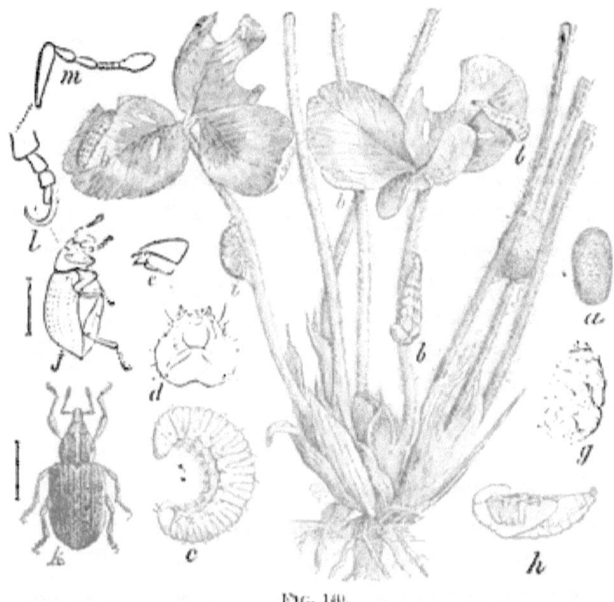

Fig. 140.

Fig. 140 gives a good idea of the insect and its work; *a*, egg; *b b b b*, larvæ; *c*, recently hatched larva; *d*, head of larva; *e*, jaws: *f*, cocoon; *g*, meshes of cocoon; *h*, pupa; *i*, weevil natural size; *j*, side view; *k*, dorsal view; *l*, tarsus and claws of beetle; *m*, antenna.

The eggs are oblong, oval, yellow, 1 m m (1-25 of an inch) long. The larva is yellowish at first, but becomes greenish-yellow with age. There is a pale rose colored dorsal line. The body is rough, length 14 m m (.65 inch). The pupa is well shown in figure and is greenish with yellow markings. It pupates in an oval yellow cocoon of coarse threads.

The imago is dark brown; sides of thorax and elytra dull yellow, with a central yellow line on the thorax. There are rows of black raised points along the inner half of the elytra, with similar dashes of muddy yellow towards the tips. The beetle is 1 c m (2-5 of an inch) long.

The female lays her 200 or 300 eggs in the clover stem, which she punctures for that purpose, in August. Dr. Riley says the eggs are pushed into crevices at the base of the plants.

The larva usually drops when approached, so that only very young ones can be found on the plants. The anus is said to aid in walking, as it emits a sticky substance, and can hold or grasp the stem of the plant. The larvæ are more active at night, but are very timid even then, and can be observed on the plants only at a distance. The larvæ feed upon the clover, and mature in about two months. The pupa state lasts ten days, so that the beetles come forth late in autumn. The cocoon may be formed among the stems of clover or just beneath the earth. The latter is probably the position where it will generally be found in the field. Some of the weevils may deposit eggs in the fall, while others may remain as imagos and lay eggs the next season. There may be two broods in a year, though Dr. Riley thinks it more likely that there is but one, and that variation in size and time of appearance is caused by retarded or accelerated development; while the larva does no inconsiderable damage, far the most is done by the mature beetle. The weevils are voracious eaters, consuming every part of the plant above the earth, and like most weevils feeding by night and hiding by day in crevices in the earth or among the stems of the plants. Like the larvæ they are very timid, and fall at the slightest jar of the plants. It feeds on all kinds of clover, red, white, and alsike. The beetles in July and August often do very serious damage, completing the work of destruction so well begun by the larvæ at an earlier date.

Dr. Riley expresses an opinion, possibly born of hope, that this insect will not spread. I have already shown how it may easily be carried far west, and as we already have seen, it surely is spreading quite rapidly, there is grave reason to fear its general spread in the Northern United States.

As we can not well use Paris green, it is probable that no better thing can be done than to plow under the clover in fields attacked in May, at which time the insects will be in the larva state, and so probably killed by this treatment. If we wait to cut for hay many of the insects would have already pupated, and so would come forth to new mischief the next year.

It is probable that the various predaceous insects will aid to diminish the numbers of this pest, and in time the parasitic insects here as well as in Europe will help to hold it in check.

Cecidomyia trifolii, Leow, Clover-leaf Midge.

Order Diptera. Family Cecidomyidae.

Loew Verhandl. Zöol. Bot., Gesell., XX., 14, p. 142, 1874.

Comstock Ann. Rept. Comm. Ag., 1879, pp. 197–199, Ill.

Lintner Rept. N. Y. Ag. Soc. 1882, p. 203, Ill.

Saunders Rept. Ontario En. Soc., Vol. XII., p. 45, 1881, Ill.

This insect is so nearly like the far more destructive clover seed midge, yet to be described, that only an expert could distinguish them the one from the other. This species has only been discovered about Washington, and unless it becomes more widely distributed, or worse still, learns the habit of its near congener, which is doing widespread and most serious harm, it will be of minor economic importance.

As will be noticed it is closely related to the well-known Hessian fly and wheat midge, so destructive to the wheat crop.

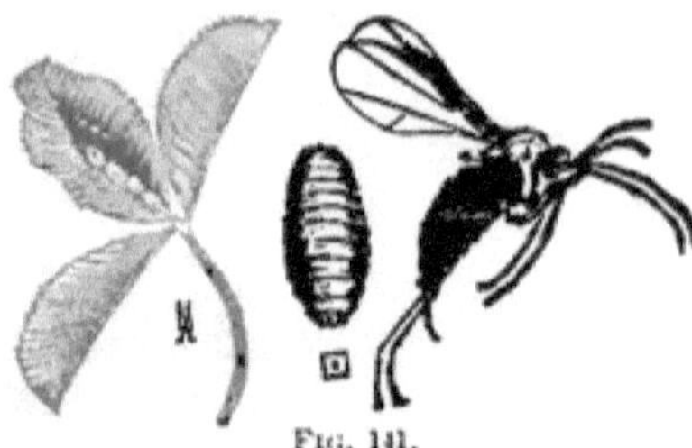

FIG. 141.

Fig. 141 shows the clover leaf concealing the larvæ; one leaf spread open exhibits the cocoons of the insect, the larva and midge or fly. The maggots are folded in the leaflets of the clover, and are at first white, but later assume an orange hue. When full grown this footless larva is 1.5 m m (.059 of an inch) long. The pupæ are enclosed in white delicate cocoons, fastened between the sides of the folded leaflets. See fig. The color of pupa is pale orange. Eyes dark, folded appendages brown. The fly is brown in color with yellowish hairs on the thorax. The female is 1.6 m m long, the male a little shorter. Except that the fly is a little smaller, and that the female has fourteen instead of sixteen joints to the antennæ, it is almost exactly like the clover seed midge, which will be more fully described and illustrated as its importance demands. The minute eggs, from two to twenty, are laid in the creases of the leaflets either of the red or white clover. In June the larva absorbs the juices of the leaflet, causing it to turn brown and to become slightly thickened, showing the tendency to form galls, which is peculiar to many Cecidomyian maggots. The irritation causes the leaflets to fold, thus forming a safe domicile for the defenceless larvæ. Late in June or early in July the flies come forth.

From the exposed condition of this insect it is very likely to become the prey of parasitic insects, and so never become very numerous. Even in considerable numbers it does no very serious harm. and unless it change its habits will never be a serious pest. We hardly need then to discuss remedies for its ravages.

Oscinis trifolii Burgess. Clover-leaf Oscinis.

Order Diptera. Family Oscinidæ.

Burgess. Ann. Rept. Comm. Agrl. 1879, p. 201.

Comstock, Ibid, pp. 200, 201.

Lintner, Rept. N. Y. Ag. Soc. 1882, p. 205.

Fitch described a species of this genus, O. tibialis, which attacks the wheat stem. See Fitch's Rept., 1st and 2d, p. 300, and for illustration Pl. 1. Fig. 5th. Dr. Riley describes Oscinis brassicæ, which attacks the cabbage, Rept. Comm. Ag. 1884, p. 322, which is fully illustrated Pl. VIII., Fig. 5th, which cut would answer in a general way for the Clover Leaf Oscinis. There are several European species which give our friends over the sea some anxiety.

The clover Oscinis is quite like our Anthomyia in habits and general appearance. The eggs are very small and white. The larvæ greenish-white, slender, tapering towards the head. They are 1.7 m m long. The puparium is shorter, oblong, and of a brown color. The fly is yellow, with the dorsal surface of its abdomen and thorax black. It is quite hairy. The length is 1.3 m m, about .05 of an inch. The eggs are probably laid in May or in early June. The larva mines the leaves and stems of white clover, possibly red as well, much as the radish maggot gouges out the plant on which it feeds. Late in June the maggot crawls from its tunnels and falls to the earth, which it enters to form the puparium. The flies appear about two weeks later. There are two and may be three broods a season. If these little sappers and miners ever become so numerous as to do serious injury we will have to resort to feeding our clover down and use ensilage for winter.

49

Tortrix Sulfureana Clem. Clover (attacks grass). Tortrix flavedana Clem. Sericoris instrutana, Clem. Leaf rollers.

Order Lepidoptera. Family Tortricidæ.

Forbes, Ill. En. Report, Vol. XIV., p. 17.

Comstock, Rept. Comm. Agr. 1880, pp. 255–258.

These insects, which are closely related to the codling moth and the apple tree leaf rollers, which are so harmful to our orchards in early summer, are all found in Michigan and the other Northern States, and South even to the Gulf. While they attack all the clovers they are not confined to them, but work on many other garden and field plants. In all the species the larvæ draw the leaves about them by means of silken threads, which they spin, and when disturbed drop and hang suspended by means of a thread, which, like a spider, they can spin as needed. A more harmful leaf roller attacks the clover seed, and will be described later.

The larva of the first species is yellowish-green, the second green, the third yellow. The larvæ are about ½ of an inch (12 to 14 m m) long. The pupæ are shorter and brown in color. The moth of the first species is bright yellow, with 2 v-shaped purple bands on each front wing. The same color marks the front and outer margins of the same wings. The back, or secondary wings, are yellowish, varying to brown. It expands a little more than ½ of an inch. The second species is a little larger. The males are dark brown, with reddish-yellow markings. Hind wings reddish. Females red, with oblique obscure bands across front wings. Females expand ¾ of an inch. The males are not quite so large. The color of the moth in the third species varies from yellow to yellowish-brown. It is about the size of the sulfureana.

These are seen feeding on the rolled up leaves, which serve both for home and food in May and June, and again in August, so there are two broods a year.

I have found Paris green sure death to orchard and shade tree leaf rollers, and without doubt it would kill these that infest the clover. Its practicality however in this case is not so apparent. It is to be hoped that parasites and other enemies will prevent these leaf rollers from becoming very serious pests.

Without doubt other Tortricids will be found to attack the clover, but as all are so nearly alike in their character and habits, what has been said will apply in a general way to all of them.

Drasteria erechtea Cram. Clover Drasteria.

Order Lepidoptera, Family Noctuidæ.

Saunders, Ont. En. Report, 1881, p. 47, Ill.

Saunders, Ont. En Report, 1875, p. 36, Ill.

French, Ill. En. Rep., Vol. VII., p. 133, Ill.

Coquillett, Ill. En. Rep., Vol. X., p. 148.

Packard, Guide to Study of Insects, p. 317.

This is as common as any moth in Michigan, and the same is true in many other States. The familiar, short, jerky flight reminding us of the tiger beetles, is seen from early spring till late autumn. Though so common, and though with slight exception (it sometimes feeds on grass) the caterpillars feed exclusively on clover, yet I think the insect is not considered a foe to be dreaded. It may be that in case of crops like the clover, where plants are numbered by the million, we suffer more from insect attack than we know.

The larva is reddish-brown, marked with longitudinal lines of dark, white and pink color. When full grown it is 3 c m (1¼ inches) long. There are only three pairs of pro legs, so the caterpillar, like others of the lower Noctuids, is a geometer, or "measuring worm." It spins a loose cocoon, in which, as also in its gait, it reminds us of the true geometers. The moth is well represented in the figure. The fore wings are dusky-brown, with darker bands crossing them,

FIG. 142.

one near the base, and another, sometimes incomplete, midway between this and the outer margin, near the apex, is a quite dark patch. Dashes of dull brown are scattered along the wing. The moth expands nearly 3 c m (1¼ inches).

The caterpillars will be seen feeding on the clover all the summer long, and at the same time the moths may be started on their short journeys as we walk over the clover fields.

Colias philodice. Common Yellow Butterfly.

Order Lepidoptera. Family Papilionidæ.

Saunders, Ont. En. Report. 1881 p. 47, Ill.

French, Ill. En. Report, Vol. VII., p. 147.

Packard, Guide to Study of Insects, p. 250.

What was said of the abundance of the Clover Drasteria is even more applicable to our yellow butterfly. Few insects are more common, more widely distributed, or better known than the sulphur-yellow butterfly which gladdens the pasture and roadside, and flecks the damp places along the roadways of all our Northern States. What was said of the food, habits, and destructiveness of the Drasteria erechtea can also be said as truly of Colias philodice.

Fig. 143

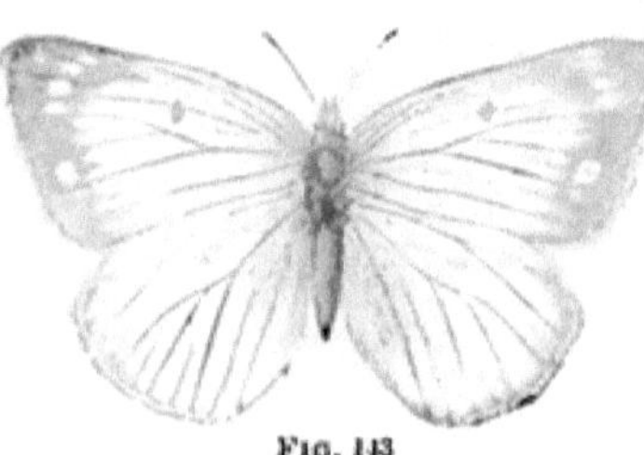

Fig. 143

Fig. 143 shows the male and female of this familiar butterfly. The eggs are long, tapering, ribbed, and though yellow at first change as the embryo develops, first to red and then to brown, just prior to hatching. The young larva is brown with a yellowish tinge. Later it changes to green. The green head has a yellowish-white stripe on each side, with a dash of red at the lower edge. The body is hairy, and when full grown the caterpillar is 2.5 c m. or one inch, long. The chrysalis, like that of our cabbage butterfly, Pieris rapæ, which it resembles, is suspended by an anal tuft and a shoulder girdle. It is pale green, tinted with yellow. On each side of the head is a dark red line, and yellow stripes are seen on the sides near the tip. The yellow, occasionally very nearly white, butterflies with wings bordered with black, sometimes gray, black, or brown are well shown in the figure. The border is narrower in the male, and encloses yellow spots in the female. A black spot is seen on the front wings of both sexes. The secondary wings are bordered with dark in both sexes, and bear an orange spot. The antennae and fringe to the wings are pink. The body is black above and paler below. The size is given in the figure.

The eggs are laid in May and August on the clover, and other leguminous plants, as peas, etc. The caterpillars are feeding from four to six weeks. The chrysalids are fastened to clover or other object, and last about twelve days. The butterflies swarm in May and again in July and August, when they are often so thick as to remind us of a snow storm.

What was said of damages and remedies in considering the last insect applies as well to this one.

Insects Attacking Clover Seed.—While the damage done to clover by some of the insects already described, especially the borers, is quite serious at times, the danger from those attacking the seed is still more formidable, and may well cause anxiety. Happily the number in this list is very limited.

Cecidomyia leguminicola, Lint. Clover Seed Midge.

Order Diptera. Family Cecidomyidæ.

Lintner, Canad. Entomol., XI., p. 44, pp. 121-124, 1879.

Lintner, Rept. N. Y. Ag. Soc., 1878, pp. 62-64.

Lintner, Rept. In. In., 1878, pp. 4-6.

Lintner, Rept. N. Y. Ag. Soc., 1880, pp. 37-41.

Lintner, Rept. Ent. Soc., Ont., 1879, pp. 28-30.

Lintner, N. Y. Ag. Soc., 1882, p. 198, Ill.

Riley, Ann. Rept. Comm. Ag., 1878, pp. 250–252, Ill.
Riley, Ann. Rept. Comm. Ag., 1884, p. 411.
Comstock, Ann. Rept. Comm.1 Ag., 879, pp. 193–197.
Saunders, Rept. Ont. En. Soc., 1881, p. 38, Ill.

This is not only one of the most alarming of our clover pests, but may be regarded as one of the most to be dreaded insects now infesting the valuable crops of the United States. It not only does very serious damage, but is spreading with great rapidity. Prof. Lintner first discovered it in a limited area in Eastern New York. Now—1885—it is known to exist in Virginia, Pennsylvania, New Jersey, Ontario, Michigan, and all through New York. The fact that the insect may remain in the seed, and thus be carried with it any distance, adds to the dangers threatened by this comparatively new pest.

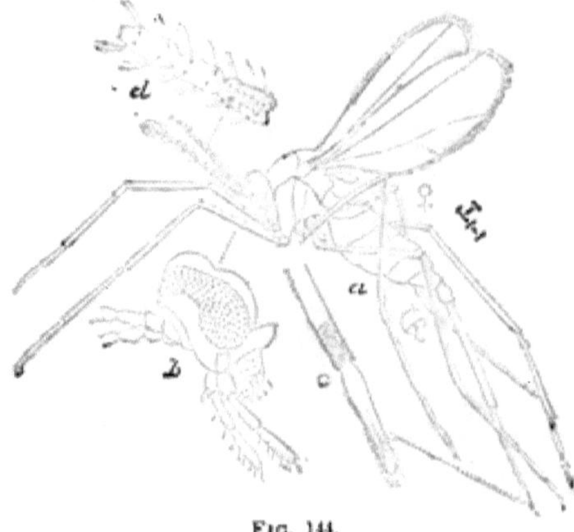

Fig. 144.

Fig. 144 *a* shows the female midge, ovipositor extended; *c*, ovipositor more magnified; *b*, head more enlarged; *d* shows greater enlargement of three joints of antennæ.

Fig. 145.

Fig. 145 *a* shows maggot or larva; *b*, head more magnified.

The eggs are oval, pale yellow, and only .025 m m (.001of an inch) long.

The larva or maggot varies from white to dark orange or orange-red. It is when full grown 2 m m (1-12 inch) long.

The pupa is orange, with brown eyes. It is found in a tough silken cocoon with more or less earth sticking to it,

As will be seen by the figures the flies resemble closely the wheat midge, C. tritici. The abdomen is red, thorax brownish-red. The antennæ are 15-jointed in the male, and 16 in the female. The wings are hairy, the palpi and ovipositor each four jointed. The male is about 1.5 m m long, the female about 3 m m. The male expands about 3.5 m m, the female 4 m m. The size varies a little. The dark scales obscure the red color, so that the flies appear dark. Underneath the color is yellowish-gray. As with the wheat midge and Hessian fly the ovipositor and clasping organs are very prominent.

The eggs are pushed, by means of the ovipositor, down into the heads of red or white clover, and lodged between the hairs that surround the separate florets. They are not glued nor placed in the florets. As with the Hessian fly the eggs may be laid singly, or in groups of two, three, four or five. As many as 50 eggs are sometimes placed in a single head of clover.

The larva affects each seed much as does the wheat midge each wheat kernel.

After absorbing the life from the seed the larva, like the maggot of the wheat midge, leaves the seed and wriggles till it escapes from the clover head and falls to the earth. Often the head of clover seems alive as a maggot pushes from nearly every seed in its effort to reach the ground. The pupa is found in its cocoon just beneath or upon the earth, under some protecting leaf, etc. The flies have been seen in New York in May, August, and quite likely some flies may issue in October. Thus there are surely two broods in New York, and possibly three. There are certainly three farther South. The larvæ will be seen full grown in the seed at the North in June, in July, and again in September. They probably pass the winter as pupæ.

Of late the larvæ have been found in seed in the market.

This is an unwelcome fact, and explains the rapid distribution of these insects.

If we can bring the second, or seed crop of clover between the two broods of the midge, the seed will be saved. Late pasturing or early cutting of the first crop will accomplish this. If the early crop be mowed a few days before timothy spikes appear, it will kill the young maggots and will bring the second crop at the desired time. Two parasites, Eurytoma funebris and Platygaster error, destroy a great many of the maggots. Already Mr. Howard finds Eurytoma funebris and Platygaster error engaged in this good work. Success to them, and may their tribe increase.

If seed is found stocked with the larvæ it should be put into a close vessel, as a jug or barrel, and bisulphide of carbon added. This will kill all the larvæ post haste. Even an open barrel, water tight, may be used by placing a buffalo robe, or other air-tight cover, over it. The fact that this insect is as far West as Michigan, and possibly as far as Illinois, makes it a matter of general interest. In the future, clover seed will be valuable.

Grapholitha interstinctana, Clem. Clover-head Caterpillar.

Order Lepidoptera. Family Tortricidae.

Comstock, Rept. Comm. Agr. 1880, p. 254.

Clemens, Proceed. Acad. Nat. Sci. Phil. 1860, p. 351.

Grote, Bull. Buffalo Soc. Vol. I., p. 92.

These caterpillars are also quite widely distributed. Grote and Comstock have taken them in New York, Grote in Pennsylvania, Comstock in Washington, and I have found them quite common in Michigan. A single larva feeds on several and often all the seeds of a single head of clover. Sometimes I would find two larvæ in a single head. The insect was quite common about Lansing last summer—1885.

The larvæ are dirty white, often greenish, 8 m m long, and spin white cocoons in the clover heads. The bodies bear many white hairs

Chrysalis light brown; 5 m m long. The anal segment bears six hooks, two dorsal, four lateral. The moths are small, brown, often nearly black, with white lines and dots marking the wings. They expand 10 m m.

There are three broods of the moth appearing respectively in early June, August and September; the larvæ appear soon after.

The same remedies which are successful against the clover-seed midge are effectual in checking the work of this moth. An Ichneumon parasite was found preying on these by Prof. Comstock. It is Phanerotoma tibialis. It is light brown, with a large dorsal yellow spot, and is 3.5 m m long.

Insects Attacking Clover Hay. Asopia costalis, Fabr. Clover Hay-worm.

Order Lepidoptera. Family Pyralidæ.
Riley, Mo. Ent. Rept. Vol. VI., p. 102, Ill.
Saunders, Ont. Ent. Rep. 1880, p. 45, Ill.
French, Ill. Ent. Rep., Vol. VII., p. 47.

This insect works on dried clover or clover hay while in the mow or stack. It is generally distributed, and scarcely a season goes by that I do not receive specimens, with request for information regarding the natural history and habits of the insect. This belongs to the same family as the bee moth and the meal moth, Pyralis farinalis, Harr., which is often very common about barns where meal is stored, and which sometimes also feeds on clover hay.

The color of the larva, Fig. 146, is dark brown, lighter beneath. The intersegmental spaces are darker than the segments, which makes the larvæ appear to be ringed. It is 18 m m (¾ of an inch) long. The cocoon is white, and 12 m m long. The chrysalis is yellow in color; length ⅓ of an inch. The imago, or moth, is purple, with a silken lustre. There are two bright yellow spots on the primary wings. The posterior wings are lighter in color than the primaries. All the wings are margined with orange, which terminates with a glossy yellow fringe. They expand about 2 c m, or .8 of an inch.

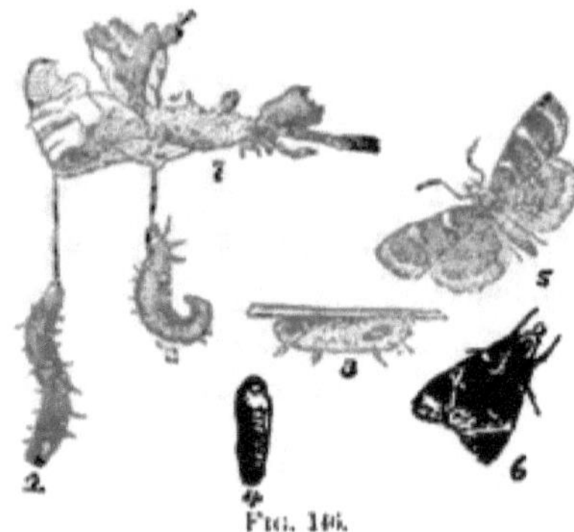

Fig. 146.

In Fig. 146, 1 and 2 shows the larvæ suspended by threads; 3 represents the cocoon; 4 the chrysalis; 5 moth with wings spread; 6 moth at rest; and 7 larva concealed in a case of silk which it has spun.

These moths are attracted by lights, and are often seen about our lamps in mid-summer. The eggs are laid on clover. The larvæ work in a silken case, and so often fairly mat the hay in one great mass. The larvæ attract attention in summer working upon the hay, but more usually in February and March, when stacks and mows of clover may be fairly alive with larvæ. These often crawl far into the stacks, where they are protected from cold, and so are sometimes said to bear a zero temperature without becoming dormant, though the truth is they have had a warm nest. I have seen them drop from a mow suspended by a silken thread, so thousands could be swept away by one stroke of a rake. They often leave stack or mow and seek some concealed place in which to pupate.

It is probably true that leaving clover hay in mow or stack year after year will promote the rapid increase of these pests. Feeding out all the hay each winter wouad be a wise precaution, or if any hay is to remain over let it be other than clover.

Of the many other species mentioned at the beginning of this article none are as yet sufficiently important to demand full description. Most work as much if not more on other plants. Some are very rare insects, and others, though common, seem not to attract any general attention by their presence.

Insects Injurious to Grass Crops.—The insects which are known to attack our grasses make even a more formidable list than those injurious to the clovers. Eighty or more species either depend wholly or in part upon our grasses for food. In the following list Hy. after the name indicates that the insect belongs to the order Hymenoptera; Lep., Lepidoptera; Dip., Diptera; Col., Coleoptera; Hom., Homoptera; Hem., Hemeptera; Or., Orthoplera. Ill. refers to Ill. Entomological Reports; Mo. to Missouri Entomological Reports; U. S. Reports, U. S. Commissioner of Agriculture; Harr., Harris Injurious Insects; Ont., Ontario Entomological Report; Pack., Packard's Guide to the Study of Insects; Streck., Strecker's Catalogue of Macrolepidoptera; Fitch, Fitch's N. Y. Reports:

Acridium Americanum. Drury, *Or.*, Ill., Vol. IX., p. 129. Mo., Vol. VIII., p. 103, Ill.

Agonoderus, all of the species, *Col. Ill.*, Vol. XII., p. 111.

Agrotis c. nigrum. Linn. *Lep.*, Ill. Vol. VII., pp. 89, 202. Ibid. Vol. X., p. 132.

Agrotis fennica, Tausch, *Lep.*, Rep. Mich. St. Board Ag. 1883, p. 423. Ont. Vol. XV., 1884, pp. 13, 15, 21, 24. Mich. Hort. Rep. 1884, p. 81.

Agrotis saucia, Hüb., *Lep.*, Ill. Vol. VII., pp. 94, 211. Ibid. Vol. X., p. 134. Mo., Vol. I., p. 74. U. S. 1881, p. 297, Ill. Harr., p. 444.

Agrotis tessellata, Harr., *Lep.*, Ill. Vol. VII., pp. 91, 206. Ibid, Vol. X., p. 133. Harr., p. 445. Ont., Vol. X., p. 39.

Amara, all the species, *Col.*, Ill. Vol. XII., p. 110.

Anisodactylus, all the species, *Col.*, Ill. Vol. XII., p. 111.

Aphis Maidis, Fitch, *Hom.*, Fitch, Vols. I. and II., p. 318. Ill. Vol. XIII., p. 46.

Arctia (Leucarctia) acraea, Sm., *Lep.*, Harr., p. 351. Pack., p. 286, Ill. Vol. VII., pp. 79, 183. Ibid., Vol. X., pp. 115, 170. Ibid. Vol. XI., p. 62.

Arctia phalerata, Harr., *Lep.*, Harr., p. 347. Ill. Vol. VII., p. 181. Ibid, Vol. X., p. 115.

Blissus leucopterus, Say, *Het.*, Harr., p. 198. Ill. Vol. VII., pp. 15, 40. Ibid, Vol. XII., p. 32. Ill. Mo., Vol. II., p. 15, Ill. Ibid., VII., p. 19.

Bryobia pratensis, Gar., *Acarina*. Ill. Vol. XIV., p. 73.

Bryobia pallida, Gar., *Acarina*, Ill. Vol. XIV., p. 74.

Calathus gregarius, Say, *Col.*, Ill. Vol. XII., p. 109.

Caloptenus bivittatus, Say, *Or.*, Ill. Vol. IX., p. 126. Mo., Vol. VII., p. 173. Ill.

Caloptenus differentialis, Thos., *Or.*, Ill. Vol. IX., p. 127. Ill. Mo., Vol. VII., p. 173. Ibid, Vol. VIII., p. 153.

Caloptenus femur-rubrum, De G., *Or.*, Ill. Vol. I., p. 99. Ibid, Vol. VII., p. 35, Ill. Harr., p. 174, Ill.

Caloptenus spretus, Tho., *Or.*, Ill. Vol. I., p. 82. Ibid, Vol. VII., p. 35, Ill. Ibid, Vol. IX., p. 121. Mo., Vol. VII., p. 121. Ibid, Vol. VIII., p. 57. Ibid, Vol. IX., p. 157. Report of U. S. Entomological Commission.

Chytolita morbidalis, Guen., *Lep.*, Ill. Vol. X., pp. 138, 182.

Coccinellidæ, *Col.*, Ill. Vol. XII., p. 116.

Cotalpa lanigera, Linn., *Col.*, Ill. Vol. XIII., p. 146, Ill. American Naturalist, 1869, pp. 186, 411. Harr., p. 24, Ill. Mo. Vol. V., p. 10.

Crambus vulgivagellus, Clem., *Lep.*, Lintner's Ent. Report, Vol. I., p. 127. Canadian Entomologist, Vol. XII., p. 17. Ibid, Vol. XIII., p. 181. Am. Nat., Vol. XV., pp. 574, 750, 914. Ont. 1881, pp. 6, 13. U. S. 1881-1882, p. 179.

Ctenucha virginica, Char., *Lep.*, Ill., Vol. X., p. 170. Lintner's En. Con., Vol. III., p. 155. Pack., p. 283.

Debis Portlandia, Fab., *Lep.*, Ill. Vol. X., p. 92. Strecker's Catalogue, p. 148.

Dichelia Sulfureana, Clem., *Lep.*, Ill., Vol. XIV., p. 17. Fernald's Catalogue, p. 21. U. S. 1880, p. 255.

Drasteria erechtea, Cram., *Lep.*, Ill. Vol. X., p. 118. Ont. 1881, p. 47, Ill. Ibid, 1875, p. 36.

Elateridæ, *Col.*, Harris, p. 55. Ill. Vol. V., p. 92. Ibid, Vol. VI., p. 21. Ibid, Vol. VII., p. 19. Ibid, Vol. XII., p. 27. Mo., Vol. II., p. 16. Fitch, Vol. X., p. 63.

Elater mancus, Say, *Col.*, Harris, p. 56.

Galerita janus, Fab., *Col.*, Ill. Vol. XII., p. 108.

Glyphina eragrostidis, Midd., *Hom.*, Ill. Vol. VIII., p. 144.

Gortyna nitela, Guenee., *Lep.*, Ill. Vol. VII., p. 100. Ibid, Vol. X., p. 151. Mo., Vol. I., p. 56. Ibid, Vol. III., p. 105. Ibid, Vol. VIII., p. 37.

Hadena devastatrix, Bruce, *Lep.*, Ill. Vol. VII., p. 216. Mo. Vol. I., p.

83. Ibid, Supplement, p. 56. Harris, p. 445. Fitch, Vols. I. and II., p. 315.

Harpalus, all of this Genus, *Col.*, Ill. Vol. XII., p. 112.

Heliophila phragmitidicola, Guenee, *Lep.*, Ill. Vol. VII., p. 224.

Heliophila Harveyi, Grote (albilinea, Hübn) *Lep.*, Mo. Vol. IX., p. 50. Ill. Vol. VII., p. 223.

Isosoma Elymi, French, *Hy.*, Ill. Vol. XI., p. 81. Canadian Entomologist, Jan., 1882.

Isosoma hordei, Harr., *Hy.*, Ill., Vol. XI., p. 75. Fitch, Vol. VI.-IX., p. 154. Harris, 553. Mich. Ag. Rept. 1884, p. 322. Mo. Vol. II., p. 92.

Lachnosterna fraterna, Harr., *Col.*, Ill. Vol. VI., p. 101. Harris, p. 32.

Lachnosterna fusca, Fröhl. *Col.*, Ill. Vol. VI., p. 97. Ibid, Vol. VII., p. 32. Mo. Vol. I., p. 156. Ill. Harris, p. 30. Fitch, Vols. I. and II., p. 248.

Lachnosterna hirticula, Knoch., *Col.*, Ill. Vol. V., p. 87. Harris, p. 32.

Lachnosterna ilicis, Knoch., *Col.*, Ill. Vol. V., p. 87.

Lachnosterna pilosicollis, Knoch., *Col.*, Harris, p. 33. Ill. Vol. V., p. 87.

Laphygma frugiperda, Guenee. *Lep.*, Ill. Vol. VII., pp. 97, 219. Ibid, Vol. X., p. 138, Vol. XIV., p. 55. Mo. Vol. II., p. 41.

Leucania pseudargyria, Guenee, *Lep.*, Ill. Vol. X., p. 139.

Leucania albilinea, Hübn., *Lep.*, Mo. Vol. IX., pp. 50-55.

Leucania unipuncta, Haw., *Lep.*, Harris, p. 627. Ill. Vol. VI., p. 56, VII., p. 101. Mo. Vol. I., p. 109. Ibid, Vol. II., p. 37. Ibid, Vol. VIII., pp. 22, 182. Ibid, Vol. II., p. 37.

Leucarctia acraea, Smith, *Lep.*, Ill. Vol. VII., p. 183. Ibid, Vol. X., p. 170, Packard, p. 286.

Limothrips poaphagus, Com., *Het.* Fernald. Grasses of Maine, p. 42. Comstock Notes on Entomology, p. 120.

Loxopeza atriventris, Say, *Col.*, Ill. Vol. XII., pp. 109, 115.

Lygus lineolaris, Beauv., *Hem.*, Ill. Vol. XIII., p. 115. Mo. Vol. II., p. 113. Harr., p. 201. U. S. Vol. 1881, p. 312.

Macrodactylus subspinosis, Fabr., *Col.*, Ill. Vol. I., p. 21. Ibid, Vol. VI., p. 103. Lintner Rept., Vol. I., p. 227. Harr., p. 35. Fitch Vol. II., p. 245. U. S. 1863, p. 567, 1867, p. 71, 1868, pp. 87, 104. Am. Entomol., Vol. I., p. 251. Mich. Pom. Report 1872, p. 667. Mich. Ag. Report 1874, p. 145.

Neonympha Canthus, Linn., *Lep.*, Ill. Vol. X., p. 91.

Neonympha eurytris, Fab., *Lep.*, Ill. Vol. X., p. 90. Strecker's Catalogue, p. 148. Harris, p. 306. Pack., p. 264.

Neonympha gemma, Hüb., *Lep.*, Ill. Vol. X., p. 91. Strecker's Catalogue Macrolepidoptera, p. 150.

Neonympha phocion, Fabr., *Lep.*, Strecker's Catalogue, p. 149. Buffalo Bulletin, Vol. II., p. 244.

Neonympha sosybius, Fabr., *Lep.*, Ill. Vol. X., p. 91. Strecker's Catalogue, p. 149. Buffalo Bulletin, Vol. II., p. 145.

Nephelodes violans, Guen., *Lep.*, Ill. Vol. VII., pp. 29, 220. Ibid. Vol. X., p. 139. Lintner's En. Report, Vol. I., p. 90. Am. Ento., Vol. III., p. 231. Am. Nat., Vol. XV., p. 575. Canadian Entomologist, Vol. VIII., p. 69. Trans. Kan. Acad. Science, Vol. IV., p. 45.

Pamphila Delaware, Edw., *Lep.*, Ill. Vol. X., p. 96.

Pamphila vitellius, Fabr., Strecker's Cat., p. 171. Proceed. Ent. Soc. Phil. II., 18, 19.

Pamphila hobomok, Harr., *Lep.*, Strecker's Cat., p. 172. Harr., p. 313. Canadian Ent., Vol. I., p. 66. Proceed. Bos. Soc. Nat. Hist., Vol. XI., p. 381. Ill. Vol. X., p. 97.

Pamphila, Iowa, Scud., *Lep.*, Strecker's Cat., p. 173.

Pamphila Mystic, Edw., *Lep.*, Strecker's Cat., p. 165. Proceed Ent. Soc. Phil., Vol. II., p. 15. Canadian Ent., Vol. I., p. 66. Packard, p. 270.

Pamphila Peckius, Kerby, *Lep.*, Ill. Vol. VII., p. 160. Ibid. Vol. X., p. 178. Harris, p. 315.

Pamphila phyleus, Dru., *Lep.*, Strecker's Cat., p. 164. Ill. Vol. X., pp. 96, 176.

Pamphila Samoset, Scud., *Lep.*, Strecker's Cat., p. 174. Proc. Ent. Soc. Phil. Vol. II., p. 507. Trans. Ent. Soc., Vol. I., p. 3.

Pamphila Sassacus, Harr., *Lep.*, Harr., p. 315. Ill. Vol. VII., p. 159. Ibid. Vol. X., p. 97. Boston Soc. Nat. Hist., Vol. II., p. 346.

Patrobus longicornis, Say, *Col.*, Ill. Vol. XII., p. 113.

Philometra serraticornis, Grote, *Lep.*, Ill., Vol. VII., p. 246.

Platynus limbatus, Say, Ill. Vol. XII., p. 109.

Plusia simplex, Guen., *Lep.*, Ill. Vol. IX., p. 48. Ibid. Vol. XI., p. 38.

Pseudoglossa lubricalis, Geyer, Ill., Vol. X., pp. 138, 182.

Pterostichus, all of the Genus. *Col.*, Ill. Vol. XII., pp. 110, 115.

Pulvinaria innumerabilis, Rath., *Hom.*, Ill. Vol. XIV., p. 103. Mich. Ag. Rep. 1883, p. 429. Ill. Am. Naturalist, Vol. XII., pp. 655-661. Proc. Dav. Ia. Acad. Sci., Vol. II., p. 293. U. S. 1884, p. 350.

Rhizobius poæ, Thom., *Hom.*, Ill. Vol. VIII., p. 166.

Satyrus alope, Fab., *Lep.*, Ill. Vol. VII., p. 156. Ibid. Vol. X., p. 92. Strecker's Cat., p. 157. Harris, p. 305.

Satyrus nephele, Kirb., *Lep.*, Ill. Vol. VII., p. 156. Ibid, Vol. X., p. 92. Vol. VI. Proceed. Ent. Soc. Phil., p. 195. Harr., p. 306. Bull. Buffalo Soc., Vol. II., p. 242.

Scelodonta pubescens, Mels., *Col.*, Ill. Vol. XIII., p. 163.

Scepsis fulvicollis, Hubn., *Lep.*, Ill., Vol. X., p. 171.

Schizoneura panicola, Thom., *Hom.*, Ill. Vol. VIII., p. 138. Ibid, Vol. XIII., pp. 42, 51.

Sciara? *Dip.*, Ill. Vol. XIII., p. 59.

Selandria? Saw Fly, *Hy.*, U. S. 1884, p. 401.

Siphonophora avenæ, Fab., *Hom.*, Fitch, Vol. VI.-IX., p. 91. Mo. Vol. II., pp. 5, 6. 10. Ill. Vol. VIII., pp. 29, 51.

Siphonophora setariæ, Thom., *Hom.*, Ill. Vol. VIII., p. 56.

Spilosoma Virginica, Fabr., *Lep.*, Harris. p. 349. Ill. Vol. IV., p. 188, Ibid, Vol. VII., pp. 80, 183, 277, 280. Ibid. Vol. X., pp. 116, 169. Packard, p. 287. Mo. Vol. III., p. 68.

Sphenophorus parvulus, Gyll., Forbes' Notes. Seen eating grass in July and August.

Sphenophorus sculptilis, Uhl. *Col.*, Lintner, En. Rep., Vol. I., p. 253. Mo. Vol. III , p. 59. U. S. 1879, p. 248. Ibid. 1880, p. 272. Ont., 1880, p. 56. Am. Nat., Vol. XV., p. 915. U. S. 1881-1882, p. 139.

Tychea panici, Thom., *Hom.*, Ill. Vol. VIII., p. 169.

In the above list I have given only such species as I know, or have good reason to believe, feed in part or wholly on grass. I have given references that the literature may be more easily investigated, though many authors referred to do not speak of the insects as enemies to our grasses, they do give habits and characters which are important.

In the list given others might very safely have been included. It is probably true that all the species of Lachnosterna—allies of our May or June beetle, the common white grub—and many species of related genera, are injurious to grasses, as they quite generally feed on the roots of these plants in the grub or larval state. It is also probable that others of the genera Agrotis, Hadena, etc., perhaps all the cut worms, are enemies of our meadows and pastures. Till within a couple of years Agrotis fennica, the Black Army Worm, was supposed to be one of the most inoffensive of these moths. Now we know that it may devastate whole meadows. Recently a Paralid, or snout moth, Crambus vulgivagellus, Clem., which has been supposed to be innoxious, did immense damage in Northern New York. The same moth is seen each year in Michigan and other States, and we do not know when it, or other species of the same genus, may not come to any locality in our Northern States to the ruination of our meadows and pastures. Another moth, Nephelodes violens, Guenee, has had a history similar to that of the Crambus, just mentioned. This moth I find while trapping moths by sugar every year here in Lansing, sometimes in great numbers. We cannot tell when it may come in devastating numbers in any locality in the United States. It is probable that several species of Elaters—spring beetles—the dreaded wire worms, are great pests to our meadows. It is quite likely that they do far more damage to grasses than is known or suspected. The same may as truly be said of the army worm moth, and other species of the Heliophila (Leucania). We note their ravages only when they come in armies. Yet I notice they are quite common every year, and as they are not usually driven by force of numbers to leave the meadows for other pasturage their blasting work, though not inconsiderable, is unnoticed. Many species of grass-hoppers, not mentioned in the above list—indeed nearly all of our locusts—are at times more or less destructive to grasses, and like the cut worms, wire worms, white grubs and army worms, work unperceived. Only when they come in swarms, as they have the past season (1875), do they attract attention.

In the above list I have not included any of the Chlorops, or Oscinis, but from the habits of the closely related Meromyza, as wheat enemies, the abundance of the flies of these genera on grass in summer, and the added fact that we often find the maggots mining in the culms makes it possible that they do more or less harm to our species of Gramineæ. It has been thought that these maggots were what caused the June grass to wither in summer, as so frequently observed. This is more likely due to species of thrips, three or more of which I have taken from the culms. Sometimes the grass withers from the attack of the stalk borer, Gortyna nitela.

I have also omitted all mention of leaf hoppers in the above list. Yet it is not improbable, indeed I think it certain, that various species of Tettigonia Heleochara and Jassus may and do often quite considerable damage to our grass crops. The larvæ of these tree or leaf hoppers are often seen enveloped in their spittle, like secretions, on our grasses, and as such insects must suck all their nutriment from the grass, they can but be quite a serious damage. Of the Hemiptera, with the exception of a few lice, plant and bark lice, the tarnished plant bug and the chinch bug, the above list speaks not. It is quite likely that other plant and bark lice, and several Heteroptera, especially of the genera Capsus and Phytocoris, may be found to work no inconsiderable harm to our grasses.

It will be seen that there are included in the list several genera of the family Carabidæ, all of which have been considered heretofore as predaceous species, and so beneficial. Prof. S. A. Forbes has well shown that many of these ground beetles are largely vegetable feeders, and that grass is the principal food of most of these species. Prof. Forbes also finds that nearly if not all of the Coccinellidæ (lady bird beetles) feed in part on the pollen of grasses. These probably do very little harm.

As was remarked in reference to the insects infesting our clovers, many in the above list live in part on other plants, and many do very little apparent harm to pasture or meadow. A detailed description will be given only of such species as are noted enemies.

51

Lachnosterna fusca, Frohl. May Beetle—White grub.

Order Coleoptera. Family Scarabæidæ.

Without doubt the White Grub, which is the larva of the common May beetle, though probably other species of this same and allied genera are much like this one in appearance and habits, is one of the very worst enemies of the grasses, as by eating off the roots whole meadows, pasture fields, and lawns are entirely ruined. Often the roots of the grass are so entirely consumed that all may be raked off, leaving the entire field as clean as a well tilled summer fallow.

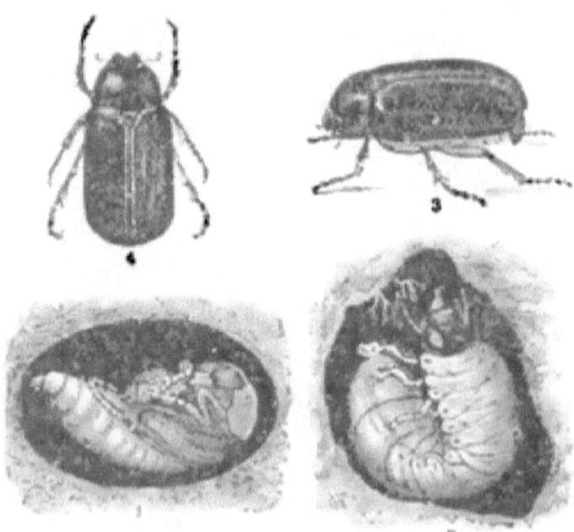

Fig. 147.

The figure (147) shows the insects in the several stages so well that little else is needed. The eggs are laid in the grass fields. The white, usually curved and wrinkled grubs with brown heads, feed for three years. The pupa is found in earthen cells, and is not different from other coleopterous pupæ.

In May and June the beetles come forth from the earth, and the females lay their eggs each to the number of from forty to sixty. It is probable that these eggs are always laid either in grass plats or where other vegetation is rank and plentiful. The beetles are nocturnal, and as is well known are attracted by lights, and so often become very annoying as they enter our rooms and houses. Sometimes the beetles so swarm in trees as to remind one of a swarm of bees. Indeed they often do no little mischief in eating the foliage of oaks and other trees during these summer love feasts. The grubs eat for three seasons.

Thus they are often found in the same grass field of varying sizes. The third spring they transform to pupæ, and in May the beetles begin to appear. It is during the second summer that they do the most harm. They are now large and sleek, and when they are very numerous, as is often the case, they sometimes do great damage, not only to grass but to our cereal crops and corn.

Fall plowing, by giving the birds and other insectivorous animals a better chance, is often practiced with excellent results in fighting these pests. Sometimes swine is turned into the meadows where they fatten on the grubs instead of on the grass which the grubs have already destroyed. Nothing is better where a field is badly infested than to turn in swine. In lawns the bare space must be spaded up and either sodded or else new grass seed sown. Rolling, which is sometimes advised, will do little or no good. I have already spoken of birds. The crow, blackbird, or purple grackle is specially serviceable. I have seen a flock of those birds clean a lawn in exceedingly quick time. Predaceous wasps and beetles also prey upon these grubs. They are also often seen to afford a pasturage for large fungous growths, which destroy them. Not only do white grubs do harm to our grasses, but they also attack corn, wheat and strawberries which are planted upon sod, and the latter when grown for a series of years in one place.

Agrotians. Cut Worms.

Order Lepidoptera. Family Noctuidæ.

Not only the real cut worms of the genera Agrotis, Hadena, and Mamestra, but many species of the same genera that do not cut off the food as do the typical cut worms, are often injurious to the grasses. From the very nature of our grasses much harm might be done, and yet, unless it were very great, go unnoticed by the practical man. It is more than likely that with the more intensive agriculture of the future, made necessary by a more dense population, note will be taken of these injuries which now are unnoticed. While we may believe that most of our cut worms attack and destroy many a spear of grass in pasture and meadow, we actually know that Agrotis fennica may utterly devastate whole meadows, as it has done in parts of Michigan for the past two years.

These gray, sober colored noctuid moths, Fig. 148, lay their scores of white eggs upon the stems of grasses. The larvæ, Fig. 149, may be dirty white or variously striped. Those which spend the day beneath the ground are more frequently light colored. They pupate in the earth. Fig. 150 shows the pupa.

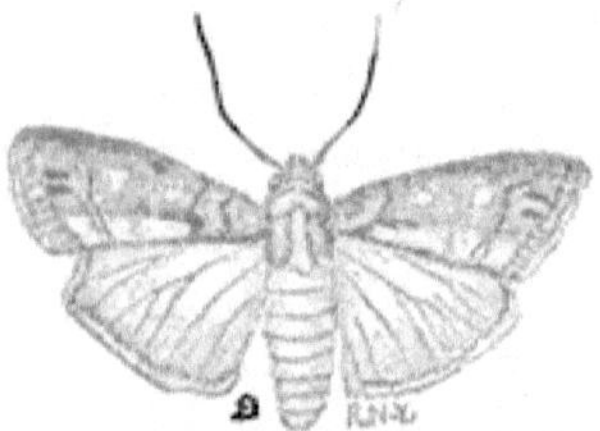

Fig. 148. Agrotis fennica and wing of Var.

The moths of different species may be found from June to October. Agrotis fennica is seen as a moth in July. It is probably true of all the species that the eggs are laid soon after the moths appear. In most cases these hatch the same season, and the larvæ become partly grown, but do their greatest mischief the following May and June. The eggs of A. fennica do not hatch till spring, when the larvæ eat ravenously and grow very rapidly. So we see that in all cases June is the dreaded month when these insects lay heavy tribute on the produce of the farmer.

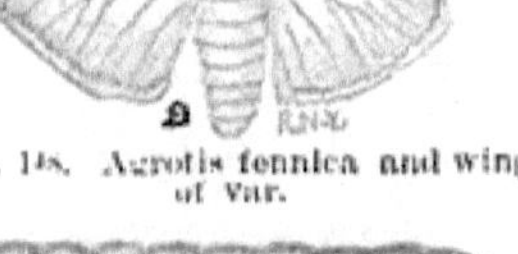

Fig. 149.

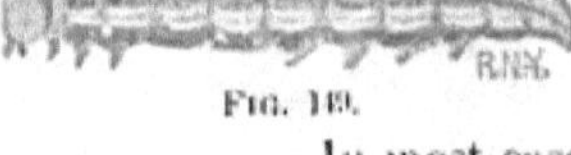

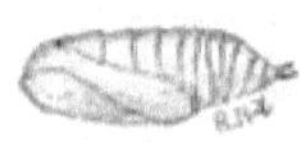

Fig. 150.

We must depend on the natural enemies very largely to overcome these injurious insects in our grass fields. The extent of the area of grass fields, the number of insects and their concealed condition makes all kinds of known warfare impracticable. When they cover a field, as did the A. fennica the meadows in Bay county, Michigan, we may adopt the same remedy as in case of the white grub, give up the fields to the swine. While we may bandage our grape-vines, fruit trees, and garden plants, and thus protect them, and while we may bait the cut worms of clean cultivated corn fields with bunches of grass poisoned with the arsenites and thus kill them, or later dig them out at a profit, none of these methods are available in the meadow.

Leucania unipuncta, Haw. Army Worm.

Order Lepidoptera. Family Noctuidae.

This in ect is so largely the prey to insect enemies, parasitic and predaceous, that it is only rarely that it does marked injury. Yet the entomologist knows that the moths are very common each year, and there can be no doubt but that it does considerable injury in our grass fields every season. It is only when its numbers, through favorable surroundings, become so immensely numerous as to make it necessary for the caterpillars to swarm forth from the meadows to get food, that we usually take note of its presence or become conscious of its power for mischief.

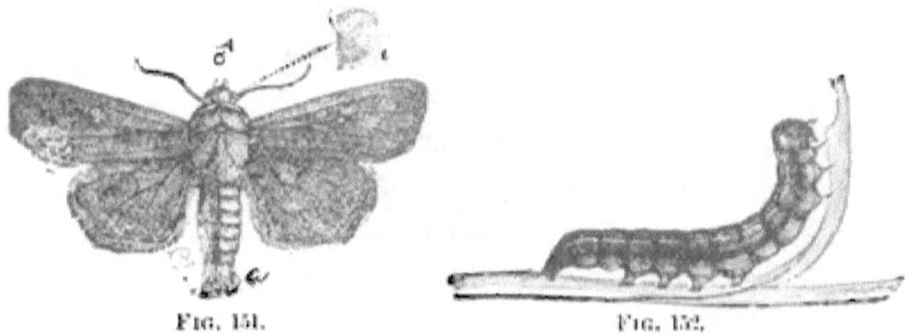

FIG. 151. FIG. 152.

The figures show well the appearance of the insect in its several stages. The moth, Fig. 151, is yellowish-brown, often with a greenish tinge with a discal white dot on each front wing, which gives the specific name.

The caterpillar, Fig. 152, is striped longitudinally with dark and light gray lines. It pupates like all noctuids in the earth.

The moths are abundant in August and September. The eggs are laid in the sheaths of the grass. The caterpillars are nearly grown in July, and then is when they devastate meadow and oat field.

Here as elsewhere, with the enemies of our grasses, we must trust largely to the other insects and birds that prey upon them. Usually this is sufficient to so reduce their numbers that their presence causes no anxiety, or even makes itself known to the farmer. When they migrate, in armies, threatening meadow and grain field, it is recommended to scatter straw, and when they become involved in its meshes to burn them, straw and all. Also to poison with the arsenites a portion of the grain or grass on the side of the field towards which the army is advancing, and thus hope by wholesale poisoning to save a portion of the crop. This has usually failed, as the army is often so large that they can devastate acres even though poisoned in this manner. The most satisfactory method yet recommended has proved to be the furrow or ditch. This is left steep on the side toward the field to be protected, and to have holes, like post holes, dug in it at frequent intervals. These holes receive the caterpillars, and the latter, by use of a convenient stick, large at one end, may be quickly killed, and so the holes made ready for a fresh lot of victims. A board fence of slight height has been made to take the place of the ditch in some cases with good results. Of course the encouragement of our insectivorous birds will aid here, as everywhere, to help solve this insect problem.

Elaters. Wire Worms.

Order Coleoptera. Family Elateridae.

The wire worms, like the white grubs, are the larvæ of beetles, live between two and three years in the earth, and by feeding upon the roots often do great injury to cereal crops, corn, and, though not so much dreaded in meadows and pastures, they are, beyond question, often quite injurious to nearly all our grasses.

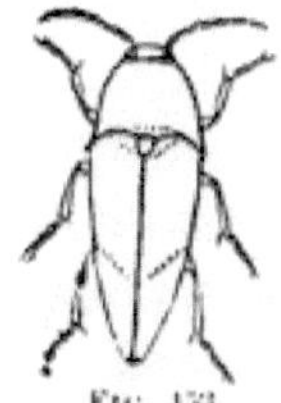
FIG. 153.

The various species of spring beetles, Fig. 153, are seen in June, and not infrequently fly into our rooms. Their long, slim form, usually brown color, and especially their habit of springing when placed on their back, which is effected by a sort of ventral spring pole arrangement, give ready means to identify the beetles. The long, cylindrical grubs, Fig. 154, with their six jointed thoracic feet, are also hard to mistake. Indeed the name wire worms is very appropriate.

FIG. 154.

As in case of white grubs the eggs are laid in meadows and pastures about the roots of grasses, where for three years the slender grubs eat and grow. While complaint is not usually made of injury to grass, yet such injury must be common and extensive. The grass blades are so countless that though numerous plants are killed they are not missed; but let the sward be plowed, and the second year plant corn, or sow oats or wheat, and if the wire worms are present—they are now rapidly approaching maturity—they often do incalculable damage, ruining, it may be, whole fields of grain. That they do not more injury the first year after plowing is not so strange. It is the habit of the grubs of this family of beetles to eat rotten or decaying wood, etc., and so it is quite likely that these wire worms, with changed habits, really prefer a diet of decaying roots for a change, especially as it may the better satisfy the cravings of the old time inherited appetite. With the exception of buckwheat, peas and beans, there is hardly a crop but what is levied upon by these insatiable wire worms. The only recommendation that our present knowledge offers to resist these terrible pests is either to summer fallow for one one year in hopes to starve the grubs, or else to sow some crop that is distasteful to them the second year after plowing the green sward.

Blissus leucopterus, Say. Chinch Bug.

Order Hemiptera. Family Lygæidæ.

This destructive bug, though very small, is often so terribly injurious that in Illinois, Iowa, Missouri and Kansas it is often the author of millions of dollar's worth of damage, and that some times in a single State. What has been said of the other insects already referred to as to damage to grass, corn and wheat applies to this as wel.. That the chinch bug is more susceptible to seasonal peculiarities—especially wet—than most insects, is well known. While in very wet years it does little damage, in dry years it sweeps as "with the besom of destruction" the great prairies of the West. That it does so little damage in Michigan, New York and the East is doubtless owing to the fact that the climate is too rigorous for it. Very likely the hibernating bug succumbs to the severity of our long, cold winters.

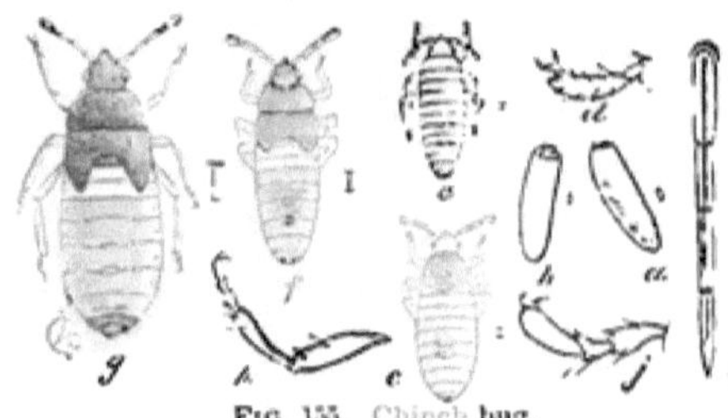

FIG. 155. Chinch bug.

This insect, Fig. 155, is hardly 4 m m long, or less than 3-20 of an inch. Its color is black with white wings marked with black spots. The bugs hibernate in winter. In May they swarm forth in nuptial flight, and soon after the egg laying begins. There are two or three broods, so from June on they will be seen in all stages. The wingless larvae, the short winged and equally active and hungry pupæ and the full fledged imago will all be seen sucking the juices from the plants at one and the same time.

Neatness in farm operations, not leaving corn-stalks and rubbish in the fields to protect and harbor the bugs in winter, is about the only remedy possible. Prof. Forbes, whose admirable researches and suggestions have been so valuable, has found that the kerosene emulsion will kill the bugs, but owing to their numbers and habits it is hardly a practical remedy.

Caloptenus—many species. Locusts, or Gray Grasshoppers.

Order Orthoptera. Family Acrididæ.

Although no grass insect is more serious in its destructiveness than the Western locust, or grasshopper, Caloptenus spretus, Thom., when it comes to make its presence felt, yet from the fact that it can never attack the vegetation except in a limited area West, and even there comes only rarely to scourge the country, it perhaps on the whole is not so serious to our forage crops as the insects already referred to. Our common, red-legged grasshopper, Caloptenus femur-rubrum, De G., with several other species, often do very serious harm in our Eastern States. Yet the fact that they come only rarely in great numbers, and then scarcely ever two years in succession, makes the insect less dreaded than it would otherwise be.

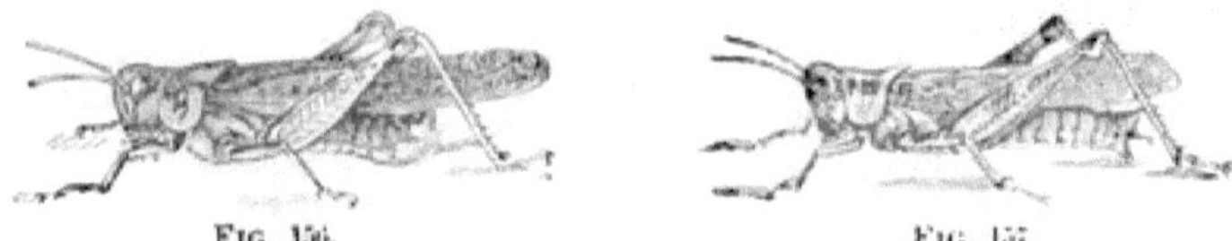

FIG. 156. FIG. 157.

Fig. 156 shows the Western locust, and Fig. 157 our red-legged locust, which insects resemble each other very closely. The principal difference is the longer wings of the C. spretus. Like the Chinch bug the transformations of these insects are incomplete. The larva in early summer, the pupa in mid-summer, and the imago in late summer all look alike, and have identical habits as to their food. They differ only in size and development of wings, which are at first wholly absent, and then appear for a time only as stubs, or mere pads. The females in August and September lay their large eggs in the ground. In all stages they are ravenous and indiscriminate feeders.

Dr. C. V. Riley, in his elaborate investigation of this insect, names and describes many mammals, birds, and other insects that destroy this pest. He also describes many mechanical appliances for the destruction of the pest. In California the past season locusts which were very common and harmful were destroyed by use of poison. Bran, sugar and arsenic were mixed and left where the insects could gain access to the mixture. Whether this can be made available in fields to protect grass, oats, etc., is yet to be decided by actual trial.

Crambus Vulgivagellus, Clem. The Vagabond Crambus.
Order Lepidoptera. Family Pyralidae.

This insect is not rare in Michigan, nor in other Northern States, yet it has rarely attracted attention as a serious pest in agriculture. In 1881 the pastures in parts of Northern New York were quite seriously damaged by this pest. It belongs to the same family as the bee moth, and to the same genus as the corn-root web worm, Crambus zeellus Fernald, which has done considerable damage in Illinois the past summer (1885).

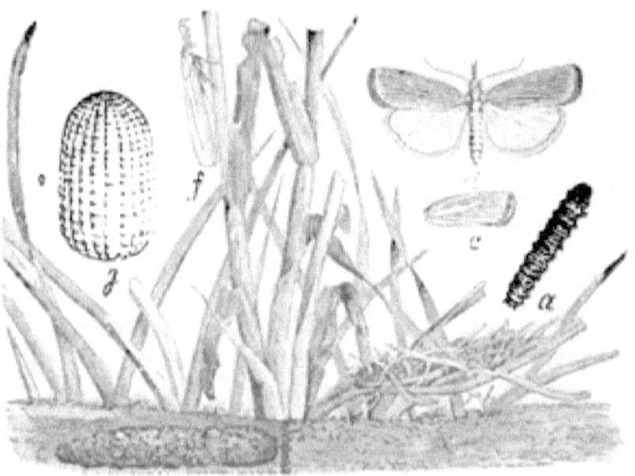

Fig. 158.

The moth, *d*, Fig. 158, expands 2.5 c m (1 inch) and like all of the species of this genus has a slender body. The front wings are of a dull yellow color. There are rows of black scales between the veins and a sub-marginal row of black dots near the outer border. The fringe has a golden reflection. The hind wings are pale yellow, with long paler fringe. The thorax and abdomen is yellow. The projecting beak—the palpi—which gives the name snout moths to this family is well marked.

The very small eggs, *g*, Fig. 158, are yellow till near hatching when they turn pink. Like the eggs of many butterflies they are ribbed, both longitudinally and transversely. The transverse ridges are less marked than the others. The eggs are .7 m m by .3 m m. The color of the caterpillar, *a*, Fig. 158, is dull green, with shining black head. There are brown tubercles along the body, each of which bears a black hair. When full grown the larva is about 2 c m (⅘ of an inch) in length. The cocoon, *b*, Fig. 158, is spun close to the earth. It is curved, attached to grass, and varies much in size. The average length is 2.25 c m, or .9 of an inch. Some cocoons are much enlarged at one end. The pale brown pupa is much the same as chrysalids of moths in general. It is 1 c m (.4 of an inch) in length.

The eggs are deposited in dry pastures and meadows in late August. They seem to be merely dropped on the ground. They hatch in a little over a week, and the young caterpillars eat sparingly, but do little harm ere they go into winter quarters. They commence to feed as soon as the grass starts in spring. The brown spots in the grass fields where all has been eaten to the very roots, which latter have not been disturbed, show to the unobservant even that a serious enemy is at hand. When very numerous whole acres are fairly mown off close to the ground. While they prefer June grass they will eat any grass, and even oats and wheat. Like the corn-root Crambus they spin a web in which they live while devastating the meadows. They feed by night, and when not feeding are concealed in a cylindrical case of pieces of grass and fecal pellets held together by silken threads. The most damage is done in May. Often the caterpillars gather in immense numbers on the trunks of trees near the ground.

They spin their cocoons late in May, which are placed upright in the ground just below the surface. They do not pupate till the first of August, and do not emerge as moths till late in the same month.

Prof. J. A. Lintner, who has given an excellent account of this pest in his 1st An. Rep., speaks of several enemies, parasitic and predaceous, which are probably what keep this pest from doing greater damage. He also suggests burning by firing the pastures. He further recommends trying a liberal application of lime, plaster, ashes, and especially gas-lime. Plowing in autumn would doubtless destroy the eggs. We may reasonably hope that we shall not have frequent attacks of this insect; possibly it will never do so much damage again. Yet it has come once, and so we may at least fear that it will again, and to be fore-warned is to be fore-armed.

Before closing it is well to state that in company with the above Prof. Lintner found a caterpillar, the larva of a moth common in Michigan and all through the North. It is Nephelodes violans. I have space only to state that it was not very injurious, though may increase and become so at any time. What has been said as to habits, and especially of remedies, in relation to the Crambus and army worms, will probably be true of this insect if it should ever become a serious enemy.

I have not space to describe more of the insects noxious to our grasses. Those described are the only ones which have given anxiety, and while the others may become more numerous and therefore harmful, they are not likely to do so. In connection with the list given above I have referred to authors who have written upon each insect, and in many cases not only are the descriptions full but excellent illustrations add to the interest and value of the treatises. It is not unlikely that new enemies will attack our forage crops; but if so they will almost certainly be like one or more of the old familiar ones, and so by studying their habits and determining their natural history we shall at once know which of the old and well tried remedies to adopt.

CHAPTER XVII.

THE FUNGI OF FORAGE PLANTS.

BY WILLIAM TRELEASE, D. SC.

Grasses afford a nidus for the development of a large number of fungi, so that they are a favorite collecting ground with students of these plants; but the greater number of species are found on dry stems and leaves, which they seize upon, as a rule, only after their death, and though the number of truly parasitic species is by no means small, there are but few that seriously injure valuable grasses. The number of noxious species on clovers and other forage plants of the pea-family is also small; hence this chapter includes a few which are of such frequent occurrence as to attract general attention.

For the most part the fungi of forage plants are directly injurious by weakening them and appropriating to themselves the food needed for making a good growth; but they likewise lower the nutritive value of the crop that is produced. In cases where seed is an object, the loss is even greater, since the yield of diseased plants is greatly lessened, while the quality of their seed is always poor. The annual loss in our meadows and pastures due to these causes cannot be stated, from the lack of reliable statistics, but in some seasons a moderate estimate places it in the millions.

Besides these direct injuries to the crops the fungi of grasses are the cause of a very considerable loss to the farmer in another way. Ergot and corn-smut have long been known to possess active medicinal and poisonous properties, and it has been demonstrated that abortion and certain diseases of the feet of cattle follow the prolonged use of ergotized hay or pasturage. How many of the smuts and other fungi of grasses possess similar or other detrimental properties is at present merely a matter of conjecture; but some of them occur in sufficient quantity to merit suspicion until they have been shown to be harmless. Students will find the principal parasites of grasses represented by actual specimens in fascicle 2 of Seymour and Earle's Economic Fungi, Cambridge, Mass., 1891.

SMUTS.

1. **Corn-smut** (*Ustilago zeæ mays, D.C.*). Forming galls, often of large size, in the leaves and other parts of Indian corn and teosinte, that are finally transformed into dusty masses of brown spores.

No fungus is more widely distributed or better known than corn-smut. Like other smuts, its germinating spores attack young plants, its mycelium or spawn making its way upward through their growing tissues without producing any evident effect until it prepares to fruit, when it increases and leads to the formation of the smut-galls, that are ultimately filled with myriads of round brown spores, each densely covered by short, sharp spines. These spores, which measure 9-13 micro-millimeters, preserve their power of germination for several years, or, in fresh barnyard manure, etc., they develop at once, multiplying indefinitely by the production of yeast-like secondary spores, each of which has the power of infecting a seedling corn plant.

Gathering and burning the smut-galls and smutty ears, while they are still green, to prevent the accumulation of spores in the soil, rotating the crop when smut has become firmly established in a field; treating seed corn with copperas-water and lime, etc., before planting; and using only old, well-rotted manure or artificial fertilizers, have all been proposed as preventives of smut.

2. **The leaf-smut of Timothy** (*Tilletia striæformis, Westd.*) Forming black, smutty lines in the leaves of Timothy and other

grasses, which are finally reduced to brown shreds, covered with dusty spores.

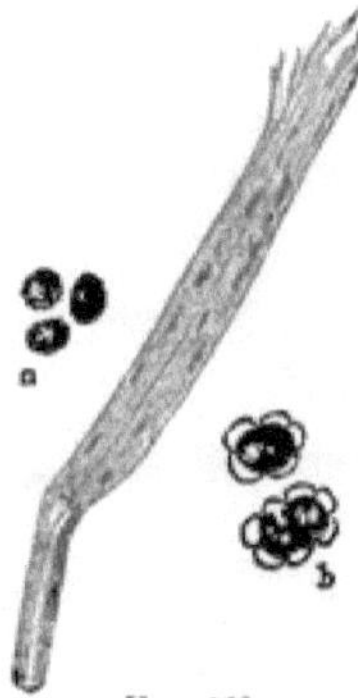

FIG. 159.

The first appearance of this disease is in the formation of lead-colored, thickened lines, about 1-64 in. wide and 1-16 to ¼ in. long, between the nerves of the leaf. The epidermis, which at first covers them and gives them their gray color, soon breaks away, revealing a powdery mass of black-brown spores, which are irregularly rounded or egg-shaped, and closely studded with short spines. They measure 10–12 micro-millimeters, and in their microscopical characters closely resemble the spores of the corn-smut.

Similar black lines are formed in the leaves of species of *Glyceria* by *Ustilago longissima* (Sow.) which has smooth brown spores, 3.5 to 7 micro-millimeters in diameter, and in the leaves of wild rye and other grasses by *Urocystis occulta* (Wall.) the dark brown opaque spores of which measure 10–20 micro-millimeters, and usually occur in clusters of 2–4, closely surrounded by masses of half-round, colorless cells of slightly greater diameter.

Ustilago hypodytes (Schl.) occurs on the stem of quack grass and other species, usually forming black smut masses inside the leaf-sheath, and *U. grandis* (Fr.) causes cat-tail-like swellings on the internodes of the reed.

The fruit of many grasses is replaced by other smut fungi, the number of which is very considerable. The commonest are: *Ustilago panici glauci* (Wall.), very abundant in autumn on pigeon grass; *U. rabenhorstiana* (Kuehn), on crab grasses and sand burs; and *U. segetum*, (P.), in oats, barley, wheat, etc.

Draining the soil well, transferring the crops to new land when they have begun to smut badly, and exercising care with respect to manure are preventive measures.

FIG. 159.—*Tilletia striaeformis* in Timothy leaf; spores at *a*; *b*. *Urocystis occulta*, spores.

RUSTS.

3 **Grass-rust,** (*Puccinia graminis, P.*) *Order Basidiomycetes, Sub-order Uredineae.*

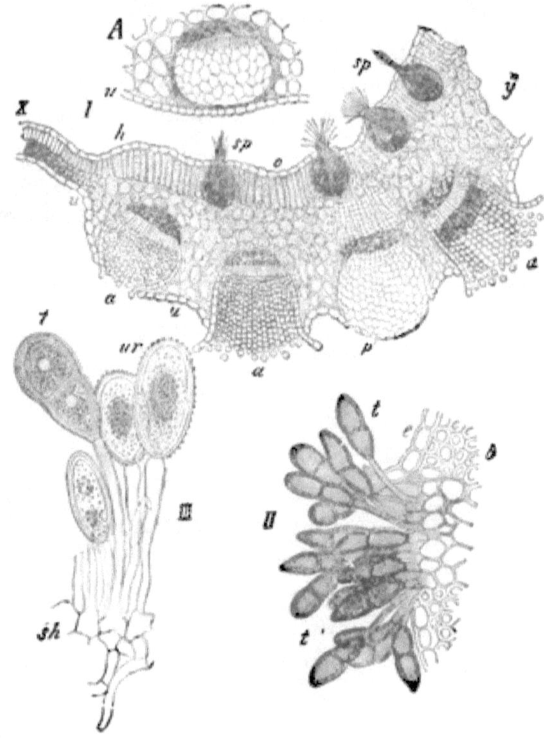

FIG. 160.—Several stages of grass-rust. *A*, young aecidium fruit; *r*, section of Barberry leaf; *a p*, aecidium fruit; *s p*, spermagonia; *II.*, a mass of teleutospores on a leaf of a grass; *III.*, three uredospores, *u r*; with one teleutospore, *t.*—(From De Bary.)

Forming orange-red, powdery spots and lines on the leaves and stems of cereals and meadow-grasses, that give place later to dead-black velvety lines.

The sheaths and culms of the smaller grasses, especially quack grasses and red-top, are very often attacked by this rust (called mildew and brand in England), which produces the same disastrous effects on them as on wheat. The red rust or uredo state appears from early spring until fall. It consists of microscopic one-celled rough spores, borne on branches of a myceli-

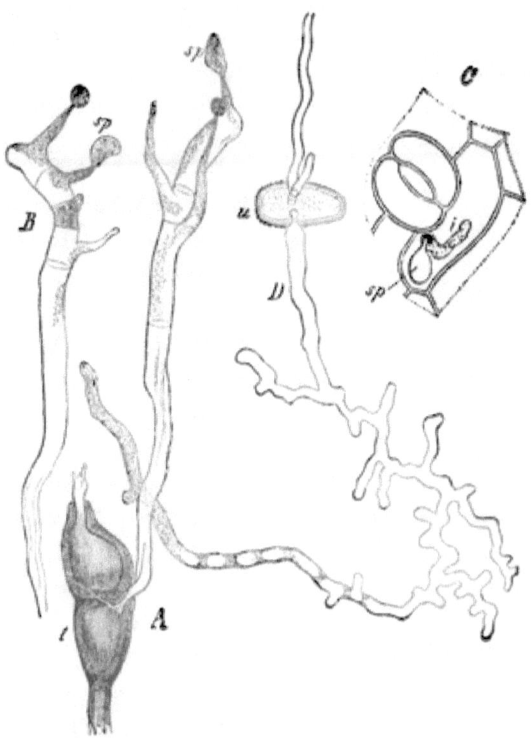

FIG. 161.—Grass-rust. *A*, germinating teleutospore, *t*; B, promycelium, with sporidia; C, *s p*, sporidium, germinating on the lower side of a leaf; *u*, a germinating uredospore.—(From DeBary.)

um that vegetates within the grass, and only appears on the surface to fruit, which propagate the disease rapidly in damp warm weather. Toward the end of the season the same mycelium bears a second form of fruit—two-celled teleutospores or winter-spores—that form dense elongated black cushions where they break through the epidermis, often covering the greater part of the dead stem. These spores germinate the next spring, when they produce secondary spores that are said by an English experimenter to infect very young grass leaves, in which they form a mycelium that quickly fruits in red rust. The winter-spores have long been held to produce a mycelium in young barberry leaves, on which the common yellow cluster-cups appear as a result,—their spores again attacking grasses and forming a mycelium that bears little of the red rust, but fruits almost exclusively by winter-spores.

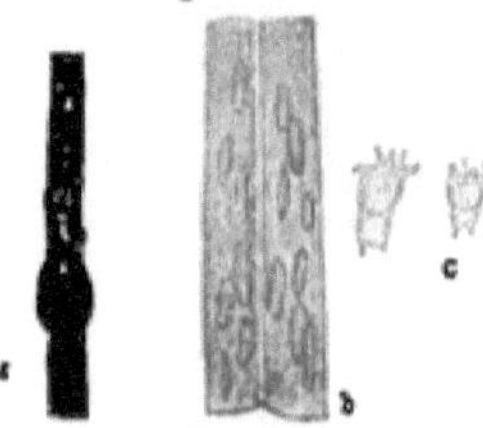

FIG. 162.

Other grasses are subject to the attacks of rust-fungi belonging to other species. *P. coronata* (Cda.), the common oat-rust, and *P. rubigo-vera*, (D. C.), the barley-rust, are not infrequent on grasses, the latter on the beautiful squirrel-tail grass. They produce smaller clusters of uredo-spores, and the cushions of teleutospores are long covered by the epidermis of the plant, and not so black. They are also more frequent on the blade than on the sheath of the leaves. *P. magnusiana* (Koern.), *P. phragmitis* (Schum.), and *P. arundinacea* (D. C.), are found on the reed. These species all have cluster-cups or æcidia on other species of plants. The rust of corn is *P. maydis* (Carrad). The tall gramma grass is infested by *P. vexans* (Farlow): *P. andropogi* (Schw.), occurs on broom-grass; *P. arundinariæ* (Schw.), on fall marsh grass; and *P. cynodontis* (Desm.), on Bermuda grass. The common rust of old witch-grass or tickle-grass is *P. emaculata* (Schw.) etc. None of these species are known to produce cluster-cups.

4. **Clover-rust** (*Uromyces trifolii*, A. & S.) Producing minute white cluster-cups, pale brown uredo-pustules and darker brown teleutospore-cushions, 1-64 in. in diameter, on the leaf-stalks and blades of clover, especially white clover.

FIG. 162.—*a. Puccinia graminis* on wheat sheath; *b. P. coronata* in oat leaf and *c* its spores.

Fig. 163.

The clover-rust bears its cluster-cups on the same plant with the other forms. They appear in early summer, in small clusters, especially on the stalk and veins of the leaves. The later cluster-cups are accompanied or followed by small round or oval pustules of rough brown uredospores, that are partly covered by the torn, lead-colored epidermis of the leaf. Both of these forms immediately reproduce a mycelium, similar to that from which they originated, in other leaves. The winter spores occur in slightly darker clusters in the fall, and germinate the following spring. They differ from the corresponding spores of *Puccinia* in being one-celled, and resemble the uredospores of the same species, except that they are somewhat darker brown, smooth, and often furnished with a blunt point at the end.

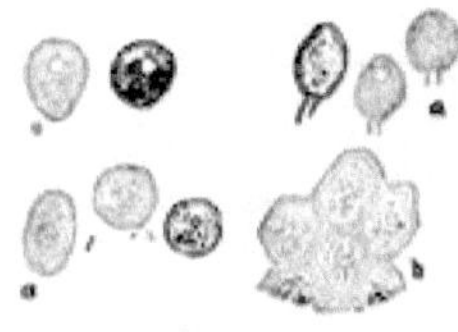
Fig. 164.

U. medicaginis falcatae (D. C.), is a related rust, found in all its stages on alfalfa and none-such, and on the wild rabbit's-foot clover and hop-clover. Its winter spores are striped by longitudinal ridges. Other species of *Uromyces* are found on different grasses. *U. dactylidis* (Otth.) occurs, in Europe, on orchard grass, the taller fescue, etc., and is represented in this country by several forms on a number of grasses. Its cluster-cups are found on the butter-cup. *U. acuminatus* (Arthur) is common on fall-marsh grass; *U. spartinae* (Farlow) on rush-salt grass; and *U. Peckianus* on the smaller salt grass (*Distichlis maritima*). These species are not known to have a cluster-cup stage.

Burning over meadows and fields covered with rusty stubble; a proper succession of crops; and the destruction of plants that

Fig. 163.—*Uromyces trifolii*, aecidium and teleutosporic sori on white clover. Fig. 164.—Uredospores, teleutospores, aecidiospores and peridial cells, of same.

serve as hosts for the cluster-cup stages of rusts. are the best methods of keeping them within check.

ASCOMYCETES.

5. **Ergot.** (*Claviceps*, Sp. *Sclerotium clavus* of authors.) Black, purple or dark gray spurs in the flowers of cereals and of various wild and cultivated grasses.

The officinal ergot, to be found in most rye-fields toward the end of summer, appears in the form of curved purple or black spurs, often an inch long and 3-16 in. in diameter, which replace the grain in one or more flowers of a spike, thus giving rise to the popular name of spurred rye, often applied to it. Spurs of the same nature, but usually shorter and stouter, are also common in the heads of wheat. Similar bodies, varying much in size, shape and color, are found in the flowers of many grasses. On the rush-salt grass they are very long and slender, and rather pale. On wild rice they are short, and even stouter than the spurs of wheat; while on smaller grasses, like red-top, Timothy, blue-grass, etc., they are much smaller, and closely resemble the pellets of mice.

These spurs are the resting form, or sclerotia, of a fungus which appears at the base of the young grain, when the grasses are in bloom. As it grows it gradually takes the place of the grain, the remains of which are pushed up at its end. While young, the fungus gives off a sweetish, ill-smelling fluid, that contains myriads of microscopic spores which are carried from plant to plant by flies, beetles and other insects that feed on the fluid, and so play an important part in spreading the disease. When the spurs have reached their growth they harden, and fall to the ground, where, as a general thing, they remain unchanged till the next spring, when each bears a number of small, stalked,

FIG. 165.—1, 2, 3, 4, Grasses affected with Ergot appearing as black spurs. (From the U. S. Dept. of Agrl.)

Fig. 165.

pink fruit-bodies, in which spores are produced at about the time when grasses are coming into bloom.

Botanists recognize several species of ergot by the form of their spring fruit; but the differences between them do not much concern their life-history, so that they need not be considered separately. From their habit of attacking only the flowers, they do not affect the general health of the grasses they grow on, while as a rule they are not abundant enough to seriously lessen the yield of seed.

FIG. 166.

Ergot has long been employed in medicine, because of its action on the uterus. That it should cause abortion when fed to stock is, therefore, not surprising. Nothing can be more misleading than the popular belief that ergot does not occur on meadow grasses in sufficient quantity to be dangerous. In examining suspected hay from several of the western States the Veterinarian of the United States Department of Agriculture found 2–6 per cent. of its entire weight to be ergot. An equally erroneous and common belief is that in pastures ergot cannot mature because the grass is so closely cropped that it cannot flower. Under close grazing most grasses produce scattering flowers, when very small, and at times nearly every one of these is ergotized.

Not long since considerable excitement was caused by the appearance of what was taken for "foot-and-mouth disease" in

FIG. 166.—Development of *Claviceps*, from Tulasne.

Kansas and other parts of the West; but on investigation it was found that the sloughing of the hoofs and other symptoms were the result of ergotism, due to the foul hay on which the cattle had been fed. Similar cases have occurred in other parts of the country, and in Europe the use of flour made from ergotized grain has occasionally given rise to epidemics of a similar nature among men. However it may be as regards abortion, ergot does not usually occur abundantly enough in closely grazed pastures to cause this trouble. It has been suggested that it may be prevented from occurring to a dangerous extent in hay by cutting grass as soon as it comes to bloom, and curing it before the ergot has matured.

Yellowish-white, irregularly rounded bodies, with a checked surface, occurring in the flowers of *Paspalum laeve*, are *Spermoedia paspali* (Fr.), the sclerotium of an entirely different fungus.

6. **Cat-tail grass fungus,** (*Epichlöe typhina*, P.). Forming a white or yellow coating around the upper leaf-sheaths of grasses.

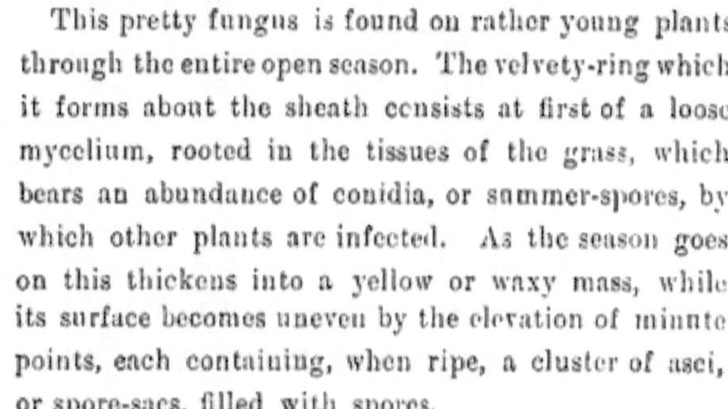

Fig. 167.

This pretty fungus is found on rather young plants through the entire open season. The velvety-ring which it forms about the sheath consists at first of a loose mycelium, rooted in the tissues of the grass, which bears an abundance of conidia, or summer-spores, by which other plants are infected. As the season goes on this thickens into a yellow or waxy mass, while its surface becomes uneven by the elevation of minute points, each containing, when ripe, a cluster of asci, or spore-sacs, filled with spores.

In Europe, meadow grasses, and especially Timothy, are some-

Fig. 167.—*Epichlöe* on sheath of grass, with enlarged fruits.

times extensively attacked by this parasite, but in America it has not been noticed to any great extent on the more valuable species—its presence being possibly overlooked in meadows because concealed by the spreading blades above.

A black fungus related to this which occurs on grass, is *Hypocrella hypoxylon* (Pk.).

7. **Black-spot disease of grass,** (*Phyllachora graminis*, P.) Coal-black spots usually under 1-32 in. wide, and 1-32–¼ in. long, on the leaves of grasses; especially conspicuous on the upper side.

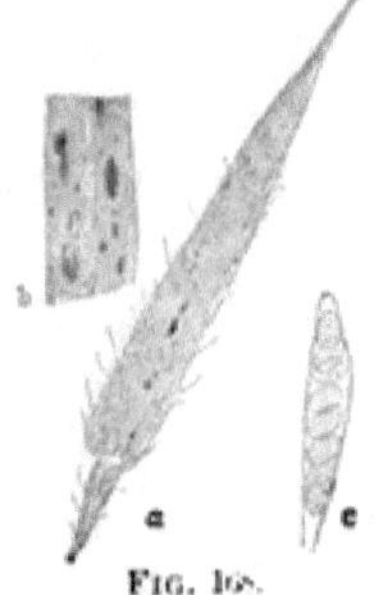

FIG. 168.

This is one of the commonest and most noticeable of grass-diseases, especially toward the end of the season, but does little harm to valuable species. It is most abundant on quack grass, hedgehog grass and the broad-leaved Panicum. The black spots are composed of dense mycelium. In them, usually after the death of the leaf, oval colorless spores are formed in asci. These spores carry the species over the winter. Smaller spores (stylospores) are produced in the same spots earlier, and serve for summer propagation. If the disease should prove troublesome, the grass may be cut early before the fungus develops, and the meadow should be burned over on the approach of cold weather to destroy the forming winter spores. (Fig. 168 illustrates the above species.)

8. **Black-spot disease of clover,** (*Phyllachora trifolii*, P.). On the lower side of clover leaves, forming at first dull-black patches, often ⅛-in. across; later occurring in the form of slightly glossy-black dots, 1-64 in. in diameter, on small whitish or pale-brown spots.

FIG. 168.—*Phyllachora graminis*. FIG. 169.—*Phyllachora trifolii*. FIG. 171.—*Polythrincium trifolii*. FIG.—170.—*Phacidium trifolii*.

Fig. 169.

In the earlier part of the season small whitish or pale-brown spots appear in the leaf, which contains the mycelium of a fungus. This fruits on the lower surface, producing numerous tufts of necklace-shaped threads, each of which ends in a 2-celled, egg-shaped conidia-spore. These tufts of threads, which, like the spores, are of a deep brown color, are packed so closely together as to completely cover the spots, though under a hand lens they can be distinguished as separate pustules. To the naked eye they appear dead-black. Later in the season similar spots are occupied by small, coal-black fruits that contain stylospores. Winter spores, produced in asci, are not known.

The conidial form of this fungus (called *Polythrincium trifolii*) is especially common on white clover, though both forms are at times found abundantly on red clover and other species. Kühn suggests growing grass with the clover as a means of lessening its injury. (See Fig. 170.)

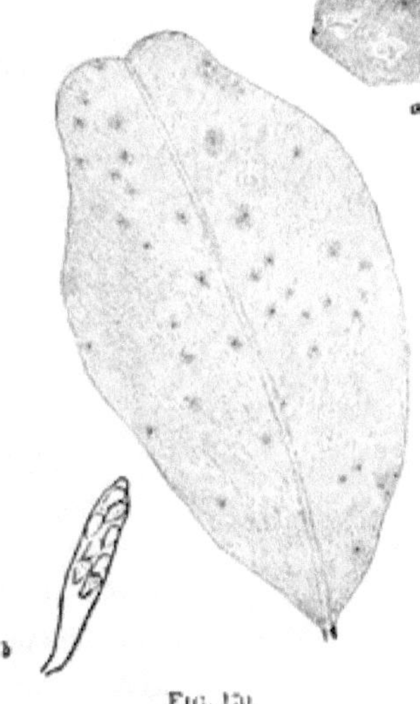

Fig. 170.

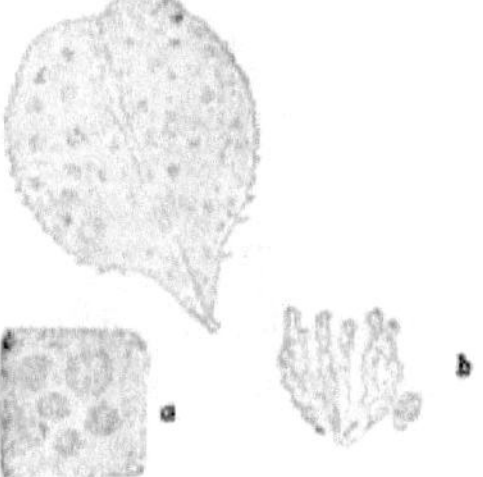

Fig. 171.

Red clover is also often marked in the summer and fall by similar but darker brown spots, bearing in the center of each a brown cup, scarcely 1.32 in. in diameter, that opens irregularly at the top and so allows the escape of its spores. This is *Phacidium* (or *Pseudopeziza*) *trifolii*, which at times does considerable damage in Europe.

9. **Violet root-fungus,** (*Leptosphæria circinans*, Fckl.). Forming a violet mold on the roots of alfalfa, red clover, etc., which soon rot, the parts above ground turning yellow and dying.

In Europe, Lucerne is subject to a disease that manifests itself by the appearance of yellow spots in the fields. These spread until the entire crop is often affected. The trouble lies in a violet-colored mold that develops on the roots of the plants, spreading from one to another through the soil, and finally producing spores by which it is apparently carried over the winter. This disease has not been recognized yet in the United States, but what is held to be a state of the root fungus—a cobwebby, white mycelium, known as the snow-mold, that covers the ground, leaves, etc., just as the snow disappears in early spring—has been noticed in great abundance at River Falls, Wis., by my friend, Professor King, so that it is not improbable that the parasitic form will soon be found. No remedies for it have been proposed, except digging ditches, as deep as the roots extend, about diseased parts of the field when it first appears to prevent it from spreading.

10. **Grass-mildew,** (*Erysiphe graminis*, D. C.). Forming a pure white, cobwebby or mealy coating on the upper side of the leaves of grasses, especially in the shade.

The German equivalent (meal-dew) of our common name for the group of fungi to which this species belongs is expressive of the appearance presented by them in their early stages, when they cover the surface of the plants they grow on with a fine, white mycelium that bears such numbers of white spores as to suggest a dusting of meal or flour.

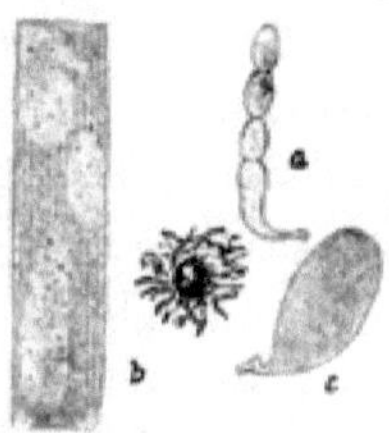

FIG. 172.

This mildew is usually found through the entire open season on grass growing in damp and shaded positions; it is especially abundant on June grass. Its cobwebby mycelium, which does not penetrate the leaves, does not at first appear to injure them but in time they succumb and dry up. Through the summer it spreads by means of its light conidia, that are easily blown about and germinate quickly while fresh, though they are unable to live through the winter. On the dead leaves, small black fruit-bodies, scarcely visible to the naked eye, are formed, in which winter-spores are produced in asci. (Fig. 172 is the illustration for the grass-mildew.)

Usually grasses do not suffer much from mildew, except in damp and shaded places. Drainage is likely to prove beneficial where it is troublesome, and infested wheat-fields have been advantageously dusted with flowers of sulphur in Europe.

11. **Sclerotium disease of clover,** (*Sclerotinia trifoliorum*, Eriks.). On clover, causing a browning of leaves or stem, which are soon covered in spots by a white mold that ultimately forms solid, wavy, black bodies, often $\frac{1}{2}$ in. long, white within.

In Europe, clovers and medicago are occasionally attacked by this fungus, which is very destructive when it occurs. The entire plant becomes filled with a mycelium which soon kills it and afterwards breaks through in places, forming black sclerotia on the various parts of the decaying plants as winter approaches. These bodies lie dormant in the soil until the following summer, when they produce fruit-bodies in the form of wavy stems, bearing brown disks or inverted cones, $\frac{1}{16}$ to $\frac{1}{4}$ in. in diameter, on their ends. When these reach the surface they shed their spores and so spread the disease.

Draining the soil well, and especially replacing clover for several years by wheat, corn, or other crops not attacked by the *Sclerotinia*, are recommended where it appears.

A large number of fungi are spoken of as imperfect fungi from the resemblance of their fruit to the conidia or stylospores of Ascomycetes. Several of these cause diseases of grasses.

FIG. 173.

The brown-spot disease of pigeon-grass, early spear-grass, and other species, is due to *Septoria graminum*, (Desm.) (Fig. 173), that form a mycelium within the plant, usually killing it in places which turn brown and are finally dotted with the minute black fruit-bodies of the parasite, within which slender colorless spores are produced. In Europe, a similar disease is also caused by a related fungus (*Dilophospora graminis*, Desm.) whose spores differ in having brush-like appendages at their ends. Both are at times destructive, but affect the cereals more than the smaller grasses. *Mastigosporium album*, (Riess), and *Scolecotrichum graminis*, (Fckl.), cause diseases of the leaves of grass in Europe. The last named appeared on orchard grass in great abundance about Madison, Wisconsin, in 1886. *Hadrotrichum phragmitis*, (Fckl.), forms small, dark-brown pustules on leaves of the reed, resembling those of a rust-fungus, even under a hand-lens. The gray-spot disease of crab-grass is due to *Pyricularia grisea* (Cke.), another imperfect form that bears pear-shaped conidia on threads that protrude through the stomata of the gray spots.

Sporobolus indicus, a grass of the Southern States, somewhat esteemed for pasturage while young, is often called "black-seed grass" or "smut-grass" from the fact that its flowering parts are generally covered by the dark-brown fruit of *Helminthosporium ravenelii*, (Curt.), that is often so abundant as to form a

Fig. 173.—*Septoria graminis.*

dense, spongy mass. Wire grass is subject to similar attacks from *H. nodosum*, (B. & C.), and other species of the same genus occur on different grasses in a similar manner.

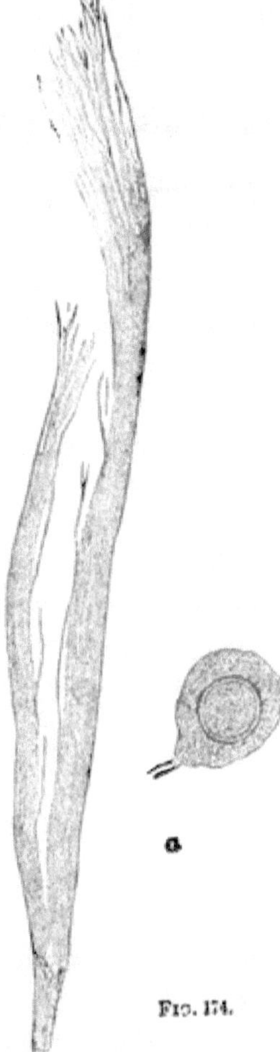

FIG. 174.

PERONOSPOREAE.

12. **Grass-peronospora.** (*Peronospora graminicola*, Sacc.) In the leaves of Hungarian grass and pigeon grass, ultimately filling them with a snuff-brown, powdery mass. (See Fig. 174.)

Hungarian grass (*Setaria italica*) is sometimes attacked by a parasite clearly related to the notorious potato blight, which forms a mycelium in the leaves of the grass, in the cells of which it lives. Branches of this emerge sparsely through the stomata and bear colorless conidia which spread the disease. Later in the season these spores are replaced by winter spores (oöspores) that originate on branches of the mycelium within the leaf by a process of fertilization. These spores are contained in thick-walled, brown envelopes, and presumably infect new plants in the spring. So far, this disease has not proved seriously destructive, though the leaves attacked are reduced to mere shreds when the winter spores are ripe. The flower-clusters of pigeon grass are greatly changed by the fungus, according to Dr. Halsted.

FIG. 174.—Shredded leaf of Hungarian grass and oöspore of *Peronospora graminicola*.

13. **Clover peronospora,** (*Peronospora trifoliorum,* DeBary). A dirty white or purple-brown mold, often completely covering the lower surface of the leaves of clover, alfalfa, none-such, etc. (See Fig. 175.)

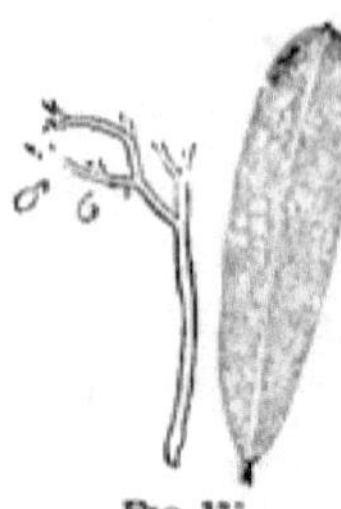

FIG. 175.

The life history of this species is quite similar to that of the last, though they differ greatly in appearance. The leaves that it occurs on are paler than the others, and the threads that escape through their stomata and bear conidia are so numerous and bushy as to form a dense coating on their under side. Oöspores are produced in smaller numbers than in the last species, and, as they are thin-walled and nearly colorless, they are only to be found after careful microscopical examination. Another species of the same genus (*P. vicial,* Berk.) is found on the leaves of vetches and of the pea.

14. **Seedling-rot,** (*Pythium debaryanum,* Hesse). Causing young plants of clover, millet, corn, and many other species to rot close to the ground or "damp-off," as it is called in greenhouses.

Several species of *Pythium* attack living plants. The present species is said to be widely distributed in garden soil in Europe and causes serious trouble by attacking seedling plants. It can be recognized by its effects on the plants, which quickly decay near the ground. They contain a delicate, colorless mycelium that fruits on the surface of the decaying parts, when these are kept damp, producing conidia, swarm-spores, and oöspores.

15. **Fairy-Ring Fungi.**—Bright green circles, several feet in diameter, closely surrounded by a narrow strip of dead or dying grass, are frequently seen in lawns or pastures, and are commonly called "fairy-rings." They are caused by several species of toadstools (the commonest is *Marasmius oreades*) that spread a short distance outward every year, their mycelium destroying

the grass in the roots of which it grows, and so causing the brown ring, on which an abundant crop of toad-stool fruits forms in the fall, which by their decay enrich the soil so that it produces a ranker vegetation the next season.

An appearance which may be called false fairy-rings is occasionally produced by *Physarum cinereum*, one of the slime-molds, on the leaves and stems of grasses. This fungus grows unnoticed on decaying matter in the ground, often creeping out in a regular manner from its starting point until a more or less perfect circle six or seven feet in diameter is formed, when it suddenly appears upon the plants it has grown under, and produces its dusty, ash-colored fruit in such abundance as to attract attention from a distance. From its mode of life, it does little if any harm to the grass, further than to make a little of it unpalatable to animals.

FIG. 175.

16. Root-gall fungi of clover, etc.—Galls which vary in size and shape, according to the species examined, are always found in greater or less number on the roots of normally grown leguminosae. They are caused by one or more microscopic fungi, sometimes referred to the genus *Rhizobium*, widely distributed in the soil where leguminous plants are grown. The fungus penetrates the tender rootlets, especially through the root hairs, giving rise to the development of galls, but its presence, unlike that of most gall-producing species, appears to be beneficial to the plants attacked, since it seems now fairly demonstrated that the assimilation of atmospheric nitrogen, which has long been attributed to leguminous crops, is effected by them through the agency of these fungi.

DEBRIS.

After the house is finished, the debris often contains a few choice brick and some stone that did not seem to exactly fit in anywhere. There is a barrel or so of good mortar, half a load of sand, a little nice lumber, a bunch and a half of shingles, and one of lath. There are remnants of nails and screws, paint, oil, putty, glass, and wall-paper. Some of these are as good as any employed in constructing the building. The most worthless fragments are carted away and covered up or burned.

So in writing a lecture, a story, or a book, there will often be more or less surplus materials. A change in the plan, perhaps, will make it seem best to leave out some things for want of a suitable place to use them.

I once supposed the following quotations among many other things would certainly find a place in the former pages, either as headings to chapter or paragraph or in some other place. A few were thus used, but most were left over. Here are some of the remnants:

"Go to grass."

"All flesh is grass."—*Isaiah.*

"The staff of life."—Said of wheat.

"Let the earth bring forth grass."—*Leviticus.*

"Sweet fields arrayed in living green."

"Grass is rather a good savings bank."—*Joseph Harris.*

"Grass is the pivotal crop of American agriculture."—*Geo. Geddes.*

"Grass is king among the crops of the earth."—*Alex. Hyde.*

"The grasses are the foundation of all agriculture."

"He maketh me to lie down in green pastures."—*23d Psalm.*

"A water meadow is the triumph of agricultural art."—*Pusey in Jour. Roy. Ag. Soc., 1849.*

"Farmers pay too little atttention to their pastures."—*N. H. Agrl. Rept.*

"The cheapest manure a farmer can use is clover seed."—*American Proverb.*

"No grass, no cattle; no cattle, no manure; no manure, no crops."—*Belgian Proverb.*

"Then learn to toil and gaily sing,
All flesh is grass, and grass is king."

—*Missouri Agrl. Rept.*

"The term grass is only another name for beef, mutton, bread and clothing."

"Feed your land before it is hungry; rest it before it is weary; weed it before it is foul."—*English Farmer.*

"One year's seeding
Is seven years' weeding."

"He who makes two blades of grass grow where only one grew before, is a great public benefactor."—*Dean Swift, in about* [illegible]

"And the ripe harvest of new-mown hay
Gives it a sweet and wholesome odor."

—*Colley Cibber.*

"The melancholy days are come, the saddest of the year,
Of wailing winds, and naked woods, and meadows brown and sear."

—*Bryant.*

"Plants do not grow where they like best, but where other plants will let them."—*Dean Herbert.*

"How doth the little busy bee
Improve each shining hour,
By carrying pollen day by day
To fertilize each flower."

"And he gave it for his opinion that whoever could make two ears of corn or two blades of grass to grow upon a spot of ground where only one grew before, would deserve better of mankind, and do more essential service to his country, than the whole race of politicians put together."—*Gulliver's Travels.*

"But of all sorts of vegetation, the grasses seem to be most neglected; neither the farmer nor the grazier seem to distinguish the annual from the perennial, the hardy from the tender, nor the succulent and nutritive from the dry and juiceless. The study of grasses would be of great consequence to a northerly and grazing kingdom."—*White's Nat. His. of Selbourne.*

BIBLIOGRAPHY.

No attempt has been made to render this list of authors complete, yet it contains the leading authorities which have furnished the greatest help in preparing this volume.

Agricultural Gazette (English), 1880.

Agricul. Reports, U. S., for 1879, '80, '81, '82, '83, '84.

Am. Agriculturist. Short notes, 1870 and later.

Am. Jour. Sci. Numerous short articles and notes.

The American Naturalist, several volumes, Phila.

Trans. Lin. Soc. The Morphology of the Flowers of Grasses, by Geo. Bentham. Hand-book of the British Flora.

Bentham and Hooker. Genera Plantarum, vol. 3, London.

Dr. C. E. Bessey. Botany for schools and colleges, N. Y.

Botanical Gazette, Ind. Numerous notes.

Robert Brown's Miscellaneous Botanical Works, 2 vols. Ray. Soc., London.

Robert Brown, Campst. Manual of Botany. Edin. and London.

Prof. James Buckman. Prize Essay. Jour. Roy. Agrl. Soc., 1854.

Bulletin of the Torrey Bot. Club, N. Y.

William Carruthers, Consulting Botanist, Jour. Roy. Agrl. Soc. His annual reports for some years.

Carter on Laying Down Land to Grass. A pamphlet. Eng.

A. W. Cheever, in N. H. Agrl. Report, 1875.

The Clover Leaf, 1880, '81, '82, '83, '84. Birdsell Mnfg. Co., South Bend, Ind.

Rept. Conn. Board of Agrl., 1868 and later. Numerous valuable notes.

The Country Gentleman. Many good articles, from 1870 to 1886.

C. Darwin. Cross and Self-fertilization of Plants, and Power of Movements in Plants.

P. Duchartre. Elements de Botanique. Paris.

M. J. Duval-Jouve. Histotaxie des Feuilles des Graminees, in Annales des Sciences Naturelles. Paris.

Encyclopædia Britannica. Article on Grasses, by H. T.

Morgan Evans. Jour. Roy. Agrl. Soc., 1876.

Prof. C. H. Fernald. The Grasses of Maine.

C. L. Flint. Grasses and Forage Plants. Boston.

The Garden. Vols. 4 and 8, Ornamental Grasses; vol. 8, Wild Grasses for Bouquets. London.

Gardener's Chronicle. Fertilization of the Flowers. March, 1874, Feb., 1875.

Botanical Text-Book. A. Gray and G. L. Goodale.

Manual of Botany. A. Gray.

Wm. Gorrie. Articles in Morton's Cyclo. of Agrl.

Prof. J. Stanton Gould. Grasses and their Culture. N. Y. Agrl. Rept., 1869.

Prof. J. S. Gould. Lecture on Grasses. Maine Agrl. Rept., 1872.

E. Hackel. Monographia Festucarum Europaearum.

Joseph Harris. Value of an Analysis of Grasses. N. Y. Agrl. Rept., 1865.

J. Henderson. Hand-book of the Grasses. New York.

A. Henfrey. An Element. Course of Bot. London.

J. D. Hooker. The Student's Flora of the British Islands.

Rev. C. W. Howard, of S. C. A Manual of the Cultivation of the Grasses and Forage Plants.

James Hunter, a pamphlet, Eng. Permanent Pasture Grasses.

Alex. Hyde. Twelve lectures on agriculture before the Lowell Institute, Boston.

Indiana Farmer.

Jour. Roy. Agrl. Soc. of Eng. Many valuable papers in many volumes, notably for 1854, '56, '58, '59, '60, '61, '66, '69, '72, '74, '75, '76, '77, '82.

J. B. Killebrew. The Grasses and Forage Plants of Tenn.

I. A. Lapham in Wis. Agrl. Rept., p. 409, 1853.

Lawes and Gilbert. Philosoph. Transactions. London.

Lawes and Gilbert. Treatment of Pastures. Jour. Roy. Agrl. Soc., 1858, 1859.

Dr. J. Lindley. The Vegetable Kingdom.

J. Lindley. Many articles in Norton's Cyclopedia of Agricul.

The Treasury of Botany. Lindley and Moore.

E. J. Lowe. British Grasses.

Maine Agrl. Rept. Discussions and notes, 1870, '71, '72, '76, '81.

Maout and Decaisne. Translated by Hooker. Descrip. and Analyt. Bot.

Dr. Maxwell T. Masters. Plant Life on the Farm. London.

Dr. L. D. Morse, in Missouri Agrl. Rept., p. 214 1868.

Michigan Board of Agriculture, 1871, '75, '77, '78, '80, '81, '82, '85. Reports and lectures by W. J. Beal.

Crops of the Farm, by J. C. Morton and others, London.

The Fertilization of Flowers. Prof. Hermann Müller.

Nat. Live Stock Jour., 1872, '73, '81.

N. H. Agrl. Value of Quack Grass. p. 142. 1853.

Prof. J. R. Page, University of Virginia. Report for 1879–80.

Penn. Agrl. Rept., 1881.

Dr. D. L. Phares, of Miss. The Farmer's Book of Grasses and other Forage Plants.

M. Plues. British Grasses.

The Prairie Farmer, 1869 and later.

Prantl and Vines. Text-book of Botany. Phila.

The Press, Phila. Pa., 1884 and later.

Proceedings of the Association for the Advancement of Science. Articles by W. J. Beal.

Proceedings of the Soc. for the Promotion of Agrl. Sci., vols. 1, 2, 3. Contributions by W. J. Beal.

The Rural New Yorker, 34 Park Row, New York. Many articles by able writers, especially for July, 1885.

A Text-book of Botany. J. Sachs.

James Sanderson. Grass with or without a Crop. Trans. of Highland Soc., 1863.

Prof. N. S. Shaler. Science, p. 186. March, 1883.

G. Sinclair's Hortus Gramineus Woburnensis, 1826, London.

Sowerby and Johnson. The Grasses of Great Britain.

Prof. L. Stockbridge. Management of Pastures. Maine Agrl. Rept., 1876, '81.

Sutton & Sons, Eng. Permanent Pastures. A pamphlet.

J. J. Thomas. Prize Essay. N. Y. Agrl. Rept., 1843.

Dr. Geo. Thurber. Geolog. Sur. Cal. Botany, vol. 2.

The Tribune, N. Y., 1870 and later.

C. B. Trinius. Species Graminum, 3 vols.

Dr. A. Voelcker. Jour. Roy. Agrl. Soc., 1866, 1874.

R. Warington. The Chemistry of the Farm. London.

Webb & Sons, Eng. Permanent Pastures. A pamphlet.

J. C. Wheeler & Sons, Eng. Book on Grasses. A pamphlet.

Botanist and Florist. A. Wood.

INDEX.

PAGE

Dentate, toothed.
Deyeuxia Canadensis. 179
Diadelphous, filaments combining in two sets.
Digitately, palmately, fingered.
Diœcious, unisexual, the two sexes borne on different plants........................ 38
Distichlis, in Pacific slope..... 83
Distichous, placed in two vertical rows....................36, 64
Distribution of seeds.........100, 101
Dock, narrow, Rumex.......... 223
Dodder......................... 221
Dodge, J. R., most valuable wild grasses................ 81
Seed of new grasses......... 301
Dog-grass....................... 167
Dolichos 364
Dorsal, on the back of.......... 36
Downing, A. J., on lawns...... 309
Drainage, advantages of........ 240
Improves the quality of grasses 281
Drasteria erechtea 387
Drupes 62
Drying grass, effect of.......... 228
Paper 11
Duke of Bedford............... 199
Duval-Jouve, on tortion of leaves 30
Dying seeds.................... 206
Dysart, Hon. S., on saving seed of Timothy................. 106

Early cultivation of grasses.... 197
Eel-grass....................... 65
Egyptian millet or grass........ 171
Elaters 408
Elements most useful to plants. 329
Elliott, Jared, early cultivation of Timothy and Fowl Meadow grass.............. 199
Elliptical, oval or oblong with regularly rounded ends.
Elymus villosus................ 200

PAGE

Embryo, a rudimentary plant in the seed.................... 65
Of Indian corn............... 42
Emersen, R. W., on weeds...... 214
Endogenous, plants in which the fibro-vascular bundles of the stem are scattered without order 64
Endorhizal 65
Endosperm, food stored in the seed outside the embryo, 41, 42, 65
England, grasses sown in...... 201
Epicampes, leaf of............. 22
Epichtoë 423
Epidermis..............14, 15, 28, 31
Affected by climate......... 15
Of poa pratensis............. 15
Eragrostis, fertilized........... 39
Ergot 420
In agrostis.................... 213
Ericaceæ 62
Erodium 216
Erysiphe on grasses............ 426
Euphorbia...................... 223
Evaporation 27
Evergreen grass, see Arrhenatherum.
Experiments of J. B. Lawes on grass lands 273
Experiments, on seeds 208, 210

Fairy-ring fungi................ 430
Falcate 363
False red-top. 140
Families of most worth..60, 61, 62, 63, 64
Farmyard manure on grass lands 267, 269, 270
Fermentation of hay........... 298
Fertilization of the flowers of clovers........................ 324
Of grasses......37, 38, 39, 40, 41
Fertilizers, effect of.....271, 275, 288
Festuca.
Arundinacea................ 131
Bees on flowers............. 38